The Hidden Lives of Lab Animals

The Hidden Lives of Lab Animals

A VET'S VISION FOR A MORE
HUMANE FUTURE

Larry Carbone

UNIVERSITY OF CALIFORNIA PRESS

University of California Press
Oakland, California

© 2026 by Larry Carbone

Library of Congress Cataloging-in-Publication Data

Names: Carbone, Larry author
Title: The hidden lives of lab animals : a vet's vision for a more humane
    future / Larry Carbone.
Description: Oakland, California : University of California Press, [2026] |
    Includes bibliographical references and index.
Identifiers: LCCN 2025029056 (print) | LCCN 2025029057 (ebook) |
    ISBN 9780520403963 cloth | ISBN 9780520403970 ebook
Subjects: LCSH: Animal experimentation | Laboratory animals
Classification: LCC HV4915 .c369 2026 (print) | LCC HV4915 (ebook) |
    DDC 179/.4—dc23/eng/20250912
LC record available at https://lccn.loc.gov/2025029056
LC ebook record available at https://lccn.loc.gov/2025029057

Manufactured in the United States of America

GPSR Authorized Representative: Easy Access System Europe,
Mustamäe tee 50, 10621 Tallinn, Estonia, gpsr.requests@easproject
.com

34 33 32 31 30 29 28 27 26 25
10 9 8 7 6 5 4 3 2 1

publication supported by a grant from
The Community Foundation for Greater New Haven
as part of the *Urban Haven Project*

For David, with all my love and appreciation

Contents

Python

Python bivittatus

FROM THE ZOO TO THE LAB, WHAT ANIMALS WANT

"Take a break and come watch," my new boss said. "It's time for Esther's lunch."

It was my first day on my first-ever job, at the Franklin Park Zoo in Boston, fourteen years old. I'd spent the morning scraping possum poo out of cages. I dipped my sponge into a hot bucket of Vesphene, the disinfectant whose sickly sweet smell came to mean *home* to me, and scrubbed. Then I switched to the paint scraper to tackle the really stubborn stuff. No gloves in those days, so I learned how well a fingernail works for those really caked-on bits of poo. I scraped and scrubbed with my right hand while holding a long-handled wooden brush in my left, not for cleaning but for keeping the possum pressed against the wall of the cage so I could clean, clear of her fifty teeth. That day I was so very proud of how clean I got those possum cages while keeping all ten fingers out of the possums' mouths. This was my audition for another day of unpaid dirty work. How could watching Esther, whoever she was, eat her lunch compete?

Esther, it turned out, was a ten-foot-long Burmese python, who relished all the live rats she could catch, kill, and swallow. Her rats had arrived in a cardboard box, the words "Charles River Laboratories" in bold font. Where else but a lab-supply house could you buy enough healthy, live rats to sate a snake with a taste for freshly killed meals? Each time the zookeeper tossed her a rat, we would shout to the victims as though watching a horror movie: "That's Esther's back you're standing on!" As the doomed rat ventured, sniffing, near the flicking tongue, we'd yell, "Go back! Go back!" But we didn't really want the rats to go back. We wanted to watch Esther strike and then position her rat for the deadly squeeze. The rat would twitch and squeal. We watched the tail flick fast, then slower, and slower, until it stopped, and we knew the rat had lost, as the rats always did. We weren't cruel or heartless, we believed, just animal-care professionals looking after our hungry charge.

Esther did not always consume her boxful of rats, and I dutifully cleaned and fed her leftovers, the survivors. But I did not really care about the rats—or the tadpoles I caught for the turtles or the feral pigeons and wild rodents we worked to exterminate on the zoo grounds. Their lives, their welfare, their very existence: none of that mattered to me in those days. They were second-class animals in my ethical universe, and they are still second-class animals according to the US Animal Welfare Act, one of the two main laws in which laboratory animals find some protections. In fact, legally, they are not even "animals"—as I'll detail later.

My zookeeper friends and I deeply cared about—even loved—our animals. At lunch I might share my sandwich with Rufus, a woolly monkey who lived in his small cage alone, off exhibit, except on days when he traveled in our Zoo Mobile. His long brown tail would wrap around my arm as he worked the chicken salad or the gooey peanut butter in his mouth. Other days my lunchtime meant a bike ride to a nearby pond to dig up plants to decorate our turtles' home, the Reptile Pit, and catch tadpoles for them to hunt and eat. When it was my turn to bring leopard cubs or raccoon babies home for night feedings, my alarm woke me every ninety minutes to groggily prepare their formula and patiently encourage them to suckle. At home I watched reruns of an obscure TV show called *Wild Cargo*, where manly men went out to catch live animals for zoos.[1] I was no manly man, but I wanted that job! My friend and I read the zoo-industry newsletter with its listing of various zoos' surplus animals for trade or sale, imagining the elaborate cages and islands we'd create for the animals we'd purchase when we ran the zoo, him as the director and me, the zoo vet. Until that happy day arrived, we had plenty more manure to shovel.

Although *Wild Cargo*, to its credit, showed some of the collateral killing that comes with capturing live animals to send to the States for zoos and circuses (and not something I'd have thought about in those days: labs), I didn't think deeply about whatever violence ripped our baby, Aggie, from her elephant herd or Winston from his chimp family. I never wondered how my beloved Rufus came from his monkey troop in steamy South American treetops to a solitary cage in wintry Boston. I did not seriously question why we had so many zoo-born leopard and tiger cubs to hand-rear in our nursery. Were their mothers too stressed for the job in their small zoo cages? Would they have kept their cubs tucked out of sight and nurtured them, or did my bosses decide the higher priority was a zoo nursery full of the very cutest baby animals?

Nor did I know or think much about where my animals went when they outgrew life in our children's zoo. The raccoon kits I had bottle-reared would be released in parks near Boston when they were the size of stripey-tailed basketballs and their personalities changed almost overnight from cuddly to curmudgeonly. I do not know, and did not know to ask, how they might brave the Boston winter on their own. While some animals moved into indoor cages after the zoo's summer season, others went . . . somewhere. A decade after my days feeding grapes one at a time to our young charges, a professor told my vet-school class about a chimp he'd known in a lab near Boston. The chimp's name: Winston. "My" Winston? Had Winston, as I ended up doing, gone from sunny summer days in the Children's Zoo to life in an animal lab?

If forced to contemplate these questions I would have said the animals' health and happiness were our sole priority, and putting on a show for our zoo-going public was secondary. In those days I did not grasp the pressure zoo managers felt to put on a good show and the ways they would compromise the animals' health and welfare to keep the paying public coming back for more.[2] If zoo administrators were willing to compromise the animals' welfare to put on a good show for visitors, I nonetheless had important mentors who started me thinking about the ethics of how we relate to animals. My first boss, Richard Farinato, was a fierce advocate for the animals, a passion he carried throughout his life. He did not waver in calling me out on any sign of indifference or ignorance, such as thinking Esther's leftover rats were somehow lesser creatures than the exotic animals on display. Even my lunch came under scrutiny from the main chimp care-giver, who was the first vegetarian I ever met. By the time I traded life in the zoo for life in a dorm, I was coming to see that every contact with animals is an ethical encounter. I started seeing that caging zoo animals has its dark side, as did my chicken-salad sandwich.

My poo-scraping skills came in handy while I worked my way through my undergraduate years at Cornell. I took the worst job of my life, week-end help at the university's poultry farm. My eyes stinging from the ammonia, I walked the aisles of the egg barns collecting and labeling eggs from hens crammed into small wire cages three tiers high. In the broiler-chicken barns, hundreds of technically cage-free chickens huddled tightly together in an indoor enclosure so crowded their wood-chip bedding was invisible. They hobbled around on arthritic legs, their swollen joints barely able to carry their weight in the short weeks they were alive.

Cornell is an unusual campus, a posh Ivy League university that none-theless has its own poultry, swine, sheep, cow, fish, and horse farms. It operated its own Ivy League mink farm too, back then, though no longer. The university runs these farms as research sites so that its scientists can study how various diets and housing systems affect animals—and the meat, milk, and eggs we take from them. Industrialized farming was and remains the industry standard, so our Ivy League chickens crowded tightly together as they modeled life down on the factory farm. If the birds weren't pecking me, they were pecking one another. This was my first encounter with research animals, and I left it as quickly as I could for a job flipping burgers and frying eggs in the dining hall, not because the birds' suffering so distressed me, I confess, but because winter was coming, and I had no car to get me to bird barns on snowy mornings.

I took whatever animal courses I could fit in my schedule and found whatever animal jobs I could. One summer I bicycled around goldenrod meadows, catching bombardier beetles and orb-weaving spiders for an animal-behavior lab. There scientists pitted beetle against spider in gladi-atorial combat, studying how the beetles sprayed chemicals at the attack-ing spiders to defend themselves. Animal testing takes many strange forms. The scientists who ran the chicken farm and the beetle coliseum had sole discretion over how they treated the animals. The US Animal

Welfare Act, passed ten years earlier, in 1966, excluded bugs and birds, as it still does, and state animal-cruelty laws carved out exceptions for lab animals. Spiders and beetles and factory-farmed Cornell chickens certainly found no anticruelty protections.

Between college and vet school I applied for work as an animal-care technician at Cornell's research-animal facility, or vivarium. During my interview I toured the Vet Research Tower, which housed hundreds of mice, rats, rabbits, dogs, sheep, and cats, all of them objects of research in the professors' laboratories. The halls were suspiciously clean. Eerily quiet. I smelled no animal smells. Staff in coveralls mopped floors rather than play with the animals, who had numbers instead of names. This was not the zoo; this did not feel like a place any animal could call home.

I took the job and passed through the looking glass to a world where animals were not quite as they seemed. The longer I stayed, the curiouser the animals got. Hundreds of caged mice shared a room; some of them—the "nudes"—were perfectly bald but for their whiskers, and others had large tumors on their flanks. Cats lived in sealed glass-fronted cubicles with filtered air in and out; I could see them meowing but could not hear them through the glass. Pregnant sheep stood in narrow cages that kept them from walking or even turning around, as a jumble of wires and catheters snaked their way from the sheep's fetus and womb to machines that measured hormones and recorded contractions. Several rooms contained steel rabbit cages, much too small for even a single hop. Some rabbit cages housed woodchucks instead of rabbits, an even tighter squeeze. I saw dogs too, but not dogs like most people know. These Cornell dogs had fragile, easily torn skin—a disease called Ehlers-Danlos syndrome in human patients—from which the scientists sometimes took biopsy samples to study their abnormal collagen. I met the hundreds of beagles in the school's other research kennels, along with Labrador retrievers with hip dysplasia, German shepherd dogs with cardiac arrythmias, and golden retrievers with muscular dystrophy.

Scientists intentionally selected sick and abnormal animals as breeding stock, recapitulating human diseases in animals that might model them in the lab. At the zoo I had been able to convince myself that the animals were there solely for their own safety and well-being, not to serve any human interests. I could harbor no such illusion in the lab vivarium. Though some of our animals—cats and dogs in particular—might be with us for years, we knew most would die young in some experiment or other.

The senior animal-care technicians teaching me my new duties had a clear sense of ownership of "their" animals, one of them with *her* rabbits, another with *her* cats, a third with *his* dogs. They called the individual animals, though never the mice or rats, by name rather than number and in our lunchroom swapped stories about their quirks and preferences with affection. Their guardianship of their animal wards included a watchful eye on the scientists' staff. Were the research staff gentle with the animals? Did they stay with the animals as they woke up from sedation? Did they come quickly when the caregivers called them about a health problem? Did they kill them—*euthanize* and *sacrifice* were the words I learned to use, cloaking reality in professional jargon—promptly when they got seriously ill? No matter how sweet or funny particular animals might be, the techs knew their bodies belonged to science and that they would say good-bye sooner rather than later.

I had not expected to work in the animal labs long, only until my then boyfriend finished his vet degree. I ended up staying for more than forty years—as animal caregiver, vet assistant, student worker, and laboratory animal veterinarian, first at the vet school at Cornell and then at the University of California–San Francisco medical school. My job as a lab vet behind the looking glass was a curious twist on what people expect a vet to do. Most vets perform clinical work that includes preventive medicine, such as giving vaccine shots to kittens, flea-prevention medicines for dogs, or wormers for horses. Clinical work also includes diagnosing and treating sick and injured animals. Companion-animal vets typically focus on the individual dog, cat, python, and so on. Farm vets may treat either the individual patient or a whole herd. Either way, animals are their patients, and their owners, guardians, or farmers are the vet's clients.

My work as a lab vet has included plenty of preventive and therapeutic medicine, much like the vet in a local clinic. When animal caregivers or scientists, my clients, said their mouse or monkey was sick, I examined and treated the animal, my patient. To prevent illness I vaccinated cats or monitored mouse and monkey colonies for assorted viruses and parasites. Vet care for individual dogs, cats, rabbits, and even monkeys is similar in the lab to the private vet–practice version. I examine the animal patient, take a medical history, run some bloodwork, and prescribe my treatment.

Mouse medicine rarely treats the individual animal as a patient. Mouse vets focus on the colony. A sick mouse we kill upon discovery in the mouse room, so we can quickly diagnose a problem that could spread to other mice. Preventive health care verges on science fiction in the mouse room. Some mouse strains have virtually no immune system, genetically programmed to lack the lymphocytes and other white blood cells that fight off infections. To keep them alive, vets house them in sterilized cages with filtered air and autoclaved food, directing staff to don face masks and sterile gloves if they must handle them. Most labs extend this infection control to mice who should have a more normal immune capacity, excluding not just the bugs like fur mites and hepatitis viruses that can make the mice sick but also a range of bacteria and viruses that may (but usually won't) make the mice a bit sick and may (but usually won't) affect the experiments. These cages and their paraphernalia, such as cage-cleaning robots, cost tens of thousands of dollars. Necessary or not, this approach to housing what should be healthy mice is so expensive as to lock us into keeping the systems running, at top dollar, with consequences for mouse welfare and for mouse usefulness in experiments.

Dog-and-cat vets serve a range of clients. Expert dog trainers, shockingly negligent owners, and devoted caregivers who would rather go hungry than see their cat in distress all may come through the door on any given day. Many like or even love their pet and will do what they feel they must for their animal's well-being, but with limits on the time and money they will devote. So too in my practice: some of my scientist clients were oblivious to their animals' suffering and sloppy about following the rules, while others remained alert, contacting me with questions, for innovative ways to better their animal care. Some of my favorites were clinician-researchers who saw human patients by day, running an animal lab as best they could in their "spare" time, driven by their desire to cure the diseases that still bedeviled the patients under their care. Some of those were expert scientists as well as physicians. Others, whose surgical skills and clinical prowess I do not question, had no idea how to formulate a hypothesis or design a workable animal experiment. As I explained the logistics of setting up a lab and the questions the ethics committee would have, some decided against shopping for some mice or monkeys or other animals, sticking to the work of caring for their human patients that they knew best.

As I started work as a lab vet, classmates who'd gone into private practice faced the anguish that the best medical care they had to offer could cost thousands of dollars. That's a steep price tag, especially since no vet can guarantee a perfect outcome. Through the 1970s and 1980s, many universities reorganized their animal-care programs, putting all the animals under the oversight of a campus head veterinarian. At Cornell that meant that animals in psychology, zoology, agriculture, and human-nutrition research would answer to the head lab vet, based in the veterinary college. Campuses differ in how they fund animal care, but a common arrangement is for a scientist's grant to directly pay for buying animals, plus a daily fee to pay for animal care. Cornell's first campus-wide vet, the man who eventually hired me, refused to charge researchers for health and welfare efforts, such as vet exams, animal medicines, animal-welfare enrichments, or vets' time spent in developing animal protocols that a committee would approve. So, as my friends in practice bemoaned the agony of seeing animals suffering when owners could not or would not consent to expensive care, I was able to simply prescribe a clinical course of action, knowing that the researchers would not need to pay for my work out of their grants. The tranche of money supporting my work came from what the National Institutes of Health refers to as indirect costs, or overhead, as well as my wealthy university's endowment.

As a lab vet, I worried less about whether my clients, the scientists, would put financial limits on my care of their animals. The biggest difference between my work and a pet vet's practice is that my patients were in the lab to serve a purpose in a way that pets are not. Dogs and cats in modern homes have very little work to do. If clients bring their dog to a vet because the dog no longer chases tennis balls, the vet's task is to figure out why and to cure or manage whatever health issue may be lurking. The dog's welfare is the goal, not the clients' need to get the dog "back to work" promptly returning thrown balls. In one sense, then, my work in the lab was more like a horse or cow vet's job, returning the animal to good health, not necessarily so the animal can feel better but so the animal can work better, be that work racing, making milk, or, in the case of my patients, producing scientific data. The animals must also be "fit for purpose" for whatever work we are expecting of them. I need to be careful with my treatments, knowing that antibiotics or immune drugs like cortisone and prednisone could bollix an infectious disease experiment. Vets and scientists may try not to convert the animals into well-functioning machines, not to reduce them to furry test tubes, but, when animal welfare and research data do conflict, the needs of the research often take precedence. This is especially true in labs that require sick animals to most accurately mimic sick people in the search for cures.

As a lab vet, clinical care of research animals was a fraction of my duties, balancing the goals of keeping animals' bodies functioning well for the experimenters' needs and hoping that that also would safeguard the animal patients' well-being. I helped scientists design their animal experiments and trained them on how to carry them out. In my role of training researchers how to run their experiments, I emphasized how animal illness, pain, and distress are often as bad for the experiment as they are for the animal. I was the animal-welfare cop too, auditing scientists' labs to check that they were following their approved animal protocols with a modicum of competence. I had my seat on the campus animal ethics committee overseeing the scientists' animal labs, and, when Animal Welfare Act inspectors from the US Department of Agriculture showed up at our door, I escorted them through labs and animal-housing rooms. I crafted our annual institutional reports to regulatory agencies as well as Freedom of Information requests from watchdog activist groups, aiming to report all the legally required information without unnecessarily divulging anything more. Working as I did on large university campuses, I shared those tasks with a team of other vets and veterinary technicians. At small mouse-only labs, a single vet does all those jobs, sometimes just showing up once a month to check in. Solo practitioner or part of a large group, all of us lab vets do the same balancing act. As one lab vet reported in a 2004 article on the worst jobs in science, a vet who takes up lab work goes "from someone who makes sick animals healthy to someone who makes healthy animals sick."[3]

From the day I started work in the animal labs, I knew the work would be ethically fraught. I'd known of zoo colleagues who'd taken jobs in animal labs; it was not something anyone wanted to talk about or think about. One day, a year after I'd launched my lab vet career fresh out of vet school, Bernard Rollin gave an electrifying talk on my campus. A philosopher who'd started the first ethics courses for US vet students, he put in words the discomfort I'd been carrying in working in a place where virtually no animals' lives ended well. I wanted to be that man! I wanted to be a veterinary ethicist. We chatted, and he offered me the opportunity to do graduate studies in moral philosophy with him in Colorado. Before signing on and heading west, I explored some other programs—no one was offering a real degree in vet ethics, but a few folks doing work who touched on the issues I wanted to solve were receptive to me cobbling together a degree with them.[4]

As tempted as I was to hang up my stethoscope and scrubs and bury myself in the philosophy books, I stayed put, in the Cornell labs. I'd spent years training to be a vet and was not ready to set practice aside just yet. Beyond that, as more and more folks were weighing in on animal ethics, I saw a gap I might fill. Philosophers and social scientists, even Rollin, who rolled up his sleeves to spend time with scientists, vets, and ranchers, wrote about animals and ethics from one or more steps removed. They did not really seem to *know* animals the way I wanted to. On the other hand, vets and other animal people writing to defend or condemn animal labs rarely seemed to really grapple with the work of the ethicists, like Bernie Rollin, Peter Singer, Andrew Rowan, or the others who shaped my perceptions. I wanted to amplify my work beyond my own cases, to write articles and my first book, which might help animal people in the labs grapple with the ethics of animal testing. My compromise was to do some post–vet graduate studies moonlighting as a PhD candidate in Cornell's History and Philosophy of Science and Technology Program while working my day job as a mouse and monkey vet. I wanted the vet expertise to really know everything I could about mouse and monkey pain and how to diagnose, treat, and prevent it. More, I wanted the experience, ethical and clinical, of examining my mouse or monkey patients, holding their lives in my hands, doing my best to understand what they needed in the moment—more pain meds, perhaps some antibiotics, or, often, euthanasia to end their suffering—and taking responsibility for my clinical choices.

A lab vet is one of the stranger professions a person could choose. Plenty of kids want to be a veterinarian when they grow up; none aspire to become the vet in an animal laboratory—same with me, except that is exactly where I ended up. Few readers of this book will ever have visited an animal lab or likely wanted to. Those who do want to are the animal activists, looking for material for exposés or just a chance to be an independent watchdog. We on the inside work diligently to exclude those people. Transparency is low. Readers may have heard from scientists that animal research is vitally necessary for the medical progress they want. They have also heard the animal activists' claims that scientists waste public funds running useless animal tests behind closed doors. Useful information beyond the accusations and counter-accusations of animal rights "extremists" battling "cruel" scientists is scarce. I offer here an authoritative, balanced view that neither glorifies nor condemns scientists who choose to use animals in their labs.

This book is highly personal. I take readers on the path I have taken working in animal labs, primarily in two large, top-tier universities. Vets at smaller colleges or in private companies will have different stories, as will the scientists whose animals we vets care for. Scientists and vets in countries with stricter or laxer laws will similarly have different experiences to share. For this reason I've gone beyond my own stories in visits to other labs in the United States and abroad, meeting with countless vets, scientists, and animal activists, and I have spent hours poring through historical documents.

My conclusion is that animal labs are still necessary if we want continued medical progress. They are more essential than animal activists portray but less so than research defenders claim. I know that animals do still suffer in those labs, though the activists exaggerate and the researchers downplay the amount and type of suffering. Even in a society that has decided that sensitive animals will suffer for human benefit, most people believe we should set limits to that suffering, and I am a witness that we are not meeting that goal. Thus I call for us to continue the welfare improvements that vets and behaviorists champion, support scientists' efforts to validate nonanimal replacements, and find better ways to weed out animal experiments that do not justify harming animals. The decision to allow animal experiments is an ethical one, and so I call for opening up ethics committees and government policymaking to a range of voices of people across the spectrum of values and worldviews, from the most ardent scientists to the most vocal animal advocates.

Animals are in labs in service to humans, but my job has been to be in the labs for the animals. And so, I write first and foremost about the animals, titling the chapters for animal species I have known or worked with. The various animals will lead you into the issues I cover in each chapter, just as "Python" tracks my personal story from a teenage zookeeper to a senior lab veterinarian, ethicist, and author. Follow the animals to learn about what animal experiments are, to consider their necessity, to envision a better world for the animals, and, finally, to find ideas for joining the effort for more compassionate animal care, an effort that stands to bring animals better lives and simultaneously make for sounder, more reproducible science that truly enhances human, animal, and environmental health.

In the next chapter, "Woodchuck," I show nonscientist readers the wide range of animal research, starting with some curious experiments I encountered as a young vet, in which scientists modeled how heavy alcohol consumption can affect the course of human hepatitis infections. This chapter covers the basics of what an animal experiment is, and I highlight issues that affect the animals' welfare the most. We urge scientists to replace animals if they can, reduce their numbers to the minimum necessary, and refine experiments to be as painless as possible for the animals: the "Three Rs" of alternatives in animal research. I thus explain why labs hold such a menagerie, with mice for some experiments and frogs for others, but also the animals most likely to bring protesters to the lab, such as monkeys, dogs, and cats. I explain how statistical analysis drives the need for multiple animals in experimental and control groups. I describe why scientists strive to eradicate any differences in their study groups, such as imbibing woodchucks and teetotalers, to minimize variability that can derail some experiments and lead others to spurious, nonreproducible results. I show how the quest for a standardized woodchuck, monkey, or mouse sometimes leads to animals living impoverished, regimented laboratory lives that often, paradoxically, make them worse research subjects as well as less happy creatures.

Next, "Marmoset" tells of a young monkey I treated, sick from the monkey version of a dread human disease, multiple sclerosis. Several years and marmoset monkeys later, the lab did indeed produce a successful drug for managing symptoms, an important step in scientists' quest for an outright cure. I ask, How necessary were the marmoset experiments and the mouse studies that accompanied them? Animal experimentation is indefensible if it doesn't deliver what scientists promise. Animal rights activists can list the many known failures, while research activists list the successes, including dozens of Nobel prizes based on animal experiments. How necessary were dog experiments in the 1920s for discovering insulin as a treatment for human and canine diabetes? And even if they were necessary then, does modern science a century later still require animals? I share my belief that we do still need some animal research if we want continued cures and vaccines. But not all animal experiments are equally productive. Not all tackle important questions. In this chapter I call for broadening the representation for deciding the importance of research projects: A cure for cancer? And one for male baldness? Does basic research justify animal experiments if scientists do not have specific applications in mind for their discoveries? Scientists must not only tackle important issues; they must run experiments with a high likelihood of producing sound data that can be replicated in other labs, and, for human medical research, the animal data must translate across species from the lab bench to the human hospital bed. A cure for mouse cancer is only as valuable as its role for leading to a cure for human cancer.

Our decision to continue some animal labs obliges us as ethical people to find ways to lighten the burdens of living in cages and serving in experiments. It obliges us to accommodate animals' welfare in our laws and in our daily practices. I write how four kinds of animals have served as ambassadors, shaping how we think about animals in labs and how we set standards for their treatment. Dogs, rabbits, primates, and mice each earn a chapter on the role they play in inspiring better animal protections or, in the case of mice, eroding those protections.

In "Dog" I write how dogs have long been the standard-bearers for centuries in the crusades to abolish or reform animal labs. In the 1800s England passed its first laboratory animal protections. Dogs were the primary concern, though they brought cats and horses along on their coattails. A century later exposés of pet dog trafficking for labs triggered the US Animal Welfare Act. Dogs, as our loving companions, were the face of campaigns to stop pet theft and to stop animal shelters from sending animals to labs. More recently, they are the poster animals for laws requiring labs to find postresearch homes for animals. Dogs as joyful athletes drove a push to get animals into bigger cages and from there out for exercise. I show how sometimes dogs inspired protections that carried over for other animals, while at other times some protections accrued to them alone. Sadly, the intense concern in the 1960s on dogs as lost pets resulted in an Animal Welfare Act that focused almost exclusively on how labs acquire and house animals, with Congress forbidding the US Department of Agriculture, the enforcement agency for the act, from regulating what scientists actually do to dogs in the lab. That issue would wait twenty years, until two primate species, rhesus monkeys and baboons, featured in high-profile lab abuse exposés and made regulation of experimental methods inevitable.

A rabbit is a quieter animal, less likely to stir emotions than a dog. When Louis Pasteur was using them in his rabies experiments, they were a menu item rather than a pet, whether raised in rabbitries or hunted in the field. But in 1980 fluffy white rabbits were Easter bunnies and pets in the United States, not dinner, and a public relations campaign highlighting their use in cosmetics-safety testing put the quest for nonanimal safety tests on a solid footing. In this chapter I describe how the focus on rabbits' eyes broadened to encompass all animals in safety tests for cosmetics, medicines, and other chemicals. This chapter explores how scientists and regulators validate that an alternative nonanimal test is as reliable as the animal toxicity tests, whose reliability is woefully inadequate. Unlike animal experiments in basic science and hypothesis testing, animal safety testing is rapidly disappearing, replaced by cells, synthetic tissues, AI-driven modeling, and other nonanimal replacements. Rabbits are benefiting, along with the other common species in toxicology labs, including mice, rats, monkeys, dogs, zebra fish, and horseshoe crabs. Should you buy only "cruelty-free" cosmetics and household products that claim they are "not tested on animals"? I give a qualified "probably," explaining how complicated even that simple question is.

Scientists and their critics agree that animals in labs should experience no unnecessary suffering. As a vet overseeing animal welfare in the labs, I ask if I can recognize and measure animal suffering and happiness to guide my efforts for the animals. In "Chicken" I show how scientists in England began asking farm animals in general, with chickens first in line, "What do you want?" I describe increasingly sophisticated tests to measure animals' preferences, experiences, aversions, and even their personalities as we move our goal from preventing unnecessary suffering to fostering animal happiness. In this chapter I focus on how welfare science can lead to better housing for laboratory animals. While costly, reforms in animal housing offer the benefit that healthy, unstressed animals almost invariably become better-quality research subjects, improving whatever value we get from our animal experiments.

As a kid working in the zoo, I chafed at how visitors laughed at our monkeys and chimps, seeing them as smelly clowns at best. Television shows reinforced this prejudice, until nature documentaries—I credit Jane Goodall in particular—got us to look in chimps' eyes and stop our smirking. Their new image, as our smart cousins deserving respect, spread to other primates, lab monkeys included. In the 1980s exposés from two monkey labs spurred the US Congress to update the Animal Welfare Act in ways that dog lovers had not managed to do. The most significant change was the new requirement that all animal labs covered by the act establish in-house animal ethics committees (in the United States, we typically call them Institutional Animal Care and Use Committees). I save my critique of those committees for later chapters, and in "Chimpanzee" I highlight the 1985 law's requirements to tend to the psychological well-being of lab and zoo primates, with calls to "enrich" the impoverished, solitary housing that was all too common at the time. In 1985 Congress singled out athletic, playful dogs as the sole recipients of mandatory exercise provisions; our brainy primate cousins alone got enrichment and psychological well-being protections. As I describe what can count as a truly enriched environment, I highlight how the idea caught on—if not in the Animal Welfare Act then in self-directed programs, further encouraged by accreditation programs and the National Institutes of Health's rule book for lab animals—that all animals in labs deserve attention to their emotional well-being, not just their physical health.

Animals in labs deserve the best living situations, with room to move around, environments they can explore, and the chance for social interactions, plus space to get away from aggressive cage mates. But what happens when we take them from their cages and bring them to the lab for an experiment? A cozy cage will not spare them the pain and distress that many experiments can cause. In "Rat" I deploy the animal-welfare science from the "Chicken" chapter to tackle the question of animal pain and its close relative, distress. No one taught me in vet school just how difficult it is to recognize pain in animals—not the sharp yelp of pain if a person steps on a dog's tail but the aching throbbing of a mouse or monkey, alone in a cage through the night after a surgery experiment. In "Rat" I show how welfare scientists and pain biologists are developing sophisticated methods to assess animal pain and developing more potent, longer-acting pain medicines for lab animals' comfort. I remind scientists that untreated pain can bollix an experiment just as surely as indiscriminate use of pain medicines can and urge us to err on the side of overtreatment rather than undertreatment. I remain skeptical that we vets are good enough yet at detecting and treating pain, so I urge ethics committees to assume the worst case as they grapple over what experiments to approve or reject, and I always encourage scientists to choose nonpainful experiments, rendering moot questions over how to manage the pain they would cause.

Where dogs, primates, and rabbits were the heroes of laboratory animal–protection campaigns, "Mouse" champions the antihero. Mice and rats vastly outnumber all other mammals in US research labs, composing around 99 percent of lab mammals, and yet, alone among countries with lab animal–protection laws, the US Animal Welfare Act defines them as "not animals" and excludes them from its protections. Unlike countries with a single law and single government department overseeing laboratory animal welfare, the US system is arcane and leaves some animals out in the cold. Our Animal Welfare Act is a flagship law, first passed in 1966. Its strengths include unannounced inspections from veterinarians from the US Department of Agriculture and publicly accessible reports on how many animals a facility uses. Its gaping hole is that it excludes almost all animals, well over 99 percent, used in the United States. It defines mice, rats, and fish as "not animals" under its protections. A separate law authorizes the National Institutes of Health to oversee animal care, including the mice and the fish, but only at labs (universities yes, private companies not usually) that receive federal research dollars. It rarely sends inspectors out, relying heavily on institutions' self-reports of any animal-welfare problems or research misbehavior. That trust and collegiality is nice if you're on the inside, hoping to avoid federal repercussions when scientists or the lab are breaking the rules, but leaves animal watchdog groups suspicious.

Finally, rounding out the US system is a private organization, the Association for Assessment and Accreditation of Laboratory Animal Care International, that accredits labs that voluntarily seek its certification. It's a thorough review but raises suspicions, as it comes but once every three years, relies on vets and scientists from other labs instead of the government, and, being a private enterprise, is secretive and immune to Freedom of Information requests. In the US system, big universities operate under the two laws, plus most go through accreditation reviews. Small colleges and private biotech startups often operate under none of the three. That's the situation many mice find themselves in, but since we have no national system for counting mice, we do not know how many. I've long campaigned for letting mice be "animals" under our Animal Welfare Act and would happily eliminate the National Institutes of Health program and let the US Department of Agriculture consolidate all lab animal–welfare oversight under its system of inspectors and its publicly viewable reports.

As they trail in legal protections, mice form the cutting edge in new research technologies, as scientists manipulate their genes to create a dizzying array of engineered varieties. Much of this gene jockeying has enabled scientists to downscale their experiments, such as largely replacing dogs in diabetes labs and woodchucks in hepatitis experiments with mice. Mouse experiments are not just more efficient than those with larger animals but often also move labs away from government oversight. I caution that as go mice, so go monkeys and others, for new gene-editing technologies pioneered in mice are now available for creating new mutated monkeys genetically programmed for anxiety, depression, and other debilitating illnesses.

In "Flea" I ask which animals deserve what kinds of ethical consideration in the animal lab? All of our welfare laws, policies, and practices allow some use of animals for human benefit, while trying to set some limits. I emphasize that harming animals requires justification, but which animals? What counts as a harm? What benefits to humans are strong enough to justify harming animals? Scientists and vets can test various species for their degree of sentience (or the neurological and emotional capacity to feel pain and pleasure) to guide decisions about what animals, from fleas to mice to monkeys, merit concern. They can determine the relative severity of various experiments, finding that early-stage cancers and simple injections harm animals less than major surgeries. Ethical questions require these technical assessments but demand more. For example, pain is unquestionably a harm to animals, but we have no scientific way to decide if painless killing, ending a healthy animal's life, is a harm that requires strong justification. Scientists alone cannot judge what projects are important enough—curing baldness versus curing cancer—to allow for harming animals. Thus I call for including the widest range of values and viewpoints, noting that the animals themselves get no voice, in ethics committees' decisions on what scientific projects to permit and which to reject because they cause too much pain and distress for too little potential benefit. Current practices in most countries, and especially in the United States, fall short of this communal ethical process.

At their best, laws reflect social ethics, setting standards of practice and methods for enforcing compliance. In "Rhesus Monkey" I describe the patchwork system we have in the United States, where two federal laws plus a voluntary program for accreditation mandate some welfare protections for some animals in some laboratories. Most countries have a system of regulated self-regulation, in which scientists, vets, and ethics committees determine how institutions will care for their research animals, with occasional self-reports and occasional inspections to verify legal compliance. Animal rights groups strive to gain insider information as outside watchdogs, pressuring government agencies to act when they detect evidence of malfeasance in the labs. From my experience as a vet practitioner under the US system, I describe the challenges I faced, in which vets shoulder competing and sometimes contradictory roles: we oversee scientists' animal use, manage the animal facilities, and promote animal experiments with the minimum of outside disruption, whether from activists or regulators.

As I started work on this book, I visited the zoo where I'd first watched Esther lunching on her rats fifty years earlier. At our reunion coworkers and I debated just which oak tree we were sitting under; so much had changed through multiple renovations. Zoo work too has changed, incorporating welfare science in a greater awareness of animals' mental and emotional health, not just for cuddly babies but for all stages of the animals' lives. In "Gorilla" I review the parallel evolution of animal care in zoos and labs, looking for ideas to bring into the research facilities.

Believing as I do that animal research remains important for medical and scientific progress and that animals suffer far more than is justifiable in labs, I offer several proposals to weed out low-value science; broaden the range of voices in ethical decisions; promote happier, longer lives for lab animals; and simultaneously decrease their suffering while reducing the negative effects that animal stress has on scientists' experiments. I suggest ways our laws should change, but, having little hope for serious legal reforms in the United States any time soon, if anything, I fear backsliding. People can and must go beyond what laws require, whether as a show of institutional responsibility to shareholder, consumer, and benefactors' sensibilities or a genuine urge to do good by the animals from whom we demand so much. I offer suggestions for scientists, institutions, vets, and my reading public, outsiders to laboratory life, to rise to the challenge of better science through better animal care.

Laboratory animals are hidden from public view. Few people see their pain or distress or occasional happiness, and few appreciate in any detail how much we benefit from them. After forty years working with them, my knowledge too remains limited, though that work has given me the chance to watch, learn, and participate. As I show some of the realities of these animals' world, I share my visions for what we must do to make their world and ours a more humane place.

Woodchuck

Marmota monax

FASHIONING ANIMALS INTO MODELS

I was not prepared to find woodchucks day drinking in the service of science when I started my working life in animal research. I had taken plenty of biology lab classes as a college student, where I charted the behaviors of honeybees and field voles, dissected embalmed sharks and cats, and seined Ithaca's gorges for fish and caddisflies. I had done my work-study stint in Cornell's poultry farm and biked the goldenrod fields collecting spiders and beetles, never considering that those chickens and bugs in the farm and classrooms had anything in common with mice and monkeys in medical labs.

Despite my newly printed biology diploma, I knew almost nothing about the numbers or kinds of animals researchers press into service when I applied for a job caring for Cornell's lab animals. I did not know why some scientists used rats, some dogs, some sheep, some woodchucks, when their common goal was medical discoveries for human health. Like anyone who has encountered animal rights messaging and images, I may have envisioned scientists force-feeding rats poison to see how many died, dripping eye makeup into rabbits' eyes till they went blind, or shocking monkeys in cages just to see what would happen.

One goal of this book is to give readers a robust view of the many ways and reasons scientists study animals, in the field, on farms, and in laboratories. These facts may convince some readers of the absolute necessity of animal experiments, while others will decide animal labs are all useless, cruel, or both. My personal conclusion is that we should be doing fewer animal experiments and that we can do them with less pain and suffering for the animals. This chapter is my primer to give all readers a common set of facts on which they can make their judgments. Only with this detailed knowledge of the wide scope of animal research can we sort out which experiments to allow or phase out and brainstorm practices to refine whatever animal tests that we keep.

My life as a technician in Cornell's animal labs came with a steep learning curve. Most of the work was easy to master: I cleaned mouse cages, I cleaned dog cages, I cleaned rabbit cages, I cleaned more mouse cages. Vivarium designers favored efficiency and infection control over the animals' enjoyment or comfort, and so our animals lived in steel, concrete, or plastic boxes, much easier to clean than the dirt or grass pens at the zoo. At the zoo we transformed animals' cages into homes, giving them pools to splash in, logs to sleep in, tadpoles to hunt and eat. But our lab mice got no chance to burrow, or dig, or build a proper nest. None of the animals did. They lived in barren but superclean cages, a thin layer of sawdust or the like to soak up their urine. Scientists and vets believed that any departures from strict hygiene and uniform treatment of the animals could derail the experiments.

I was already pretty good at handling the species I encountered at the vivarium, though some, like the Ehlers-Danlos dogs with their hyperfragile skin, took a special touch, lest a quick grab of a squirmy pup's scruff might tear their skin. Handling woodchucks in lab cages was not all that different from grappling with the possums and porcupines from my zoo days: I would shoo them into a corner and keep their back to me and their teeth away while I scraped out their dung and bedding. Woodchucks, large rodents about the size of a cat, have always populated the pastures around Cornell's agricultural campus. With their buck teeth and pudgy physique, woodchucks—you can also call them groundhogs, whistle-pigs, or, more formally, *Marmota monax*—are vaguely comical. You may see them if you walk a field on a sunny morning, sitting up on their haunches, holding some plant, perhaps a farmer's carrot, in their two front paws, while those big front teeth gnaw the vegetable down to nothing. They freeze stock-still if a person approaches, then with a sharp whistle dive down into their burrow. I find them cute as can be, but I'm not a vegetable farmer whose crops they are raiding or a dairy herder whose cows can break their legs when they misstep into a woodchuck hole. Nor would scientists want them in their lab if mice could do the job. Woodchucks have formidable teeth—leather gauntlets and a syringe of ketamine are the standard tools for a simple blood sample—and most lab cages will not contain them. It's much easier to work with mice.

Woodchucks are hibernators. Alone among lab animals, they have their own holiday—Groundhog Day, February 2—when their Pennsylvania representative, Punxsutawney Phil, interrupts his hibernation to check for his shadow and determine if winter has passed. Their unique physiology, with annual cycles of weight loss and hormonal swings, led some Pennsylvania scientists in the 1970s to round up some local woodchucks and launch a breeding colony for lab studies. Who knew what valuable insights into human health woodchucks might provide? The researchers focused on metabolic and heart disease, not infectious disease or cancer, as they followed their chucks from birth to death, with an autopsy for every animal who died on their watch.

Luckily—for the scientists if not for the woodchucks—the colony animals had a high incidence of liver disease and even liver cancer, similar to those in human patients with hepatitis B virus. They discovered that woodchucks harbored their own version of the human hep B virus.[1] In the late 1970s Cornell scientists imported some of Punxsutawney Phil's neighbors to breed as a lab colony, not to study hibernation but to get inside those livers and see what they could learn, hopeful that chucks could be the long-awaited animal model that would help them conquer human hep B.

Woodchucks were fruitful and multiplied in Cornell labs through the 1980s and 1990s, topping out at a thousand animals a year. Visiting one of our many woodchuck rooms one day, I saw animals with bottles instead of food bowls, with notes that the smoothies in the chucks' drinking bottles held measured doses of alcohol and were to be handled only by the researchers. The curious idea of tipsy woodchucks lodged in my memory for years, though the team never published peer-reviewed papers on this work. I tracked down the graduate dissertation documenting this research, all that exists as the only confirmation I had not made this up.

The tippling woodchucks' story shows how scientists define a hypothesis, map out an animal experiment, exclude whatever variability and biases they can control, and settle on the number of animals they believe will solve their problem. It illustrates too how various types of animals come and go as the species of choice in any field of research, sometimes because of scientific developments; sometimes because shifting societal values or legal restrictions push scientists away from one animal in search of others; sometimes, as with the woodchucks, because of a combination of the two, with the right dose of serendipity. Hepatitis researchers had a clear mission in mind with their experiments: to use woodchucks as models to improve vaccines and care of human hepatitis patients. Other experiments I'll describe focus on basic biology with no particular disease in the crosshairs. Some focus on domestic or wild animal health. Many animal experiments do not touch on health or disease at all.

The scientists serving up the alcohol-laced smoothies were veterinarians, but healthier woodchucks were not their motivation. They did not necessarily care about the effects of woodchucks' drinking habits on woodchuck health, at least not directly. They already knew that both heavy alcohol consumption and chronic viral hepatitis can lead to liver cancer in human patients and suspected that fuller epidemiological work would show the two insults synergizing in humans. The researchers chose woodchucks as surrogates—as *models*—of human patients, infecting them with the woodchuck version of the human hepatitis B virus and giving them alcohol-laced smoothies as their sole diet. They wanted to see if these two factors together would more quickly lead to liver cancer in the animals and, if so, to figure out how. They were confident that much of what was true of woodchuck hepatitis patients would be true of—that is, would *translate* to—human hepatitis patients.

The woodchucks served in a *controlled* experiment, one of the most common modern approaches to human and animal research. In a controlled animal experiment, scientists seek to standardize everything in the animals' lives except for the one or two variables that they are examining. Here four treatment groups were included: woodchucks without alcohol and without the virus; with alcohol and without the virus; without alcohol and with the virus; and with alcohol and with the virus. Everything else in the chucks' lives—time of infection, type of caging, day length, housing temperature, schedule of blood samples, age and sex of the chucks—must be fully brought under the scientists' control so that the effects of the alcohol could stand out clearly against any background variability, or the noise in the system. Thus the researchers' note to us animal-care and vet staff: do not touch these bottles and jeopardize our control over the chucks' alcohol intake.

An animal experiment has many moving parts. Get them wrong and problems await. Scientists must choose the right animal model of the correct sex, strain, and age. They must use enough animals for statistical analysis; one-off anecdotes are interesting but rarely produce conclusive information. Scientists must cut through extraneous variables and sources of bias, from health issues in the animals to incompetent experimenters to uncontrolled factors in the animals' environments such as food, light, temperature, noise, smells, pheromones, social environments, interactions with caregivers, and opportunities for running, climbing, or burrowing. When they get these things wrong, animals can suffer. When animals suffer, often the data they produce will suffer too, incapable of replication in other animal labs and impossible to translate reliably to knowledge about human health and disease.

Animals offer a chance to control an experiment in a way that human subjects often thwart. Some years earlier Italian researchers had tried to study how alcohol consumption affects liver health in human hepatitis patients over three years of light, heavy, or no drinking. They couldn't randomly assign subjects to study or control groups and so had no control over who got hepatitis. They could not have forced their volunteers to overindulge or abstain, so they had to rely on subjects' self-reports of their drinking, something humans are notoriously dishonest about.[2] By contrast, the chucks drank their alcohol smoothies: we gave them no choice. They carried the virus or not, because the scientists intentionally infected them, then housed them in isolation cages to prevent its spread. The Cornell vet scientists saw a chance to answer questions the Italians were asking, with a fraction of the number of subjects and no fudging of their alcohol intake. They deployed a couple of dozen captive woodchucks compared with over five hundred free-ranging Italian human volunteers, hoping that their tight control of confounding variables would give them clearer results than the human studies could produce.

The human hepatitis B virus deserves medical researchers' attention. It is a global killer that spreads via shared needles, blood transfusions, and sexual contact. Mothers can infect their babies at the time of birth. The immediate, acute disease usually produces mild effects, but some 10 percent of patients go on to develop chronic hepatitis, and some fraction of patients with chronic disease develop liver cancer. Multiply that by the several million hep B patients around the world, especially in rural Asia and Africa, and hep B looms as a major killer, in the league of malaria, HIV, and polio. As immunologists in the 1970s pursued a vaccine to prevent infection, virologists researched medicines to cure infections that slip past the vaccines, and oncologists sought treatments for the millions of chronic patients on track to liver carcinoma. If infecting woodchucks and forcing them to imbibe could help alleviate this suffering, then scientists were going to keep them drinking.

In infectious disease research, scientists cast about for an animal species that they can successfully infect and in whom disease will progress as it does in people. For example, scientists where I once worked used rabbits as their model for studies of MRSA, the antibiotic-resistant staph bacteria that causes serious skin, lung, blood, bone, and other infections in people. These dangerous bacteria can infect a wide range of animals, and rabbits, rodents, dogs, and sheep have all been put to service over the years. Viruses, on the other hand, tend to be fussier than bacteria about the species they will infect, and so whenever scientists identify a new virus—hepatitis B or C, HIV/AIDS, Zika, SARS, COVID-19, name the next emergent epidemic—the race is on to find an animal species susceptible to infection. Even before they have identified what new virus is causing an emerging epidemic, scientists start pumping patient samples into mice, rabbits, guinea pigs, and rats, trying to find a lab animal they can enlist for their experiments. Common lab animals, even monkeys, proved immune to human hepatitis B infection (and, similarly, to HIV some twenty years later).[3] Had anyone tried, they'd have found woodchucks resistant to the human hep B too. Yet chucks, with their own virus, a doppelganger of the human version, came to reign as the animal model of choice in hep B research through the 1980s and 1990s.

Woodchucks and their hepatitis virus were exciting because scientists had spent decades without a practical hep B model, relying on scarce and expensive chimps and ethically questionable human experiments. Scientists suspected that two similar diseases, serum hepatitis and infectious hepatitis, were the result of viruses, but they did not know what viruses. They began injecting and feeding material from hepatitis patients into a range of animals, but they always came up empty-handed. No serum hepatitis plus no infectious hepatitis added up to no animal model. Even chimps, our closest cousins, failed to contract infection in the initial attempts.

Facing this apparent dead end, scientists mounted increasingly ambitious hepatitis experiments drafting the one animal species reliably capable of sustaining a human hepatitis infection: humans. In the 1940s and 1950s, researchers tried quashing some hepatitis outbreaks at summer camps and in the military by jabbing patients with gamma globulin collected from recovered patients. Success! Blood transfusions were becoming a more common procedure in those days of World War II, but they risked spreading hepatitis. So scientists tried various ways to inactivate the viruses they were sure must be lurking, comparing the effects of administering their sanitized plasma versus unsanitized control plasma into human "volunteers." In some 1950s experiments, those "volunteers" were prison inmates who received blood the researchers knew carried an active hepatitis virus.[4]

In 1947 the Nuremberg Code, an international response to Nazis' medical experiments on concentration camp prisoners, introduced the ethical principle that researchers should experiment on human subjects only once they had amassed animal data. With no enforcement mechanism in the Nuremberg Code and no animals who could provide preliminary data, Saul Krugman in the 1950s took the prior decade's human experiments to a new, and now notorious, level. Krugman went to the Willowbrook Institute, where hepatitis ran rampant among the institutionalized intellectually handicapped children. Initially, Krugman tried immunizing incoming kids with convalescent patients' serum to see if this shielded them from the hepatitis infections he was sure they'd eventually contract at Willowbrook. It did. Doctors inject new vaccines or placebos and wait to see protection in the face of a naturally spreading virus. That's slow and gives them no control over whether both groups face equal exposure to the virus. So Krugman took another step, inoculating incoming children upon their arrival at Willowbrook with infectious material from infected kids.

Sure enough, the inoculated children developed a transient hepatitis, experiencing a few days of vomiting, jaundice, and loss of appetite. None got severely sick. Krugman's success in deliberately infecting the children allowed him to develop a home-brewed vaccine that gave longer-lasting immunity than his predecessors had gotten from gamma globulin shots. He demonstrated his vaccine's efficacy by inoculating yet more kids and giving controls a placebo, then once again feeding them infectious material, such as feces from infected Willowbrook kids in chocolate milk.[5]

In 1932, well before the doctors at Willowbrook started infecting children with hepatitis in the 1950s, government scientists at Tuskegee Institute started following a cohort of local Black sharecroppers with untreated syphilis to chart the course of the disease. Alexander Fleming's discovery of penicillin in the 1940s revolutionized syphilis treatment, but not in Alabama, where the Tuskegee study continued for twenty more years. The doctors never told their volunteers that they had a cure for them, leaving them to suffer with their fatal disease and to spread it to their wives. The exposé of the Tuskegee abuses, not the Willowbrook experiments, led to reform in the United States through the National Research Act of 1974 and the Belmont Report of 1979. The Willowbrook studies of the 1970s mostly escaped scrutiny while in progress, but growing public awareness of human research increasingly was making studies like Krugman's difficult to launch.[6]

As Krugman was finishing up his human studies in New York, scientists were still trying to get chimps to sustain a human-origin hep B infection. In 1961 Friedrich Deinhardt and his colleagues published their success at infecting fifty recently captured chimps in the Belgian Congo, using human material from Krugman's Willowbrook freezer. Chimps became the model species for hep B research, but they are expensive and scarce— large, strong, hard-to-care-for animals, slow to reproduce, all of them in those days wild cargo from African jungles—nothing so easy as buying a box of rats from Charles River. Most chimp studies used a tiny handful of animals, sometimes just two or three, rarely adequate for a good statistical analysis, thereby making errors likely.

Despite the challenges of these experiments, the combination of human and chimp research successfully identified the hepatitis B virus, produced the first scalable hep B vaccine with reagents from human patients' blood, and launched human clinical trials for the vaccines doctors still use today. Only after millions of shots of a vaccine developed via chimpanzee and human subjects had already gone into patients' arms did the Philadelphia woodchucks and their newly discovered woodchuck hepatitis virus arrive on the scene.

To the extent that a vaccine is the ultimate goal in most infectious disease research such as hepatitis B, the woodchuck researchers were already scooped before they even started. But though physicians already had their vaccine in hand for their human patients, Cornell imported infected chucks and set to work breeding new generations for future experiments. But the experiments were for what, exactly, if the moon-shot goal of a hepatitis vaccine had already been reached? Scientists always want to learn more. Every success or failure in the lab raises challenging new questions, and those questions often lead to yet more animal experiments. Possible advances from the world of hepatitis research still lingered. For example, an improved vaccine that wouldn't need three doses over the span of six months and wouldn't require refrigeration would have much better uptake in the poor, rural areas of Africa and Asia, where hep B wreaks havoc. Medicines that block hepatitis liver infections from exploding into fatal liver cancer in woodchucks might similarly prevent cancer in humans already infected with the hepatitis B virus. Treatments that keep alcohol consumption from accelerating liver disease in infected chucks might do the same trick for human patients. Through a scientist's eyes, every new discovery raises dozens of new questions to investigate. To a critic's eyes, when those dozens of questions call for further experiments on thousands of animals, the obvious question is when does it ever end.

Woodchucks do not fit the profile of a typical laboratory animal; they are not dogs, mice, monkeys, or guinea pigs. But the chucks' misfortune of harboring a cancer-causing virus so similar to its human counterpart offered scientists a unique opportunity. For most experiments, mice or rats are the workhorse species who carry the burden of modeling our diseases and testing our therapies for us. Many modern labs house only those two species. In my Cornell years, and later at the University of California, I met other anomalous lab animals, beyond the commonplace mouse and monkey, pressed into service for some unique aspect of their biology that might yield some useful info about humans. For cardiologists the

Watanabe breed of rabbits develops humanlike atherosclerosis when they eat a diet rich in fat and cholesterol. Ophthalmologists had their eyes on the ground squirrels that zip around Bay Area fields, as they have color vision and lots of cone cells in their retinas, more like a person than most rodents. Cornell's colony of golden retrievers included healthy females carrying a gene that caused muscular dystrophy in their male pups, similar to Duchenne's muscular dystrophy, in which apparently healthy women may give birth to affected sons.

Two otherwise near-identical cousin species of voles, a type of small field rodent, differ in their social behavior. Like nature's own experimental-versus-control group paradigm, prairie voles tend to practice monogamy while male meadow voles mate and then scurry away. Through a psychologist's lens, what an enticing pair of animals to study, dissecting the changes in their brain chemistry brought on by having had sex (and subsequently being decapitated to give psychologists their brains to analyze). This is not something researchers could readily accomplish with human volunteers, but it leads to noninvasive MRI studies of human brain chemistry to explore the biological bases of human attachment and the pathology of the inability to form attachments. Scientists look for and capitalize on such anomalous similarities between humans and nonhumans.

The drugs Ozempic and Wegovy (both licensed versions of the same drug, semaglutide) became blockbuster drugs from 2017 on, for their quiet utility for treating human diabetes as well as for their headliner success as weight-loss drugs. They did not spring fully formed from a brilliant scientist's brain into doctors' waiting rooms. Rather, animals were enlisted from the initial discovery of a hormone in Gila monster lizards' venomous saliva, through assays in hamster ovary–cell culture, to experiments in genetically modified diabetic mice. When both mice and human volunteers lost weight with the diabetes treatment under development, scientists knew they might have a safe and effective medicine for both diabetes and obesity. But not so fast: licensure from the Food and Drug Administration required fetal and adult safety testing first. Thus pregnant and nonpregnant rats, mice, rabbits, and long-tailed macaque monkeys, listed as endangered in 2022, all played a role.[7] That's quite a menagerie in every dose of Ozempic.

The hamsters, monkeys, and others played their part in the Ozempic story for their various similarities to humans. But even the myriad differences between humans and other animals can tantalize. Gila monsters' venomous saliva contains a (nonvenomous) hormone quite similar to our human hormone, GLP-1, which regulates insulin and sugar in our bodies. Plenty of animals have similar hormones that could be put to work as medicines for human diabetes patients, but for that their effects are so short-lived they would require multiple doses per day. The key difference that makes the Gila monster hormone so useful is how much longer it stays active in the body. The lizard's hormone, exendin, was similar enough to the human GLP-1 hormone to intrigue scientists, but its difference from the human's secured its place (and its synthesized derivatives, no need to harvest directly from scarce Gila monsters) in medical care.

Time and again animals' differences from us pave their way into our medical research labs. A Cornell lab kept tanks full of lampreys, a primitive and generally unappealing parasitic fish. Lampreys abound in Cayuga Lake's waters in upstate New York. As young larvae, they attach to larger fish via a circular mouth of rasping teeth, sucking out juices from their host and leaving large circular wounds, to the disgust of local anglers fishing for lake bass. In Cornell labs high above Cayuga's waters, bass swam in their aquariums, seemingly oblivious to the half dozen inches-long wormy lampreys attached to and slowly digesting them. Rare among vertebrates, larval lampreys can regenerate their spinal cord nerves after an injury in a way no human quadriplegic patient can do. And so the scientists imported lampreys into the lab to sever their spines and study how they regain function as they heal, with hopes of bringing that knowledge to the trauma center of the local emergency room. Chimps—luckily for them—struck out as models in HIV labs in part because their HIV infection does not progress to the AIDS disease. Rather than a disqualifier, though, this difference to humans could have held a key to improving the human outcomes in HIV. Shouldn't we want to know why they have this immunity? Or why elephants, bats, and naked mole rats are all so cancer resistant? To the lab insider's eye, these "negative models" hold just as much promise as the animals we choose for their similarities to us. Sometimes we experiment on animals because they are similar to us; sometimes we experiment on them because they are different from us.

As a rule of thumb, scientists will choose the simplest and cheapest test system they trust (and that they believe journal editors, Food and Drug

Administration regulators, grant reviewers, and other peers will trust) to reliably test their hypothesis. If worms and woodchucks and people all have chromosomes that change with age, researchers study the worms. If a lab can induce multiple sclerosis (MS) in both mice and monkeys, the mice will be the animal of choice unless differences in their MS rule them out. If horseshoe crabs can test for bacterial contaminants in medicines as well as rabbits do, a lab will go with horseshoe crabs, at least until the Food and Drug Administration regulators accept that a totally nonanimal assay is at least as good. If children, chimps, and chucks all contract hepatitis that becomes liver cancer while mice are immune, the woodchucks look very attractive.

I cannot name a field of medical research that relies on a single animal species as its model of choice. Cardiology is a useful case study. Sometimes small size is an advantage. Mice and rats are small and cheap compared to dogs or monkeys, and once surgeons have mastered rat open-chest surgery, they can perform a range of experiments. Even smaller, zebra fish larvae are so tiny as to be near-transparent. Under a microscope scientists can watch the heart start beating within a day of fertilization. They can manipulate genes and monitor how the mutations disrupt the fish's heart development, looking for information relevant to human babies with heart defects. This is not possible with larger animals or animals whose embryos are tucked inside a hard-shelled egg or a mother's uterus. Small size has its advantages, but cardiac surgeons developing new stents, valve replacements, or other heart surgeries want an animal with a more human-sized heart, making dogs and pigs common subjects in cardiac surgeons' labs.

Size is hardly the only criterion by which to pick an animal species for the lab. Aging monkeys are as prone to heart health challenges as we are, but they are expensive, in short supply, and require specialized facilities to house. Dogs and pigs are common in cardiology labs, but in situations where both are biologically human enough to answer the research question, the pig will typically get the nod, for nonbiological reasons. While lab dogs and lab pigs rival each other for intelligence and sensitivity and benefit from similar protections in our federal lab animal–welfare laws, public opinion does not see them equally. Who wants the public opprobrium and scrutiny, the risk of becoming an animal activists' target, that comes with being a dog experimenter? Choose a pig instead. Among the small-size candidates for modeling affairs of the heart, mice and rats also hold a non-biological advantage over comparably sized hamsters or guinea pigs with a similarly rodent biology. They remain excluded from the protections of the US Animal Welfare Act, as I describe in the chapter "Mouse," and invisible to government inspections. Thus research administrators and lab vets will caution their scientists against buying a hamster if they can get their science done with a mouse.

Once an animal species has found a place in a discipline's labs—cats somehow came to be the go-to animals for brain and spinal cord studies for decades while rabbits predominated in toxicology safety tests—their use as models acquires a certain self-perpetuating momentum. Scientists build on the knowledge and materials of their discipline. A detailed atlas of the regions of a cat's central nervous system or standardized scoring sheet for rabbit-eye injury further propels that species's role. Replacing cats with rats, or rabbits with guinea pigs, requires repeating much of the initial work to tailor it for a different animal. Woodchucks had minimal history in labs, so the Cornell scientists needed to invest substantial time, money, and animal lives into making them into a usable model.

The Cornell folks had developed their woodchuck program enough by the mid-1980s to be able to launch the alcohol experiments. They had the tools and knowledge to run the studies, but in the currency of academic science, those experiments turned out to be valueless and invisible: an obscure and hard-to-find PhD dissertation (I wrote one of those myself, I confess, though that one killed trees, not animals), resulting in no peer-reviewed journal articles and no one citing the work in their papers. Alcohol appeared to depress some aspects of immune system function, but the researchers could not say whether these perturbations were suffi-cient to affect the course of the liver disease in the woodchucks. Five (of six, or about 83 percent) animals in the alcohol group and five (of seven, or about 71 percent) drinking virgin smoothies developed liver cancer, but the author cautioned that various factors, including "the small number of animals used," left him unable to say that alcohol increased liver cancer risk in woodchucks, let alone in people.

I carry at least some of the blame for experiments that failed through poor planning to produce robust useful data. Our Cornell animal ethics committee came to life in 1986 in response to new federal mandates, and I joined it as a graduate student representative. That same year the chucks in the first round of alcohol versus virus were born in the local pastures and promptly strayed into the researchers' field traps. In hindsight I wish our committee had gotten enough details on this set of studies to see flaws that would lower its scientific value.

The word is *underpowered*. For scientists to succeed in an experiment with only six or seven subjects in a cohort, they need tight control over all the variables they can identify. They need to isolate one variable per experiment: alcohol versus placebo. Instead, the woodchuck folks included two sexes of woodchucks. They bought infected lab-reared woodchucks and compared them to uninfected wild-caught juveniles. One cohort had Philadelphia genes; the others hailed from New York. Any of this variability could muddy the effects of alcohol versus placebo. A cancer incidence of 71 versus 83 percent is a fairly small relative risk. For a statistician to buy into the claim that alcohol—not genes, sex, age, or prior history in the lab or running free—accounts for an increase in cancer, they would want either more subjects, less variability, or both.

Animal ethics principles call on scientists to use the smallest number of animals necessary to achieve their scientific aims. When they use too few, whether to meet an ethics committee's real or perceived demands or to save time and money, they publish questionable data without statistical value, if they get to publish at all. In people, age or sex differences can result in different susceptibility to hep B and its aftermath. If that's true in woodchucks, a small sample size of mixed sexes or ages might have too much noise in the system, too much small variation among the animals, to detect the "signal" of different rates of infection or cancer. And what a waste of animals' lives that is, reducing the number of animals in an experiment so far that the smallish number used are wasted. Ironically, on a per-experiment basis, scientists often should be subjecting more animals to experimentation, not fewer, if animals must be used, to ensure robust conclusions.

As a general rule of thumb, the more variability within a scientist's research cohort, the more animal or human subjects the researchers need. Through the years scientists have worked with vets to develop inbred animals as closely related as a room full of identical twins. They have eradicated every possible virus and parasite, housing animals in sterilized plastic boxes with filtered air and water. The hope is that if scientists eradicate and control all sources of variability and accurately report how they've done so, others can replicate their experiments, verify their findings as true and reproducible, and build on them. This rigorous exclusion of all variability can be powerful, allowing scientists to run experiments with a dozen or so mice that might take hundreds of human subjects to accomplish. A downside of this approach is that scientists may develop exquisite data sets about a particular age, sex, and genotype of mice all in the same laboratory that may be so polished and precise as to not even replicate in other strains or in other scientists' hands.

The pursuit to banish all variability in an animal experiment causes other mischief too, both for the quality of animal data and the quality of animal lives. Immunologists are finding that a mouse who never encounters a germ never develops a mature immune system; such mice may be useful as models of people who've similarly lived in a germfree bubble, but not so useful for the rest of us. Meanwhile, the types of sterilized cages that such protected environments allow tend to deprive mice of many activities that mice prefer. Yet these sterile cages, so expensive to buy and maintain, are standard issue for most lab mice, whether germ control is relevant to the experiments or not. Just as pediatricians are now worrying that an overly clean childhood results in kids immunologically unprepared when infections do come their way, mouse immunologists worry that pet-store mice with their associated germs might make for better science, more translatable to humans, than the overly clean mice in sheltered, sterile cages that they've been studying for decades.

I've known neuroscientists to push back against any sort of cage enrichment for their animals, fearing that a rich and varied environment would skew "natural" brain development. This means even just a bit of paper that mice could build a cozy nest with, any sort of treats or dietary variety, or any sort of social life get denied to the mice in their labs. Again, as the mice suffer this deprivation, that itself can skew brain development, leaving scientists at risk of studying psychopathologies when they want to study normal brains. Finally, on this theme, as I show in the chapter "Rat," scientists sometimes resist ethics committee and vets' urging to give their animals better pain medicines, again from a conviction that pain medicines, more than pain itself, affect animals' bodies and make them inferior research subjects, when the opposite is often equally likely. Nonetheless, this need to control extraneous variability from animal experiments continues as an important challenge for scientists pursuing the cleanest data possible and for welfare specialists trying to enliven confined animals' days.

I've spent most of my career in university labs, where scientists aim to discover new information, test hypotheses, and be the first to publish new findings in professional journals. This is the sort of work that begets Nobel Prizes, and in my work at the University of California–San Francisco we had our share of animal researchers landing those honors.[8] This is the animal work I know best, but scientists also deploy animals for more mundane purposes, especially the final rounds of safety and efficacy testing of drugs before they move on to human clinical trials. Animals also serve for safety-testing cosmetics and industrial chemicals and household cleaners and agricultural pesticides. None of this testing will garner anyone their Nobel, but the Food and Drug Administration requires it for new drugs, the EPA requires it for pesticides, and even the Department of Commerce, regulating interstate trucking, wants to know just how hazardous the materials it authorizes for transport might be.

Lab insiders save the phrase "animal testing" for safety and efficacy screening of drugs and compounds before they go to human volunteers or straight to market. We distinguish testing from "research," perhaps to elevate its importance in public discourse or to emphasize that this is an endeavor whose final outcomes are unknown. I don't worry about the semantics. I see a spectrum, from basic biomedical studies asking what a certain cell or molecule does in the body without a specific disease in mind that this knowledge might vanquish, through experiments like the woodchuck work asking basic questions about a specific disease, on to testing whatever vaccine or medicine this basic research contributed to before human volunteers take up the work of final clinical trials. Wherever a study lies on this basic-to-applied spectrum, its justification lies in goals of improving human health and safety.

We don't know in the United States how many animals scientists use in various basic research and testing labs. I've published my estimate that the United States produced more than one hundred million mice and rats for labs in 2018. No amount of burrowing through data could tell me how many of those bred animals actually produced data, nor what kinds of labs they served in; the low-transparency systems in the United States simply do not collect those kinds of details. Two government agencies oversee lab animals in the United States: the National Institutes of Health and the US Department of Agriculture. They collect some basic annual counts lab by lab but do not tally them and make national totals available. By contrast, many European countries have signed transparency pacts to inform the public of animal use. The European Union collects total animal numbers, reporting them in aggregate and by the type of use. Currently, basic research uses less than half of all lab animals, while the later stages of drug development (the European Union labels this "translational and applied research") and safety testing to meet regulatory requirements together slightly exceed the numbers in basic labs.[9] The European Union likewise tracks animals that labs breed without ever actually using, whether labs are culling animals because they bred more than they could use or, more often, because breeding for specific genetic traits invariably leads to numerous offspring with the wrong genes who then have no place in experiments.

Research on the spectrum from basic science to regulatory testing aims to serve animal health as well as human. Scientists research new vaccines for dogs or investigate the effects of health and welfare improvements on milk production in cows. In the field, vets and immunologists tackle diseases of wild animals, such as transmissible cancers in Tasmanian devils, scabies mange in wombats, or fungal epidemics wiping out bats and frogs around the globe. For some of this work, researchers study the species of interest, while for much of it mice or other animals serve as models of larger animals. Mice, for instance, figure into more lab experiments aimed at curing Tasmanian devil facial tumor disease than do the scarce devils themselves, carnivores who would happily consume the mice drafted for their medical studies.

Scientists defend the value of "basic" research, experiments they conduct with no practical application in mind. Living as I have in the world of medical research, I have a hard time believing that things we might learn in animal labs about cells, proteins, DNA, brains, and all things in biology have zero potential for eventual deployment in conquering some disease or other. Field zoology research may offer truer examples of non–goal-oriented research. Jacanas are fascinating birds who live in the tropics around the world. Technically, they are wading birds, but their widespread toes allow them to walk on lily pads rather than doing much wading. Their other notable trait: most species are polyandrous, which is to say that females maintain harems of several males who tend to whatever eggs the boss jacana lays. They thus have attracted ornithologists throughout the years, who go beyond simply charting the birds' behaviors and run interventions and experiments such as removing eggs or causing other disturbances and charting how the birds react. The work offers little practical information that I can see. The birds are abundant enough not to be a big conservation concern, and so far no one has made a strong case that we must study jacanas' female-dominated societies to better understand human behavior. This strikes me as about as non–goal-oriented a line of research as a scientist can do, and so when scientists kill birds for these studies, as I describe later, it becomes a good case illustration of the value of cost-benefit calculations in animal experiments.

Basic, non–goal-oriented research has a stubborn habit of seeking applications beyond the "wow" of inherently interesting discovery such as the jacanas' peculiar arrangement. Ecologists have been studying the natural history of US wild animals for many years, not just the charismatic large herbivores and predators but humble rodents as well. Biologists trap, mark, release, and retrap small voles, deer mice, and other critters. Only occasionally do scientists in their papers say why they launched their study, but, given rodents' crucial place in food cycles and their menace to crops and orchards, even basic studies of their behavior might have some practical application. Field biologists suspected that, unique among small rodents including the varied species called voles, one species, *Microtus ochrogaster*, the prairie vole, a short-eared mouse-size rodent barely distinguishable to the casual observer from their cousin, the meadow vole, might enjoy monogamous-pair bonds. In capture-recapture experiments, the same female and male would show up in traps together at greater frequency than what they saw with other species. Intriguing? Sure. Useful knowledge for managing vole populations on farms? Maybe. But the real buzz around voles, the reasons labs house tens of thousands of them is the possibility that whatever makes them monogamous when so few other rodents see the appeal might work similarly in humans. Scientists are studying how oxytocin, vasopressin, and other brain chemicals serve to foster social bonding in these animals and, by extension, in humans. A monogamy pill? Maybe not, but surely brain chemicals have a role in human social and antisocial behavior, and maybe, just maybe, findings in voles can somehow help human psychiatric patients with assorted deficits in attachment and sociality.[10] Divorce lawyers need not panic; psychiatrists may see fruit from this animal model, but it shows so clearly why scientists are so resistant to accusations that basic science with no immediately obvious payout does not merit the privilege of using animals in labs.

Not all animals in science serve to generate new information, either for their own species's benefit or as models of humans. The EU-Norway data reveal that approximately 15 percent of animals in labs serve as sources of medicines, monoclonal antibodies, and other products. Premarin is the acronym-y name for the *pregnant mare urine* that finds its way into menopause treatments. Horseshoe crabs' blue blood yields cells that screen medical fluids for bacterial toxins. And just as scientists are finding non-animal replacements, especially in safety testing, that can reduce animal use in some animal labs, a new animal arrives on the horizon. Early primate-to-human transplants—a baboon heart for Baby Fae in 1984, baboon blood to cure Jeff Getty's AIDS in 1995—failed. Scientists are now growing genetically modified pigs, engineered to harmonize with a human immune system and, if their kidneys or hearts are going into a human patient, to stay small enough to fit in a human chest or abdomen. Early efforts have not been resounding successes but show enough promise that we should expect these animal uses to increase through the next decade.

A particularly unpleasant, and thankfully uncommon, role for animals is to feed the nasty pests and vectors that spread human and nonhuman infections. In Lyme disease research, our Cornell lab enrolled rabbits to feed larger ticks and deer mice to feed larval ticks (all the while hoping that jumpy little deer mice would not escape with ticks attached and start a local epidemic in a town where Lyme disease had yet to establish itself). Anthony Fauci, formerly of the National Institutes of Health, has come under fire for his support of using dogs to feed sandflies in experiments to kill sandflies or to develop immune treatments of the leishmaniasis parasite they spread. Anesthetizing the host mammal is of limited value, as drugs in the deer mouse's or dog's blood might affect the parasites and skew the experiments. Much better, when feasible, is to feed the little parasites on blood from slaughterhouse cows, as two of my Cornell vet professors did, feeding fleas in special chambers that let them poke their proboscis through a wax paper membrane rather than a cat's skin.

Animals come and go as the model of choice, with bigger animals often supplanted by mice, and mice by nonanimal alternatives. Today the only woodchucks on the Cornell campus are the occasional varmints munching and digging in the horse and cow pastures. The breeding barns and woodchuck labs are shuttered. Some labs around the country still use the occasional woodchuck, now obtained from a commercial outfit in Idaho. Woodchucks had their day as the world increasingly frowned on experiments on institutionalized humans and as the logistics and cost of chimpanzee testing assured that chimps could never remain the model of choice for hep B experiments. Woodchucks have passed the baton to a range of genetically engineered mice, to "humanized" mice (mice who carry human-origin liver and immune cells), and to cultured cells for hep B experiments. Mice and cultured cells offer advantages in terms of the experiments that scientists are now able to conduct, working more closely with the human virus in human cells than a woodchuck model using hep B's virus cousin.

As a tech and then a vet, I learned why our facilities had such a varied range of animals and such a large number too. For work on the ethics committee, I had to understand how scientists design their experiments. As a lab vet in basic university science, I rarely saw the findings of "my" researchers' discoveries being translated from animal models to human patients. If anything, I had a ringside seat to see the failures, whether from planning an experiment with too few woodchucks to assigning undertrained students to perform complex procedures alone, without a mentor or assistant at hand, to not communicating to lab staff the procedures that the ethics committee approved. I should be the biggest cynic about animal experiments. And I would be, but for the amazing discoveries I've seen my clients, my researchers, bring to light from their animal experiments. This juxtaposition, the dismal failures and the exciting discoveries at the same campus, convince me that we must find better ways to separate the good animal experiments from the bad, well before the researcher gets out the mouse catalog and starts shopping for animals.

To eliminate the worst animal experiments and improve on any that may remain, these basics of animal modeling are crucial. Scientists and the ethics committees who oversee their work must consider what kinds of animals they will enroll and how many. They must search for new non-animal alternatives to replace animals. They must consider whether efforts to limit variability within research cohorts are also perversely working against the animals' comfort and well-being. They must seek to replace, reduce, and refine their animal use, the "Three Rs" rubric enshrined in most current lab animal protections. And then they must go further, asking what purposes a given experiment may serve, and decide if the goals truly merit harming animals.

Exciting discoveries in animal labs are just step one. Whether in tissue cultures, mice, chimps, or chucks, scientists' discoveries about what hepatitis viruses do to liver cells in labs are only of use if they really and truly reflect what happens in human patients. And so the crucial question to which I devote the next chapter is how reliably do any of these findings in animal labs *translate* into important discoveries for human health.

Marmoset

Callithrix jacchus

SCORING THE VALUE OF ANIMAL RESEARCH

I was still quite new, maybe even my first day as a lab vet at the University of California–San Francisco, when a veterinary technician called in a panic. An animal was down on his side in his cage and looking bad. He was weak and limp; I should come quickly. She reminded me which building and which floor I should hustle to. When I got off the elevator on eleven, the sickly sweet smell alone told me I'd found my way to the marmoset colony. The little monkeys' chattering, ramped up because one of their own was in the technician's hands on the exam table, confirmed it.

The patient was a young male marmoset, a research monkey. I don't remember whether he had a name other than the number tattooed on his chest. Most animal-care staff, including the folks who best knew the marmosets, were already gone for the day, and I had never treated a marmoset before. I knew nothing of this animal's history, and I did not take the time to ask—in hindsight, my big mistake. Instead, I jumped into the standard first-response mode that I would use for any species: draw some blood for diagnostics, place an IV to run in fluids, put him on an oxygen mask. Stabilize the patient, then form a plan of action.

No one can resist a marmoset's charms. Think of a bushy-tailed squirrel with a surprisingly human face. They have white tufts of hair by the ears, like an old man's untrimmed sideburns. When you talk to them, especially if they think you're holding a mealworm behind your back, they cock their head from side to side, with eyes open wide and eyebrows twitching, as if making a show of their genuine interest in whatever you have to say. Twins are the norm in marmoset reproduction. Dad and any weaned young still with the family carry newborn babies around, presenting them to Mom for nursing. In the wild a mother, father, and several adult young defend a territory against rival marmosets. They chatter and squeal, marking everything with their pungent, smelly urine. In a lab, family groups occupy several separate cages—at the University of California–San Francisco (UCSF), about the size of a refrigerator—within a room. You can hear their squeaking and squawking from down the hall, and the smell of their urine stays in their cages despite multiple washings.

Captured in jungles throughout Central and South America, marmosets and other small monkeys found themselves in the United States and Europe by the thousands through the 1950s and 1960s, until the 1975 international Convention on Trade in Endangered Species put limits on trapping them for zoos, the pet trade, and lab studies. Much of this midcentury traffic in the wide range of monkey species was a bit of a fishing expedition: get them into the labs, learn how to breed and care for them, and, surely, their usefulness for medical research would reveal itself. Microbiologists found them useful in infection research, such as when chimps proved susceptible to human hepatitis viruses or when owl monkeys took a lead role in malaria labs. Primates being our brainy cousins, neurobiologists have also been keen to bank on their neurological similarities to humans for experiments.

There I was in the marmoset room. The only lab monkeys I'd known before were some baboons in the Cornell labs, robust forty-pound animals who dwarfed a one-pound marmoset's mass. Tiny patients are a challenge: try to hear their hearts, count their respiration, or hit a vein. I could complement my baboon experience with some know-how treating marmo-sized animals, like rats and small kittens. Best though, I knew I could trust an experienced vet tech to help me with the case or, really, to let me help her.

Were I your cat's vet struggling to place a catheter in a tiny vein during an emergency, I would ask for his medical history as I worked. Did he eat something weird recently? Was he out running in the street? Were there any prior episodes of collapsing? Is he on any medications? But as lab vets, the medical history we most need is to know what experiments the researchers are conducting. Can the experiment make the animal sick? Are there medicines a vet would prescribe that could help the animal but kill the experiment? When I worked at Cornell, this was easy; as a member of the ethics committee there, I knew the details of every animal-research project on our campus. New to the San Francisco campus, I lacked this crucial knowledge. In my panic I neglected to ask, and the little monkey died in my hands as we were trying to run his IV fluids.

Had I been familiar with the marmoset project, had I asked the right questions of the research staffer watching me struggle to establish the IV, I would have known that this animal had the marmoset version of human multiple sclerosis, in which symptoms wax and wane. A down marmoset may well recover some function for a few days before the next episode, so intravenous fluids and other heroics were likely unnecessary to get the animal through the night, and my struggle to administer them may well have put the patient over the edge. Had the animal survived the night, that extra time would have allowed the scientists to run a final MRI and collect whatever fresh tissues they needed. That's cold comfort for the little monkey, who would have died—that is, whom the scientists would have killed—eventually either way. Still, any potential justification for hurting monkeys in multiple sclerosis (MS) research is lost if the animals die prematurely, in their vets' hands or alone in their cages, their tissues and samples lost to the experiment.

Struggling to save this animal's life, I did not stop to ask how necessary the project was. I worked in a world in which we all shared a basic belief that animal experiments, including those riddled with pain and suffering, must be useful and necessary. Forty years in ethics committee meetings, and I recall no conversations about whether animal labs are necessary in the grand scheme and few serious debates about whether any of the thousands of animal-use applications we reviewed one by one were necessary. We could have asked, "Necessary for what?" but almost never did. What would happen to humankind, to medical progress, if we shut down this animal experiment or that animal test or shuttered the whole lot of animal labs entirely? Our ethics committee chair reminded us that our legal mandate was to focus on minimizing any harm to the animals. To critique a scientist's claim that a particular project would make good and necessary use of the animals was taking a step beyond. It would risk slowing the scientists' work or threatening their publications, so crucial to any professor's academic survival. Our questioning could jeopardize their National Institutes of Health (NIH) grants or allow scientists at more laissez-faire campuses to beat out our researchers, outcomes our university administrators certainly would not tolerate.

This chapter is my attempt to step enough outside of that milieu to listen to those who question science insiders' presumption that we need animal laboratories. I need to ask, Are animal experiments useful or misleading or sometimes one and sometimes the other? When they are useful, are they actually necessary? Even if they once were the only route to important knowledge, could we now replace them with new ways of doing science? When animal experiments are of marginal usefulness, should we abandon some animal testing or instead figure out ways to improve it? We need this information to be able to ask, in a later chapter, even if animal labs prove to be necessary, are we justified to harm animals in labs in ways we (no longer) would harm humans? *Necessary* is not an absolute; the question is "What is this necessary *for*?" And for whom? In the context of medical research and testing, *necessary* means that without animal experiments we would seriously cripple progress toward curing or eliminating serious diseases. We might unleash unsafe and or ineffective medicines into doctors' clinics or miss discovering an important medicine altogether.

Animal labs exist in the context of broader questions of what counts as medical progress. Our species will survive without good treatments for multiple sclerosis—we've made it this far, after all—but millions of MS patients are suffering and will continue suffering as they wait for good, affordable cures to become available. Nor will animal labs solve the health-care inequities that keep established and new treatments from reaching all patients who could benefit; if they could, the United States, with its enormous investment in animal testing, should lead the world in health-care outcomes. More animal data will not convince people to use the knowledge scientists have produced so far, be that the COVID or measles vaccines that many patients avoid or healthy habits such as eating better, exercising, or cleaning up environmental messes that lead to cancers, asthma, and other diseases. Animal research will not produce medicines without serious side effects; all medicines are potentially unsafe in some people, especially at the high doses that diseases such as cancer and auto-immune illnesses require. New knowledge emerging from animal labs may add some days to the lives of people who can afford it or dramatically improve the lives of others, but the new treatments alone are useless unless doctors prescribe them and patients can and will use them.

Medical research faces several challenges, not just in the animal labs but throughout the enterprise. Scientists and their critics alike write of a crisis of replicability, or reproducibility, in which a scientist's findings, even though peer-reviewed and published in the top journals, fail to find corroboration in other scientists' hands. How can labs build on scientists' data and bring that knowledge to a useful medical treatment if even the initial scientist is unable to get the same results a second time? Animal researchers are not alone in facing this challenge, which also plagues human-subjects experiments and in vitro studies (those we perform in cells and tissues outside a human or animal body).

We need clear expectations on what counts as significant success in medical science, with its pastiche of animal tests, nonanimal alternatives, and human volunteers. Measles and polio vaccines can eradicate those viruses from communities. COVID and flu vaccines cannot match that feat but are valuable in preventing the worst of those infections. Total cures for diabetes, HIV, and most cancers have remained elusive, but the right medicines can convert these afflictions' threats of early death into manageable chronic conditions. In drug safety testing, no animal lab or nonanimal replacement can promise total safety for all patients every-where who might use it. Instead, the tests tell us that a new medicine looks safe enough to go to human clinical trials, which themselves can tell us only that it looks safe enough for broader use, not that it will never cause any patients harm.[1]

To help human patients suffering from a devastating illness, my univer-sity's scientists are seeking a total cure, certainly, but along the way patients need treatments to slow progression and manage flare-ups. Researchers believed it necessary to replicate multiple sclerosis in monkeys. Multiple sclerosis is an autoimmune disease in which the patient's immune system attacks the fatty material, myelin, that encases and protects nerves in the brain and the body. It's a terrible disease in people, waxing and waning in severity but ultimately progressing to paralysis, blindness, pain, and cognitive dysfunction in the worst-affected patients. No one yet knows the inciting cause, or etiology, of human MS. Genes play a role but don't tell the full story. Do viruses or other infections trigger it? Environmental toxins? Vitamin deficiencies? Might it have multiple causes, all converging on a similar suite of nerve lesions and symptoms?

Though the origins of the human disease are unclear, scientists have ways to trick a monkey or mouse immune system into attacking the lab subject's myelin. The animal version of MS, experimental autoimmune encephalomyelitis, is similarly debilitating. First the tail goes limp and then the lower limbs, progressing to paralysis of all four legs, with the animal on the floor of the cage, unable to get up. My little marmoset patient would have experienced this stressful, progressive loss of function, with whatever nerve pain might or might not accompany it, for weeks before I met him. Steroids and other medicines might provide some relief, but in the lab the drugs I might prescribe that might affect the course of the disease under study are often off the table, yet another uncontrolled variable clouding the experiment and its data.

Some five years after that little monkey died in my hands, the researchers moved their marmoset colony from my campus to another institution. I never heard an explanation for the move, but I was glad of it. I'd seen enough marmosets with experimental autoimmune encephalomyelitis, with no obvious benefits that I could discern from their studies. I'd had enough of the restrictions on my vet care, the constant concern that anything I did to make the animals feel better would distort the disease the scientists were trying to study.

Another decade passed and then, seventeen years after my first marmoset patient, Genentech, the biotech company not far from our San Francisco campus (and now a division of the Swiss pharmaceutical giant Roche), announced its newly approved drug, ocrelizumab, for managing some forms of MS and slowing disease progression. Neither a cure nor a way to prevent the mysterious onset of this disease, ocrelizumab nonetheless provides welcome relief and allows many MS patients to live more active lives. Marmosets were integral to this breakthrough.

When newspapers report medical breakthroughs, I always look for two things: if they quoted any UCSF scientists I might know and if they mentioned any animals whose lives contributed to the breakthrough. I learned that Dr. Stephen Hauser, whose UCSF monkeys I'd vetted so many years earlier, was the head of the team that developed and tested Genentech's new drug. UCSF's write-up touted that one of our own had developed this great new drug, describing Hauser's insight that mouse models of MS were inaccurate and that he needed a "new animal model." That "new animal model" so important to the progress, the marmoset monkeys, got no mention in UCSF's press release.[2]

Marmosets, the unnamed "new animal model," became the often-hidden monkey of choice for MS experiments. They are much smaller than rhesus monkeys, the most common lab monkey, which makes maintaining a marmoset colony cheaper. Beyond that, UCSF scientists in the mid-1990s also saw a potential advantage in marmos' odd propensity for bearing twins whose immune systems partially fuse. This quirk of biology initially launched marmos rather than other monkeys as an MS model, with scientists thinking the immune-system peculiarity might play into their experiments. In the long run this immunological quirk was irrelevant to the experiments, but by then marmosets had become the go-to monkeys for MS experiments.[3] By the time I got to UCSF in 1999, the marmoset colony stood at a hundred or so animals, as breeding pairs raised their twins, who became the research subjects.

Cortisone sometimes helps human patients with MS flare-ups, but we vets were forbidden from administering any such treatments for fear that they would alter the course of the disease under investigation and interfere with what the scientists were trying to study. Forbidden to treat any sick animals, my involvement was limited mostly to my role on the animal ethics committee and occasionally meeting the research tech to discuss whether an animal's suffering was severe enough to require that the researchers euthanize them or whether they could keep the sick monkey a few days longer so they could schedule a final MRI and sample collection.

To find a treatment for this truly devastating disease in their human patients, these doctors went to great lengths to create a terribly debilitating disease in animals. The decades-long timeline and the colony's move from my campus blocked me from following its trajectory, if any, to a successful treatment. It challenged me to answer the vital question that could justify or damn animal labs: Does animal research deliver what it promises? An affirmative answer invites more questions about ethics; just because we benefit, does that justify harming animals? But a negative answer, a determination that we don't get anything from the animal tests, has a clear result: we should be shutting down the labs.

Two vocal camps, animal rights activists and research-defense societies, look at the data and come to opposite conclusions about the usefulness of animal labs. In truth, the opposite camps tend to start with opposite conclusions, already convinced that animal tests are necessary or useless and finding confirmation of their biases in the available cases. I've networked with animal protectionists through the years, doing my best to hear their arguments and to share my insider information on how ethics committees and regulators operate, not as a mole but as someone who believes in transparency and dialogue. Time and again I've had to stop the conversation and say, "You know that's my university's information that I'm not at liberty to share." Despite my efforts to hear animal protectionists' challenges to the work "my" scientists do, I freely admit that I inhabit their world of animal labs and that that world shapes my perceptions and beliefs.

Inside the world of medical research, scientists and vets are confident that animal labs produce important results and often view animal rights activists who claim otherwise as naive at best and deceitful and even dangerous at worst. We can point to thousands of medical advances that passed through animal labs on their way to human medical care and conclude almost reflexively that animal research must work. While we admit that not every animal experiment will have important or immediate outcomes, animal researchers, and the vets supporting their work, remain convinced that in the long run animal research saves lives. That's a claim that needs examination; the fact scientists used animals in going from the lab to the clinic does not prove they needed those animals then or, with newer technologies, would need those animals going forward.

When I had my knee replacements a few years back, I thought of what animals I should thank, posthumously, for the surgery that so improved my quality of life. The anesthetics, painkillers, antiseptics, and antibiotics all came through animal labs that used rodents, dogs, or other animals, while rats served to assure that the sutures for the final skin closure were all compatible with living tissue. Blood from horseshoe crabs served to screen these medicines and devices for bacterial toxins that, if present, could have sent me into endotoxic shock. Knee replacements were not the specific target of these widely used developments for surgeries of all kinds, but, nonetheless, advancements aimed elsewhere certainly benefited those getting such replacements. As for my new titanium knees, rats were likely most to thank for assuring that the titanium and the glue attaching metal to bone would endure inside a body—that is, would be biocompatible—without the body's immune system rejecting it. Lab rats weigh less than a pound, so goats and sheep and dogs tested how well metal glued onto bone would hold up with the pressure of human weight. Even the processes surgeons tried and ultimately rejected, such as filling an arthritic joint with fat or synthetic cushioning compounds, were tested on animals, sparing me exposure to ineffective treatments and potential associated negative effects.

I took a clinical vet job at the University of California–San Francisco in 1999, and within a year I was heading up animal-welfare compliance for that medical school's labs. I learned to steer clear of the animal rights protesters picketing quietly outside every Wednesday at lunchtime, but I'd occasionally take one of their pamphlets as I headed to my office. One day the pamphlets told people to call our ethics committee chair, or me, as the welfare-compliance vet, to stop the monkey labs. They did not call for bigger cages in the monkey labs or more pain medicines or more toys; they demanded a full stop, and they demanded it now. Back in my office, my phone held the one threatening message I've—thankfully—ever received, from a person thousands of miles away claiming to have plans to come to San Francisco. That call, despite its low credibility, gave me one tiny taste of what many researchers, especially those with monkey labs, have experienced. It has certainly shaded my thoughts on openness and dialogue, both on the risks involved and, conversely, on the total necessity for mutual understanding.

The protesters at our doors evinced no doubts that the scientists on campus were wrong to think animal testing could be useful and doubly wrong to think it could be ethical. The scientists inside—hundreds of them—showed no evident doubt that their animal experiments were vital and justified. I wondered whether either group could be completely wrong in their assumptions. In this chapter I want to convey how hard it is even for us research insiders to hear the arguments about the utility versus futility of animal labs and know whom to believe. Was I right to slip past the protesters, with no more engagement than a quick "thank you" as I took their brochures? Or, given what I know firsthand about the animals inside, should I have joined them?

I never joined the protesters, but I could not write them off either. By their lights the failings of animal testing are so evident and alternatives are so readily available that scientists' persistence in running animal labs is perplexing at best, or lazy, deceitful, and disastrous for human health and animal welfare at worst. They cite animal experiments that greenlit drugs that are either useless or toxic in people, drawing on the cases that made the news. Data that never appeared in newspapers, locked away at private pharmaceutical firms, could add to their story, all those drugs in development that never got beyond the first phase of human clinical trials, despite how well they worked in the animal labs.

Each of these camps can list examples of successes or failures. Successful new medicines travel a path that weaves through animal tests, computer modeling, chemistry labs, experiments on human cells, and clinical research with human volunteers. Unsuccessful, even tragically fatal medicines travel the same path. Along with their lists of successes and failures, each camp has its own foundational stories that everyone learns with their initiation into that group. Two stories have achieved almost mythical status: dog labs brought us insulin for treating diabetes, while rats let the drug thalidomide slip into clinical use, with tragic results for expectant mothers and their babies.

In the 1920s Frederick Banting in Toronto, with his assistant Charles Best, anesthetized lab dogs and surgically removed their pancreases. As Banting expected, he induced diabetes. He then treated some of them successfully with insulin he'd purified from yet more dogs' pancreases. Within just two years, he treated his first human diabetic patient, fourteen-year-old Leonard Thompson, using insulin from slaughterhouse cows, a much more abundant source than dogs would be. Banting's insulin treatments came a couple of years too late for my paternal grandfather, who died of diabetes in 1924, but my *nonno*'s contemporaries (and assorted aunts and cousins) benefited, as insulin quickly came into widespread use by the early 1930s. For research defenders, Banting's Nobel-winning success is a foundational story of successful animal research. And not only do human physicians have insulin for their patients; we vets too use insulin to treat our diabetic patients, be they pet dogs and cats or apes and other zoo animals.

On the other hand, say the animal activists, consider thalidomide. After some testing on rodents in the 1950s, the German company Chemie Grünenthal marketed thalidomide for anxiety and morning sickness in pregnant women. Thousands of women using the drug subsequently gave birth to babies with severe limb defects or other health problems; others lost their pregnancies or saw their children die shortly after birth despite animal studies showing it was safe.[4] Animal testing had missed this risk: Can it ever be trusted?

Both stories are compelling. But both stories have alternate readings. Banting and other researchers got the idea to remove dog pancreases because autopsies of human diabetics had revealed diseased pancreases. In the decades before Leonard Thompson's successful insulin treatment, long before Nuremberg, Tuskegee, and Willowbrook abuses led doctors to test their ideas on animals before people, doctors had experimented with a variety of pancreas-based treatments in dogs *and* in human diabetic patients.

Two Scottish doctors, John Rennie and Thomas Fraser, bypassed dogs and tried treating five human patients with pancreas extracts from monkfish they got at the Aberdeen Fish Market. They chose fish because a difference in pancreas anatomy suggested their islet cells might be a good source for what we now know as insulin. They may have used too small a dose. They gave it orally rather than injecting, not knowing that insulin is rarely effective by that route. They chose subjects far advanced in their diabetes. Others had tried using cow pancreatic juices for human patients. Had these human experiments succeeded, the insulin story would not likely have become such a powerful exemplar of the power of animal labs. Yes, their insulin came from animals, and, yes, other diabetes scientists before them had done some animal studies, but the success in Aberdeen would challenge the simple myth of Banting and Best, that dog experiments were irreplaceable.

As for thalidomide, had scientists and regulators shared animal activists' interpretation of the facts, the thalidomide tragedy would have ended our reliance on animal labs. Instead, toxicologists maintained that animal safety testing works, but only if we launch even more kinds of tests in even more kinds of animals. Chemie Grünenthal had not tested their morning sickness pills in pregnant animals, and, after thalidomide's removal from the market, scientists who did found similar birth defects in rabbits and, at higher doses, in rodents. Thalidomide's lessons led to changes in regulatory requirements that were the precise opposite of what research critics would have put in place, such as using both rodents and nonrodents (usually rabbits, dogs, or primates) should something slip through rodent testing with undetected safety concerns and, especially for drugs intended for use in pregnancy, running tests in pregnant animals, often monkeys.

Banting won his Nobel prize in 1923. Chemie Grünenthal shelved thalidomide in 1961. Much of the research that resulted in my titanium knees is similarly decades old. What do these stories, or even the recent release of ocrelizumab in 2017 for multiple sclerosis, tell us about the use or necessity of animal labs now or looking to the future? Past successes do not dictate future necessities. Columbus proved to the Western world that lands and people existed beyond European shores. He did not prove that wooden sailing ships would forever after be essential tools for exploring the world around us. In the fast-developing world of animal testing and its alternatives, even my 2022 knees and the newest MS treatments are ancient history. We want to know if animal testing is necessary for moving forward.

In theory, animal research *should* work, given the many genetic overlaps and similarities among animals, including us humans; what's true of one animal will often be true of a different species, including humans. My vet school professors taught us dog anatomy and medicine as the archetype from which we could extrapolate to other animals, while also teaching us some species-specific differences a cat, horse, or cow vet should know. The first hedgehog I ever cared for, Spike, had inflamed skin and was losing quills. No one knew much about hedgehog medicine in those days, so I pretended she was a wee but prickly dog and gingerly plucked a few of her spines to examine under a scope. And there they were: mange mites. Hedgehog scabies! I crossed my fingers, hedged my bets, and gave her the then-new wonder drug for killing animal parasites, ivermectin, at a dog dose that I scaled down for her small size. She survived, I'm happy to report, and got her quills back.

Veterinary practice depends on the likelihood that what's true of our best-studied species is often true of the weird and wild species as well. Medical research carries this further, trusting that what's true of nonhumans is often enough true of people. Ivermectin works great for dog and hedgehog mange and for deworming horses. But it earned its founders the Nobel Prize for what came next, not the veterinary work, but its success in treating human parasitic diseases, most notably *Onchocerca* worms, whose larvae cause a disease in Africa called river blindness.

The counterpoint: in theory, animal research should *not* work, given the myriad, idiosyncratic, often unpredictable differences among species. Even among dogs, vets had found that ivermectin could kill collie dogs and—a quirky anomaly I'd never have predicted—some kinds of tortoises. My faith that Spike would be more like a bulldog than a collie or a tortoise in her response to ivermectin could well have killed her. Critics point out the potentially fatal consequences of ignoring the species differences that pop up unpredictably, making extrapolation from nonhuman lab animals to human patients a risky gamble.[5] Animal research's critics point out that had scientists done early studies of penicillin in guinea pigs or of aspirin in cats, they may have shelved them as too toxic to try in people. In a similar vein, were ivermectin as toxic in beagles, the staple breed of dog labs, as it is in collies, it too may have died an early death instead of becoming the ubiquitous, successful parasite treatment that it is.

Every day brings new facts to the lists of similarities and differences among human and nonhuman animals. Whether the stories we read fascinate us with how unique animals are or endear us with how our dogs or cats think just like we do, they are irrelevant to medical research unless they go further. We know lab animals are simultaneously similar to us and different. What we need to know, what could justify harming them in experiments, is whether our findings in animal labs can reliably predict information about people. To be useful, animal researchers must do better than a coin toss and better too than the available alternatives in labs and in human clinics.

My decision to treat Spike the hedgehog with ivermectin was a coin toss, hoping without prior evidence she would more resemble beagles and most other dog breeds than she would a collie. I trusted—really, I guessed—she would handle ivermectin as well as snakes and aquatic turtles do and not succumb as a land tortoise might. Indeed, we trust that ivermectin will demolish some animals, such as scabies mites and *Onchocerca* worms, so that other animals—the dogs, humans, and hedge-hogs—can clear infections. But the coin I tossed one day in deciding to prescribe Spike's treatment is not representative of how scientists bring medicines from animal labs to human patients.

In Banting's diabetes lab and Hauser's multiple sclerosis studies, researchers already had human data in hand, from autopsies and clinical patients, that led them to focus on the pancreas in diabetics and myelin in MS. Banting could have deployed human pancreas cells in cultures to watch how they release insulin as sugar levels rise around them, but such culture systems came decades after Banting first treated young Leonard with insulin. Modern multiple sclerosis researchers have a greater range of nonanimal options. Epidemiologists explore the degree to which genes, vitamins, smoking, and viruses influence the onset of MS in a way animals or cells cannot tell us, though then the animals and cells serve to explore the mechanisms of those factors. Rodent- and human-origin cells, col-lected from nerve tissue or grown from stem cells have their roles and their limits. Modern CT and MRI studies can map spinal cord regrowth in treated human patients. Animal studies persist in MS labs, where mouse and marmoset experiments explore responses to potential treatments to block the inflammation that destroys the myelin or to test methods for regrowing myelin. As a final preclinical step before humans reenter the scene in clinical trials, pregnant monkeys (typically rhesus monkeys or related macaques) screen the safety of monoclonal antibody drugs like ocrelizumab for pregnant patients. None of these varied research projects will change the course of MS care on its own, as scientists compile the range of human, animal, and nonanimal data to explore different aspects of this complicated disease.

Poring through scientists' papers on MS and other diseases that rely, though never exclusively, on animal studies has led me to a most unsatis-fying conclusion: animal experiments sometimes work and sometimes fail. Moreover, scientists' reliance on heretofore useful animal studies is in constant flux as new nonanimal tests don't just allow us to stop the animal tests but also allow insights and knowledge that the animal tests were not capable of delivering. Picture animal labs as part of a house of cards or a

Jenga tower in which animal tests, patient exams and autopsies, cells and chemicals in flasks, epidemiology reviews, and human clinical trials combine to develop a medicine and score it as safe and effective enough for human use. Looking down from the tower, could we have removed all the foundational animal tests and still have built a successful outcome? Could we have selectively removed some? Which ones? Did we have nonanimal blocks available at the time that could have substituted? Looking forward, can we keep building the tower with new and improved nonanimal methods and never put another animal in a lab cage?

Take Ozempic, for example. It started with a suspicion in the 1980s that Gila monsters' venom might yield a better hormone for regulating blood sugar than other diabetes treatments. Now artificial-intelligence modeling might better envision a chemical scientists could synthesize that could perform Ozempic's incursions on diabetes and obesity and spare the lizards. Cultured hamster cells measured its activity without harming any animals. Diabetic mice complemented that work, but how truly necessary were they? The mice, as well as human volunteers, revealed the unexpected side effect of how Ozempic tells the brain it's not hungry, adding obesity treatment to its portfolio as a diabetes treatment. Hamster ovary cells in cultures could not have discovered this effect. Pregnant monkeys served in the final stages of safety testing. Thalidomide fifty years earlier showed the importance of safety-testing drugs that pregnant patients may need, but the conservation of endangered species, financial costs of running monkey experiments, and animal-welfare needs all converge to show the urgency of replacing such studies, not just with nonmonkey but with nonanimal alternatives. Those replacements are coming soon, but they're not here yet.

Laboratory animals deserve constant expert reevaluation of the relative value of animal and nonanimal tests in every field of biology. Rarely can single scientists perform such an evaluation, as their expertise lies in one of the many fields of animal and other research; their knowledge of related fields is limited, and their biases hard to completely avoid. What I want to see are up-to-date, multidisciplinary, in-depth analyses of the necessity of animal labs. I'd hoped for precisely this when the US Congress in 2021 directed the NIH to fund a National Academies of Science, Engineering, and Medicine panel to evaluate the use of lab primates in NIH's research labs. The NASEM panel expanded their remit to cover all NIH-funded primate research, whether inside the government labs or in universities and primate centers that the NIH supports; I was happy to see that. They put together a panel of leading scientists, many of them with monkeys in their own labs but others with expertise in nonanimal alternatives. They balanced public hearings and private meetings, then ran their report past a further set of experts, including, to their credit, the primate expert at the animal-protection organization the Humane Society of the United States.

Monkeys populate every page of this report. Look for baboons in organ-transplant labs. Japanese snow monkeys serve as obesity models, while Latin American owl monkeys serve in vaccine development. Marmosets are on the rise, as the most common monkey target of genetic-engineering experiments. A table lists twelve varieties of monkeys and their special uses in NIH-funded research, with most of them somehow figuring in either infectious disease or brain and nerve studies. Rhesus macaque monkeys predominate, reflecting their numbers in the NIH-funded primate-center breeding ranches throughout the United States. The report serves also as a primer on nonanimal tests. For example, they introduce brain organoids, tiny three-dimensional cell clusters grown from human stem cells, that mimic some of the complexity of a human brain. They herald artificial-intelligence computer simulations and hybrid cell-silicon "organs on a chip," in which human cells' activity in a culture system can be recorded.

After close to two years of work, the NASEM panel concluded that primates "have contributed to numerous human health advances . . . [and] continue to be vital to the nation's ability to public health emergencies." They championed new alternative in vitro and in silico methodologies but warned that, in the coming years, those nonanimal tests would increasingly complement live lab monkeys but could not yet fully replace them. Simply put: monkey labs are still necessary for continued medical progress.[6] I know and trust some of the panelists, and I trusted their conclusions. But longtime lab insiders, myself included, can be quick to read "rhesus macaques are used to investigate Alzheimer's" as saying that critical examination demonstrates that rhesus monkey experiments produce crucial, translatable data not obtainable via human or other nonanimal research. "Monkeys have contributed" or "monkeys are used" implies, but hardly proves, that monkeys have been necessary and irreplaceable in the past or, more crucially, that they are now. Even less do such general statements testify that a particular experiment a scientist submits to the ethics committee will serve necessity.

We need experts to analyze one another's animal experiments, but this well-funded national effort fell short in two important ways. Its goals were both too narrow and too broad. First, the NASEM effort was too narrow in scope, given that the NIH likely funds less than half the primate experiments in the United States (the panel itself admitted it had no way of knowing how much). The report should also have covered the use of tens of thousands of monkeys in private pharmaceutical research and development. The NASEM panel describes what scientists see as a monkey shortage, a confluence of decreased exports out of Asia, limited breeding in the NIH-funded and private primate-breeding centers in the United States, and growing demand for primates for brain research and infectious disease studies. In the face of public and private labs' competing demands, our nation could use an overall plan for prioritizing what labs should get primates for what purposes. Certainly, the monkeys do not know whether private or public funds are paying for their time in the labs. The White Coat Waste Project lodged complaints about monkey research at the NIH, and Congress responded by commissioning this narrowly tailored report. Congress followed the White Coat Waste's focus on framing animal-welfare issues in terms of wasted taxpayer money and in the process squandered the opportunity to usefully review the bigger picture of how our country uses monkeys in labs.

On the other hand, the NASEM remit was too broad to convince me of the necessity of using monkeys in any specific application. Yes, monkeys and humans share brain regions that nonprimates simply do not have, and scientists do use monkeys in brain research; that's hardly proof that studying these regions in primates yields necessary knowledge. As one example, the report covers treatments for parkinsonism, including deep-brain stimulation in which a neurosurgeon feeds a small electrode down into the brain and gives mild electric pulses that appear to improve human symptoms in a variety of neurological conditions. The report notes this treatment as a development that was "enabled" through primate research. What we need is not just the historical record of how scientists did study rats, monkeys, and human patients to develop this treatment but an evaluation of what might have happened without the animals in the mix. More important, now that better human MRIs, cellular methods, and the like are available, we need to know how truly necessary any work with animals will be going forward. This panel was not constituted to answer such detailed questions for the myriad diseases in which primates serve as lab models, but those are the details an ethics committee or the NIH-funding reviewers need to have to accept that animal tests are the best available tool for a specific disease under study. The answer will vary not just with the disease but with the specific aspect of the disease scientists are trying to understand, be it resistance to infection, ability of drugs to block cellular damage, fine-tuning the dose to try in people, or predicting safety in pediatric, pregnant, geriatric, and all other patients.

Animal activists complained that the panel was stacked with scientists, several of them NIH-funded primate researchers. The White Coat Waste group, whose activities led to the panel's existence, certainly had no seat at the table. I should not have been surprised then when a congressional subcommittee in 2025 invited the White Coat Waste Project as witnesses to a hearing it titled "Transgender Rats and Poisoned Puppies: Oversight of Taxpayer-Funded Animal Cruelty." Three beagles were seated in the chamber but no scientists, including no one from the congressionally mandated primate review that had so recently concluded and made its report to Congress and the public. No one offered evidence to support or refute witnesses' claims that scientists have produced no cancer cures or that any apparent successes from animal labs were just "dumb luck." Overshadowed by the heated rhetoric and claims of cruelty, waste, and scientists' twisted minds was the small spark of agreement with the primate panel's scientists, that scientists should use animals only when non-animal alternatives cannot meet the needs. And asked when that day might arrive, one witness told Congress it could be forty years or more. These two groups, the primate scientists and the committee's witnesses largely talked past each other, when what we need is political support to find nonanimal tests and a way to evaluate which experiments are truly productive and which need to be retired.[7] Unsurprisingly, the two efforts staked out opposite claims about federally funded primate centers. The primate panel called for expanding and supporting them, ramping up US-based rhesus monkey breeding to end trafficking in endangered monkeys from Asia. They see increasing needs for primates in a world awaiting the next pandemic. White Coat Waste would shut them all down, as wasteful, cruel, and unproductive of any useful knowledge.

As one modest example of the sorts of granular reviews I want to see, the NIH supports a multicenter consortium of experts investigating the human inflammatory response to injury. One of the group's first projects was to focus on molecular and genetic changes in mice cells as they respond to trauma and burns. If mouse and human cells react similarly, perhaps treatments that work in model mouse labs could translate to humans. In reviewing published studies, the panel found that, regardless of the injury, burn, or physical trauma, the molecular and genetic changes in people were similar among human patients, funneled through a common pathway of inflammation. That should be a tempting target for therapeutic interventions that could span a range of injuries. And mice? It was essentially a coin toss how well mice matched the various molecular reactions that the human patients shared and whether treatments developed in mice would likely be useful in humans for these particular injuries—an expensive and time-consuming coin toss, causing plenty of mouse illness in the labs. The consortium authors do not call for an end to all animal testing and even have some suggestions for running better mouse studies of inflammatory diseases. Mostly, however, they call for a move from mouse labs for this specialized area of critical-care medicine to studies in human cells and human patients.[8] Imagine such a rigorous, granular, evidence-based review in every specialty and subspecialty. And when and if such panels find that a particular animal use fails to deliver information that can translate to the human or veterinary clinic, and no obvious fixes are evident, scientists should step away from that model. Funders should stop investing in it, and ethics committees should be empowered to just say no.

When Banting and his colleagues were running their dog and human diabetes experiments, the world's collective scientific knowledge of diabetes was quite modest. Now we face an explosion of information, with the disadvantage that it's near-impossible to corral it all and the advantage that we, and AI, now can systematically analyze the data we have, sourcing numerous published studies in a field to separate the wheat from the chaff. In their primer on using systematic reviews (and their cousin, meta-analysis) to improve animal use, Benjamin Ineichen and his colleagues show the power of animal experiments that more closely model human clinical scenarios, such as studying stroke medicines in older animals with comorbidities such as obesity rather than the young, healthy mice of most stroke studies. In a hindsight look at why probiotics helped mice with pancreatitis but not humans, systematic analysis showed up the failures of the collective animal research, such as using many different probiotics and administering them before scientists induced the condition in mice instead of as a treatment during a pancreatic attack in people. Yes, no scientist should start an animal experiment without having scoured the literature for this big-picture view of the usefulness of animals for their project.[9]

I am heartened by efforts I see around the world to improve on animal modeling as we work toward replacing animals altogether. The United Kingdom often takes the lead, as it has been doing since England passed its 1876 Animals Act. The Royal Society for the Prevention of Cruelty to Animals has at times assembled scientists, working along with groups such as the Universities Federation for Animal Welfare and the National Centre for the Replacement, Refinement and Reduction of Animals in Research, to convene groups of specialized scientists, animal-welfare scientists, and animal advocates to focus on specific issues, such as use of animals in sepsis research, to produce guidelines for how to kill lab rodents when necessary or how to minimize pain in lab octopuses. Such efforts need scientists' expertise, and I believe they likewise need advocates' skeptical eyes and the credibility that can come with them.

Skeptics, whether animal activists and other outsiders or specialists and insiders in other fields, are ill-poised to do such reviews as so little of the necessary information is visible. University researchers publish their studies in publicly accessible journals. If a drug company picks up on a promising potential drug, they have no obligation to publish successes or failures. Drug companies keep their data in their own hands and are free to publish what they choose to publish. Genentech and Roche opted to publish the data on their successful phase 2 and phase 3 human clinical trials with ocrelizumab, the multiple sclerosis medicine. Neither mice nor marmosets appear in these reports, though the company does caution use in pregnant women in its "prescribing information" for physicians, based on its studies in pregnant monkeys. Thus a patient concerned about taking medicines tested in animals or, for that matter, concerned about taking medicines *not* tested in animals can sometimes get some glimpse of how animals in the final stages of drug development may have provided some safety data. These sources still would not tell that patient how irreplaceable the monkeys were for safety assessment, nor what role they or mice or other animals played early on in research and development.[10] Had the clinical trials failed, would Genentech have shared that info? Would animal-research skeptics have that data point to add to their argument that animal studies are largely futile? I do doubt it. More invisible than the marmosets or woodchucks whose lives may have led to current treatments are the animals scientists used to develop drugs that ultimately failed to secure approval for human use.

Before the Food and Drug Administration (FDA) approves human clinical trials for an investigational new drug, it reviews a company's preclinical data, compiled from the range of cell, chemical, and human experiments and, almost invariably, animal testing. Attrition, which is a drug's failure at any stage of the human clinical trials, is an approximate reflection of the value of the animal tests. The FDA provides estimates of attrition in the various phases of human clinical trials, with a final attrition rate of over 90 percent.[11] In nine cases out of ten, animal experiments, along with the human cells, computer simulations, and other methods, oversold the safety and usefulness of a novel medicine.

A 10 percent success rate still leaves open the question of whether a drug's success actually required all the animals it used. People will differ on whether a 10 percent success rate is too low to allow continued animal use. Some animal liberationists would decry any animal suffering, even with a 100 percent success rate. Desperate parents of a dying child might believe that an even tinier shot at a cure is worth a try. Between these extremes just about everyone can agree that 70 percent would be better than 7 percent. Drug companies certainly would prefer a greater return on their years of expensive research. Animal advocates too certainly want whatever animals we use to have died for something of value.

In the face of this disappointing failure rate, we as a society have options, starting with the opposite stances of shutting down the animal labs entirely or shrugging our shoulders and keeping on running experiments the way people continue buying lottery tickets despite far worse than a ten in a hundred chance of hitting a jackpot. We must find ways to weed out the experiments least likely to produce useful information about important questions, factoring not just time and money but also animal lives and suffering into our calculus. And if we keep some animal labs running, scientists can use animals more wisely for better science outcomes and work with animal-welfare specialists to lower the potential for animal suffering, as I show in coming chapters. The path forward requires figuring out why success rates are as low as they are. How do medicines that eventually fail slip through the myriad basic lab studies in vivo in animals and in vitro in cells, through the various human experiments and patient data analyses, through the computer modeling, and through the final animal tests for safety and efficacy to launch a medicine or vaccine down the long, expensive, dangerous three phases of human clinical trials?

Just as the USDA does not acquire (and therefore cannot disseminate) information on the various uses of the forty thousand dogs or sixty thousand monkeys in US labs, the FDA does not post breakdowns of the success and failure rates of drugs going through the three phases of clinical trials. Most deeper analyses of drug failures come from companies willing to pool their resources and share their data in pursuit of better "hit" rates in their studies. Note that most of this proprietary industry data is not accessible to outsiders such as animal activists or to scientists at rival companies. The success and failure rates are not equal across all fields, with cancer drugs, for example, more often failing on the road from animal studies to human drug approvals than new vaccines or infection treatments.

The proffered reasons for drug failures vary among analysts. In a widely cited industry-sponsored study in 2014, Michael Hay and his colleagues suggested several reasons, including changes in FDA's standards for drug safety. In probing how many clinical trials fail because of poor study design, they bemoan the challenge of working with proprietary industry data for their analyses. For cancer in particular, they muse that better animal experiments could improve drug success, singling out "xenograft" models in which scientists grow human tumor cells in immune-deficient mice. An important later study pointed to evidence that poor human patient selection could lead to cancer drug failures. We should, for example, expect poor performance of a hepatitis drug if it targets molecules specific to hep C, but we enroll patients with hep B, hep A, and assorted nonviral liver diseases in our clinical trial. Likewise, if we see diseases such as cancer or, even more specific, brain cancer, we risk missing the usefulness of a treatment, giving it a failing grade in clinical trials, for a very specific brain cancer by enrolling too broad a cohort of cancer patients.[12] If scientists misread their animal data to suggest that a new medicine could vanquish all cancers in all patients, that medicine will join the 90 percent of drugs that fail. If they better match their human trials to what the animal studies actually support, they may instead hit the jackpot. The causes are many, but somehow scientists need to catch drugs with low efficacy or low safety as early in the process as possible, stop the animal experiments earlier, and avoid launching human clinical trials of drugs the animal experiments should have weeded out.

Clinical-trial failures suggest that some of the preclinical animal studies yielded overly optimistic predictions that success in the animals would translate to success for human patients. Successes come when the animal and other preclinical experiments studies are correct enough to launch a successful drug. In both failures and successes, scientists use animals to explore how the disease progresses, what points in the process might be susceptible to interventions, what doses and routes of administration look promising, and what safety and side effects profiles a drug will have in people. With animals so much in this mix in various ways, some animal tests will be more useful and accurate than others. Some will be more amenable to replacement with in vitro or other alternatives that may give even better results, often faster and sometimes cheaper than the animals, and allow particular facets of the animal experiments to end.

I believe that animals still play a vital role in important medical research for human health. As a vet on an ethics committee though, I want more than the big-picture assurance that some animal tests are essential some of the time. I want to know if a particular experiment a scientist has brought forth for committee review is likely to be useful and important and that, in the three years since the committee last reviewed the project (in the United States, the law requires committee rereviews every three years), no one has produced a more reliable nonanimal model that should shut down the animal lab.

As a general rule, the statistician John Ioannidis at Stanford and Ray and Jean Greek, a physician and vet who've spent decades critically examining the claims of animal researchers, argue that the more basic the science, the further removed from direct application for a particular disease, the less likely it is to produce useful information. This applies to animal and other experiments as well, though the Greeks' focus on the implications for animal experiments suggests a higher bar for approving animal experiments for very basic non–goal-directed research.[13]

Some proposals for doing better science apply both to animal studies and to medical research more broadly. Marcus Munafo and his colleagues' "Manifesto for Reproducible Science" (2017) ranges from steering clear of corporate sponsorship, refraining from publishing one-off studies that may not be replicable even in the same lab, and encouraging collaborations rather than the current competitive model that rewards whoever publishes first. They call for public preregistration of experiments so analysts can see whether researchers succeed in their goals and for scientists to publish negative findings, not just their splashy positive results. They call on scientists to consistently report the fullest description of their methods, whether for animal studies or other research; if others do not get the full details of how a scientists ran their experiments, they cannot hope to replicate and build on promising findings. The manifesto promotes the London-based National Centre for the 3Rs' "Experimental Design Assistant," an interactive program to help scientists design better animal experiments, a tool every ethics committee and scientist should acquaint themselves with.[14]

With a focus primarily on animal labs, Joseph Garner, Hanno Würbel, and other animal-welfare researchers advise on how to choose clinically relevant animal experiments that can effectively translate from the animal labs to the human clinics. Among their suggestions they urge scientists to really know the disease and, preferably, the patient population, relying on patients to help set priorities for research projects and to help scientists better appreciate that patients who have spent their life managing their diabetes are more than just pancreases. In their eyes (and mine), animal experiments that produce clinically useful information may be essential. Those that cannot, are not.[15]

I've described how scientists strive to eliminate every potential source of variability in their study animals. They choose a single genetically modified mouse strain for their experiments, with mice all the same sex and age, housed in identical plastic boxes that filter out infections. Animals suffer untreated pain or live lives of boredom and frustration in the scientists' quest to stamp out variability and any "noise" in the system. Garner and others challenge the assumption that these sheltered animals living their sterile lives will best produce useful clinical knowledge, and they suggest ways to embrace variability in their animals that better reflect the varied lives of human or veterinary patients. If a discovery is robust enough to be useful in a clinical setting for a wide variety of patients, it should be robust enough to shine through animal experiments in which animals can live rich and varied lives.[16] Vets, ethics committee members, and the scientists themselves need to learn from these writers if they are going to produce results that both improve human health and reduce animal suffering, but this requires stepping out of their respective bubbles and their narrow focus for some cross-talk among people with varied perspectives and knowledge. In my experience these are not the conversations we have in the ethics committee meetings, though what a great opportunity for scientists to learn other perspectives, meet others' concerns, and together build a better mouse lab.

As a university vet, I've wished for better ways, a crystal ball even, for knowing an experiment would not just succeed but contribute meaningful data in the service of important research questions. Failing that, ethics committees should be part of the cycle of planning, conducting, analyzing, and publishing experiments and, for experiments that do not meet the scientists' stated goals, assessing whether failure calls for changing the experiment for the next round or stepping away from the animals. If a lab's experiments consistently fall short, at some point the scientists should move on.

Lacking a crystal ball, ethics committees, scientists, and outside observers should actively monitor what scientists are producing. Welfare laws in many countries call on ethics committees or animal-welfare officers to police that scientists are obeying the rules and treating their animals well, not judging what the scientists accomplish. Thirty years after the fact, I have no records or recollection of our ethics review of Cornell's woodchucks-and-alcohol experiments. I am confident, however, that we never knew that the work, by science's standards, vanished. They did not publish, they did not run follow-up studies, and certainly no one ever cited this work. Our ethics committee should have known this and known of the study flaws (too few animals to overcome the mismatch between experimental and control groups) and demanded better planning for future woodchuck experiments, whether alcohol would be served or not. At UCSF our committee knew only that the marmoset colony was migrating to a different campus, but I don't believe we knew why. Was it just for more years of doing the same studies, or was it, as I learned from the *New York Times* rather than from our animal researcher, that they were on track with focused studies for developing a successful treatment for MS symptoms?[17] Scientists certainly reevaluate their own failures and successes to plan for a more successful next round of experiments, but just as they alone cannot rule that their studies will reap continued funding, I want others, such as the ethics committee and the campus vets, to have a voice in whether the ongoing work still deserves the license to harm animals.

Outsiders—that is, all of us whose laws allow and whose taxes fund animal labs—also deserve some transparency and a chance to monitor how scientists are delivering what they promise. I say this while recognizing that threats of violence against scientists and their families are real (even if actual violence is rare), as are campaigns to destroy scientists' careers and work if activists identify them as animal abusers. In the United Kingdom, the Home Office and the EU Directorate-General for Environment collect and post public and private lay summaries of approved animal projects. In the United States, activists use Freedom of Information requests for ethics committee records and for grants that the NIH funds, giving them some window into scientists with animal-use approvals. At this point the European and UK data lack sufficient identification to allow anyone to follow up, the way I call on US ethics committees to do, to explore what publications, if any, have come three or five or ten years down the road and to take the next step, seeing what impact the work has had in terms of patents filed or other scientists citing the work. This information could serve locally, giving ethics committees data on whether to continue approving ongoing projects, and more globally, showing trends of the kinds of research projects that succeed in having an impact.

In the chapter "Rabbit," I highlight how scientists and regulators evaluate nonanimal models in safety and toxicology tests. Beyond our ethical concern for harming animals, nonanimal tests, including new technologies for studies with human volunteers, can be more precise, more predictive, and more efficient than animal tests. They are the post–lab animal future, and some of that future is already here. For example, bacteria produce toxins that can persist in medical supplies even after sterilization kills the bacteria. Rabbits were once the mainstay for testing every batch of IV fluids for these toxins in a crude test that inoculated the rabbits and monitored for fever, sickness, or death. Expensive, slow, and cruel, the rabbit pyrogen test was ripe for replacement when scientists discovered that they could collect the blue blood (hemolymph, technically) of horseshoe crabs for an in vitro endotoxin assay that was a good-enough replacement of the rabbits to win FDA approval. Though the horseshoe crabs return to the salt marshes after their session as involuntary blood donors, the procedure is not without stress to the animals and has low-enough survival rates to threaten wild populations. Now the horseshoe crab assay is heading to extinction, hopefully just in time to let up pressures on the crab populations, pending regulatory updates that will approve or, eventually, require a version of the crucial horseshoe crab protein grown in cultured insect cells. Rabbits or horseshoe crabs were once necessary for patients to reliably survive vaccines and other medicines, but they are no longer.

Some new technologies hasten the post–lab animal future, but we are not there yet. Sometimes new technologies and new information will actually lead to more animal experiments. Cornell built a shining new Biotechnology Building in the late 1980s, and, believing that animal research looked just about obsolete in modern medicine, they reluctantly included three tiny animal rooms. As molecular biologists in some high-tech fields were moving away from vertebrate animals, others were developing the techniques to add or remove genes from mice, and suddenly the apparent steady decline of animals in labs reversed, with transgenic genetically engineered mice leading the way. A single new professor on campus quickly filled those three tiny animal rooms with his colonies of genetically modified mice—the first serious foray into that technology on my campus—and soon we were filling animal rooms in other campus buildings with these new creatures. Mouse numbers in labs went through the roof throughout the 1990s as labs scrambled to build new vivaria to house them all. Around the world uncounted millions of mice with designer genes fill laboratories to this day. Without shutting down the mouse labs, researchers proceeded to find new uses for genetically engineered zebra fishes, with us campus vets having to learn how to build and maintain large fish facilities alongside our mouse houses. Yet another advance in gene jockeying, called CRISPR, is bringing us genetically modified monkeys, pigs, ferrets, and other larger animals. These new genetic technologies hold promise for better developing nonanimal lab tests, but somehow keep creating new jobs for the animals at the same time, which means yet more ways to program animals to develop more and more illnesses in the lab.

Findings in humans, even in approved drugs with wide clinical use, provoke new animal experiments. The tranquilizer thalidomide, for example, is back on the market after its disastrous effects on human fetuses and babies. Seeking to understand how it caused the limb deformities that it did, scientists discovered, through further animal testing, that the drug shuts down the growth of blood vessels so necessary for fetal development. Through subsequent years of animal and in vitro experiments, researchers developed this devastating side effect into a successful treatment (though not a full cure) of a cancer, multiple myeloma, that depends on a robust blood flow. Choke off the tumor's blood supply with a drug like thalidomide, and the patient's prospects can improve dramatically.

Newspapers' science reports often gloss over whether scientists relied on animals in covering the latest and greatest discoveries. I always try to discern what I can about the animals behind the curtain. I imagine most people know that we should exercise if we want to stay healthy. Epidemiologists crunch human-patient data to test how exercise affects health outcomes in various diseases. Sports physiologists enlist human volunteers to tackle the question, "Do I really need to get off the couch that much to keep a healthy heart?" With my personal history of knee pain and my interest in laboratory animals, I hit on an article reporting that walking exercise can be helpful prior to a patient's onset of arthritis. I'd have signed up for such a study, even without knowing in advance whether I would be developing arthritis, but I would not have been eligible to participate. I'm the wrong species; the scientists instead used rats. They describe how they exercised them for sixty minutes on a treadmill, neither specifying nor denying if they used small electric shocks to keep them moving, a common practice in animal experiments of moderate or intense exercise (the kind of detail a lab vet knows to look for). They do describe the chemical they injected into the rats' knees to cause arthritis, but if the rats received analgesics for the electric shocks (which would likely make the shocks less likely to inspire continued running) or for their arthritis, the article is silent. Reading between some lines, I see the pain of arthritis and of electric shocks, so I ask, "What necessary information did this experiment give us that we could not already learn from human studies?" The answer, as with so many animal experiments probing what we already know from humans, is a molecular or mechanistic explanation of how exercise promotes knee health. Given the pain such studies cause, the burden should be on researchers to explain to a broadly representative ethics review why this molecular information is important.

Animal labs will one day be obsolete, as scientists learn new ways to gather data or, perhaps, if we as a society decide to forgo scientific advances that require harming animals. Animal-research skeptics have asked what might have happened had scientists never used animals? Might we still have medical science, but with some different focus and different outcomes? Would we know more about ways to prevent disease? Would we have a health system focused less on cures and more on maintaining health? Or would scientists have persisted longer with testing insulin in fourteen-year-olds, infecting children with hepatitis, forcing prison inmates to be guinea pigs? We cannot test these counterfactual alternative histories; we can only move forward.

AI and computer modeling cannot yet carry us to the bright day when we liberate all the lab animals. Human cells in a flask, micro organs on a chip, and patient-derived tumor cells are not yet capable of answering all the questions scientists must answer for us to take the current 10 percent success rate of drugs in development and push it toward 100 percent. They alone cannot yet keep us moving toward cures for diseases whose causes (multiple sclerosis) or treatments (brain cancers) continue to elude us. We are not going to shut down the labs and liberate the animals any time soon. NIH will continue funding animal experiments, though it may shift some funding priorities more toward preventive and community health than high-tech innovations. The FDA and EPA still require animal safety tests, but, as I show in "Rabbit," are hard at work to phase those out. Pharmaceutical companies will still use animals when they think they are essential in their research and development projects and for safety testing if they believe regulators (and protection from consumer lawsuits) require it.

Clearly, we need much more rigorous and detailed examinations of the utility of animal testing in each and every specialty and subspecialty of medical research. Harming animals requires justification, so animal experiments can never be justified if we aren't confident they will be useful. In coming chapters I'll propose ways to move beyond platitudes to practices that really might filter out unnecessary animal experiments and what to do about the animals those experiments will conscript.

For now, though, I want to move on to what we should do about the animals currently living in labs. How should we decide what experiments to allow, if any? On what species of animals? With what kinds of protections for their lives and welfare? In the next chapter, I call on dogs to guide that exploration. Though well below 1 percent of the animals in modern labs, dogs, in our complicated relationships with them, have driven much of the debate around how we should care for animals in our lives.

Dog

Canis lupus familiaris

THE POSTER PUPS OF ANIMAL-RESEARCH BATTLES

In my vet training, dogs were our Everyanimal. We learned canine anatomy and medicine and surgery, then learned the variations on the canine theme that make a horse a horse or a cat a cat. The vet school bought beagles from a specialized lab breeder for practice labs before unleashing us in the clinic with clients' pets. Each dog underwent two surgeries. After the first, a "spay" in which we removed her uterus and ovaries from her abdomen, we students had responsibility for all the aftercare, monitoring, and antibiotics. The following week we performed a more extensive surgery—an orthopedic procedure of some sort—at the end of which we killed our anesthetized dogs with an overdose. Though dogs were our default Everyanimal, horse and sheep surgery differ in important ways, so we also learned our skills with practice on sheep and ponies from assorted auction sales and Cornell's own farms. As with our dogs, we killed them at the end of our semester.

My peer cohort of students used dogs from a breeder rather than shelter animals that might have been abandoned or lost pets or hard-luck strays. Earlier students operated on a single dog week after week after week, but by the mid-1980s federal law shifted the burden from individual animals, who would have endured postsurgical pain multiple times in a semester, setting a limit of one survival surgery per animal. We used more dogs, but with less pain per dog than in earlier years. Students a few years behind me had to develop basic skills like using a scalpel and suturing on nonanimal mockups before they were allowed to try their hand with a living animal. Vet professors improved the pain treatments lab beagles got until, finally, pushed by some professors and students alike, most vet colleges terminated dog-surgery classes entirely, though most keep some training in livestock surgery. A vet student will now typically perform their first-ever dog surgery by neutering a shelter animal to get them ready for adoption.

In laboratory animal advocacy well beyond issues in vet training, dogs again and again stand as our Everyanimal. Though other species occasionally get the spotlight, mostly, we talk about dogs when we talk about our obligations to animals in labs. I will show you how, sometimes, with a wag of their tail, dogs get special treatment that pigs or mice should envy. Ending dog labs in vet schools did not end the use of hoof stock or mice for training students. At other times dogs inspire welfare reforms that carry other animals, cats in particular, along on their coattails. I'll also show how dogs can be our guides to understanding other animals. Living among dogs inspires us to try to see the world through a different lens, a perspective that can promote better care for multiple species.

Dogs feel pain, and they often let you know it, so our efforts to understand and treat their dog pain give us a map to extending that commitment to other equally sensitive species, even if we are less attuned to what they might be telling us. In Victorian England activists' concern for dogs' pain drove their successful push for laws to limit animal experiments and to require anesthetics in labs. In the United States, fears of pet theft and pity for dogs in pounds resulted in the US Laboratory Animal Welfare Act in 1966, with its strong emphasis on dog trafficking and its perplexing weakness on setting standards for what scientists actually do to the dogs, painful or not, once they're in the labs. In vet schools in the 1980s, battles over forcing students to kill dogs so they could learn to heal them barely ever touched on the other animals—farm species especially—whom students might kill for their training. In campaigns in the 2020s, though dogs are now far outnumbered by other animals in labs, activists are successfully pushing states to adopt requirements to retire dogs, requirements that include cats as well but none of the far more numerous other creatures. Commercial beagle farms with thousands of dogs arose as the inevitable response to laws banning pound dogs from crossing from shelters to labs, but the farms now find themselves in animal activists' crosshairs in a way that mouse vendors churning out millions of mice so far mostly avoid.

We bring what we know, and what we think we know, about dogs to our ethics and policy discussions. Dogs voice their pain, so we mandate anesthesia and a limit to painful dog experiments. Dogs love to run and play, so we argue over how much play time and exercise our Animal Welfare Act should require. We see dogs as abandoned strays in a heartless dogcatcher's wagon, caged in filthy pounds; thus over the decades we fight about whether pound dogs are the ideal lab subjects, plentiful and with no other prospects. Or should they be spared that final assault, even at the expense of creating new animal lives who will die in laboratories? Even dogs who've been kenneled all their lives as research subjects quickly learn civilized toilet habits and love a comfortable bed, so welfare advocates push for giving them a good home life when their time in the lab has ended. Dogs have worked at our side as loyal partners, in fields and in war; perhaps their greatest role could be teaming up with scientists to conquer a panoply of human diseases—unless that's a step too far, a betrayal of the pact of the millennia-old symbiosis between our species. So this we know: dogs are sensitive, playful, energetic, trainable, faithful to their people, and loyal to their home. Activists smartly take that knowledge into skirmishes over what should and should not happen in research laboratories.

Some of these doggy ideas fit well with other animals, and some less so. Maybe cats are less eager to serve, mice less in need of a good run, monkeys less suitable for a postlab life in a loving family's home. Our efforts to understand and treat dog pain give us a map to extending that commitment to other equally sensitive species, even if we are less attuned to what they might be telling us. And so some welfare advances targeting dogs accrue to other animals, while some remain for dogs only. Meanwhile cats, true to their mythical craftiness, almost always hitch a ride in whatever welfare protections—pet-theft protections, limits on the use of shelter animals, programs for post-lab adoption—that lab dogs win. Cats get no special protections of their own, and they have rarely featured in animal-welfare crusades.

My apologies to them and their fans that I do not even give them their own chapter here.

Dogs have occasionally inspired increased protections for other animals, but often they leave others as sacrificial animals to carry the burdens society wants to spare dogs. Pigs may be as smart and sensitive as dogs, but in many fields, cardiology especially, they have increasingly replaced dogs. Lab pigs often cost much less than a purpose-bred dog from a specialized breeder, but the main driver I have seen is a desire to avoid the public relations concerns and personal anguish that encourage scientists to switch species.[1] People, many scientists included, love dogs. They eat pigs.

Dogs have never been the only animals to find themselves in scientists' hands, but most such animals never spurred animal-protection campaigns. In 1600s England, for example, Robert Boyle showed that he could suck the air out of a sealed glass jar, knocking out or killing a range of insects, snails, mice, larks, and ultimately kittens, showing there was something in the air (turned out that it's oxygen) that animals need. William Harvey demonstrated that the heart, not the lungs as was the prevailing theory of the day, circulated the blood through the body. Most scientists of the day would have used dogs, but Harvey was not just a scientist but also the royal physician to King Charles I and apparently had an endless supply of deer from the king's royal hunting grounds, not a common reason for scientists choosing a model animal.

For a scientist, a docile, trusting dog was often just the right tool for the job. In the lab pigs get too big and too rambunctious too fast, while mice and rats are too small for many experiments. Sheep are skittish, while cats are ten claws and four fangs too feisty, the better to keep inexperienced scientists and their students at bay. Dogs weigh in at just the right size for experiments and demonstrations. Nineteenth-century scientists often conducted their experiments as demonstrations for one another to witness—a forerunner of the modern practice of posting one's data, methods, and statistics online to allow other scientists to see for themselves—and dogs combined availability, size, and relative ease of handling for public shows of lung function or blood transfusions.

For almost as long as scientists have been studying animals in labs, activists have rallied outside those labs in opposition, inspired at times by the cries of dogs coming from the labs and at other times by the suspicious secrecy scientists maintain. In Victorian England their efforts found sympathetic ears. Dogs were center stage, though never entirely alone as animals of concern; they also briefly shared a spotlight with French horses. The veterinary college in Alfort, just on the edge of Paris, made extensive use of horses in training veterinary students, including painful procedures performed without anesthesia. In the 1860s, in France and England both, people called to put a stop to the horse classes. British scientists, some of them dog users themselves, joined with animal activists to protest the use of live, unanesthetized horses for training French veterinary students. The French professors at the vet school appear to have continued doing as they pleased, but in London the movement to block French-style animal experimentation was growing.

A decade later, in 1874, the British Medical Association invited Valentin Magnan, a French psychiatrist, to discuss, with a live animal demonstration, his studies of how alcohol and absinthe affect the body. He strapped unanesthetized dogs to the table to administer either absinthe or alcohol. The dogs began howling and struggling from the first administration, while some of the doctors and scientists present shouted their protests. One dog suffered seizures and died from the absinthe. Thomas Joliffe Tuffnell, the president of the Irish Royal College of Surgeons, rushed forward to try to liberate the dogs. Police were summoned. A few months later, the Royal Society for the Prevention of Cruelty to Animals pursued charges as a violation of the 1849 Cruelty to Animals Act, but Magnan was well back in Paris by then, and the legal case went nowhere. Still, the Royal Society used this case as the evidence it needed that the voluntary code of conduct that the British Association for the Advancement of Science had put forth a few years earlier, calling for anesthetics for animal experiments, was ineffective.[2]

Queen Victoria quite fancied dogs and often sat with one for her portraits. She was all for abolishing vivisection in England, and dogs were her big concern. Her Majesty bestowed her royal assent on Parliament's passage of the Cruelty to Animals Act in 1876 to regulate animal use in science, the first such law in the world, establishing some norms and practices that US animal protectionists worked through the early twentieth century to import. Charles Darwin had a hand in crafting the version of the bill that eventually became law. He was keen to continue scientists' freedom to use animals but adamant that animal-welfare standards—the use of anesthetics for painful procedures in particular—must rule. His best-known statement on his ambivalence about animal experiments features, of course, a dog: "Everyone has heard of the dog suffering under vivisection who licked the hand of the operator; this man, unless he had a heart of stone, must have felt remorse to the last hour of his life" (or at least should have, but I know that most of us who remain in animal labs seek ways to keep any such lifelong remorse at bay). As Hal Herzog has noted, Darwin added a caveat in the next edition of his book: "unless the operation was fully justified by an increase in knowledge."[3]

The 1876 British law mandated that painful animal experiments include anesthesia, unless anesthesia would somehow interfere with the experiment. The law largely ruled out using live animals, even under anesthesia, for skills development such as vet-student training or for demonstrations such as Monsieur Magnan's absinthe lecture. An early draft prohibited any experiments on dogs, allowing cats and horses the special protection it gave to dogs. The final version sets limits on what scientists can do to any animal, but, as we will see repeated in later laws, gave special status to dogs, along with cats, horses, and mules, requiring that scientists use other species in their labs whenever they can.

Despite the new law, scientists continued using dogs in their labs and classrooms. The 1876 act set practices for government enforcement, but, then as now, activists claimed enforcement was toothless and so took a direct hand on their own, finding their way into animal labs and publicizing evidence of abuses that they claimed to have found. To be sure, they put the spotlight not on mice or rabbits but on dogs. In 1903 activists gained access to William Bayliss's medical school classes at the University College London and claimed he operated on dogs without the legally required chloroform anesthesia. Bayliss successfully sued the National Anti-vivisection Society for libel, claiming he had indeed anesthetized his dogs. To memorialize their efforts on behalf of Bayliss's dogs, animal activists erected a statue of a lab dog in Battersea. In the face of frequent vandalism, the dog statue acquired police protection. Controversy reached the point of violence—the Brown Dog Riots—with medical students in favor of having dogs for their classes carrying effigies of the brown dog and confronting labor unions, suffragettes, and even police. After about five years of controversy, the Battersea City Council decided that the brown dog statue must be destroyed, sending it to a blacksmith's forge.[4] Today, the statue's descendent, cast in bronze in 1985, lives on in Battersea Park, with a plaque repeating what the original dog's plinth asked: "Men and women of England, how long shall these things be?" I cannot envision any other animal who could have stirred people to such vehement actions.

In the United States, animal activists pushed unsuccessfully through-out the early twentieth century for legislation modeled on the British act to regulate what scientists did in their animal labs. In the face of their failure, they fought county by county and state by state against what they called "pound seizure," or what researchers called "pound release." Despite various arguments pro and con, both camps knew that preventing animal shelters from transferring or selling animals to labs would choke off the supply of dogs, putting animal experiments out of the financial reach of many scientists. Whereas many states initially passed laws requiring shel-ters to give up unadopted animals they would otherwise be euthanizing to labs, many later reversed this policy and expressly forbade the practice. In New York State in my student days, local pounds could not legally send dogs to labs, but we were allowed to buy former pound dogs and other "random-source" dogs from out of state. My then boss pushed Cornell to say no to such random-source dogs, though the switch to lab-reared bea-gles drove up the cost of my surgery class considerably. That move let us tell anyone who called looking for their lost pet that we could not possibly have the animal on our campus. Down the road our local SPCA was still euthanizing dogs for lack of enough adoptive homes, bringing their bodies to the Cornell incinerator, where they were cremated alongside our pur-pose-bred lab beagles.

The crusade to end pound seizures succeeded in moving many states to put shelter animals off-limits to scientists, but it fell far short of advancing reforms within labs the way the 1876 British law did. The US bans were about *which* dogs would go to labs, but they could do nothing to ensure that dogs or any other lab animals receive anesthesia or that scientists use animals solely for scientific discovery, not for vet or medical student sur-gery training. Shelters should be "havens of mercy" for the hard-luck dogs they take in, and pound-seizure bans may have reduced the numbers of shelter dogs heading to labs (though not the numbers being "put down" in dog pounds). By driving up costs, pound-seizure bans may in fact have reduced scientists' use of dogs in general. They were a boon for the commercial kennels breeding dogs, mostly beagles, for science, some of which house and produce thousands or tens of thousands of dogs for labs for our annual use (from US Department of Agriculture, or USDA, reports, more than forty thousand in 2021).[5]

The pound issue is mostly dead at this point. For most experiments scientists prefer the standardized beagle, with less of a mix of breeds and ages and uncontrolled variability than a shipment of random-source dogs can deliver. In 2009 the National Institutes of Health (NIH) commissioned an expert panel to look at the necessity of so-called random-source dogs and cats. This was yet another government review that focused solely on NIH-funded basic research, as if dogs, random source or purpose bred, would know or care who's footing the bill for their stint in the labs. They concluded that for experiments that truly required older, larger breed dogs, NIH might be able to support their production in licensed kennels, making pound dogs no longer necessary (at least in NIH-funded science).[6]

Cats have never gotten much coverage in lab animal campaigns.[7] Certainly their traditional use in neurology and brain research should have caused activists concern. They took a similarly troubled route as dogs did from streets and shelters to labs. As shelter dogs maintained center stage in campaigns to ban pound seizure and as activists and legislators succeeded in imposing limits, cats came along with the dogs. Pound-seizure bans, however, did nothing for the rabbits, rats, or monkeys who never detoured through shelters en route to science vivaria.

As the pound-release debates dragged on across the country, scientists and lab vets fought against national regulations either for sourcing dogs or for using them in labs. Animal activists had their portrayals of the doggy victims of science, so science defenders tried to reshape and reclaim the lab canine's public image. Aiming for public support to bolster their lobbying efforts, the Chicago-based National Society for Medical Research launched its Research Dog Heroes Awards. Newspapers printed the canine winners' pictures in newspapers as research defenders were testifying in state houses and in Washington that these valuable animals received excellent care in labs, with no benefit to come from expensive proposals for government oversight. Half a world away, Laika, a Russian street dog, was the ultimate canine science hero, the first cosmonaut to orbit the earth, her pictures in Soviet and global newspapers. She personified the Soviet head start in space exploration. But she got no hero's welcome; they sent her into space in *Sputnik 2* with enough food to last a couple of weeks at most and no plans (for the scientists) or hope (for Laika) that she return to earth alive. The Soviets reported that she died within the first four days when her spaceship fatally overheated. Months later Laika's long-dead corpse went up in flames as her spaceship reentered Earth's atmosphere.

Dogs, as portrayed by US and Soviet scientists, weren't hapless victims at all. Far from it. Just as heroic and loyal dogs have teamed up with people for farming and hunting and battle, lab dogs were our willing teammates in the quest to stamp out disease. This role as standard-bearer for the heroic contributions of lab animals throughout our nation really could only have been played by dogs. Guinea pigs as heroes? Cats as willing participants in our labs? Though science lobbyists succeeded for years in staving off overly restrictive regulations, I have never seen evidence that the Research Dog Heroes effort contributed much.

Animal activists and science defenders had tussled for decades. Knowing that regulation was inevitable, lab vets and some scientists preemptively launched their own code of conduct, the *Guide for Laboratory Animal Facilities and Care*, in 1963, with a voluntary accreditation program for lab vivaria (now called the Association for Assessment and Accreditation of Laboratory Animal Care International, founded in 1965), aiming to bring all labs up to standards that lab vets could endorse while avoiding government-imposed rules they would find onerous or pointlessly bureaucratic.[8] This was a complicated maneuver on the part of vets at a few elite research universities. Their goals included staving off the regulations that legislators without their expertise would impose, setting standards for vets and scientists at lower-caliber institutions, and protecting veterinary authority in labs from scientists and administrators who might not see the value of a vet-run animal vivarium. Despite these steps toward self-regulation, ninety years after the British Animals Act of 1876, the US Congress passed its first nationwide law, the Laboratory Animal Welfare Act of 1966. As always, dogs were the stars of the show.

Animal activists had been pushing for decades for federal regulations of what scientists could do with animals in their labs. Ironically, the canine stories that launched the 1966 law actually blunted such a forceful law. In 1965 Pepper, the dalmatian, caught the public's attention in the pages of *Sports Illustrated* magazine, not a common venue for stories about politics or animals but widely read. Pepper had disappeared—lost? stolen?—from a farm in Pennsylvania. Her family, the Lackavages, scoured the countryside and visited the livestock and dog auctions that the Humane Society of the United States was investigating. The Humane Society was an offshoot of the more conservative American Humane Association, formed because of an internal schism over how aggressively to work to ban pound release. The Lackavages read in the papers that a dog dealer's confiscated truckload of animals—including two dalmatians—had been impounded while the dealer went in search of a better truck that would meet state anticruelty laws. The Lackavages raced to the shelter. The dogs were back in the dealer's hands already, but the family identified Pepper from pictures the shelter staff had taken during the dogs' brief stay. Could the Lackavages visit the dealer's farm? Not without a warrant. And a photo of a dog's face—even with a dalmatian's unique array of spots—was not solid enough evidence for a warrant.

The dog dealer was located in representative Joseph Resnick's district, and Resnick tried to get the Lackavages access to the farm. No success. Pepper's story, "The Lost Pets That Stray to the Labs," ends in a laboratory in Montefiore Hospital in New York City, where a dalmatian—presumably Pepper—dies in a cardiology experiment.[9] But even before the article appeared in *Sports Illustrated*, Pepper's story had moved Resnick and his colleagues to introduce legislation making dognapping for labs a federal offense. I've never understood why *Sports Illustrated* did not include the crucial picture that identified Pepper; what a powerful image that would have been just a few years after the Disney film *101 Dalmatians* had made the breed so popular. But once her story hit the magazine stands, even without a picture, Pepper inspired a flood of letters to politicians across the country.

Months later *LIFE* magazine followed with its own dog story, with a teaser on the cover heralding, "Your Dog Is in Cruel Danger." In the article "Concentration Camps for Dogs," undercover investigators with the Humane Society brought reporters and photographers along with the police for a raid on a Maryland dog dealer's premises. Whereas Pepper had a name but no public face, *LIFE*'s specialty was photojournalism, and the graphic full-page photos of unnamed, emaciated dogs were deeply disturbing.[10] Plenty of people assumed that life and death in a lab was a horrific fate for dogs. They might trust that dogs suffer in labs for important medical breakthroughs. But there's a difference between "dogs" and "my dog," and these two articles made clear that *your* dog, just like Pepper Lackavage, could be one of the lost pets in cruel danger.

The specter of man's best friend stolen for labs unleashed a torrent of constituent letters, and Congress responded. President Lyndon Johnson signed the Animal Welfare Act into law in 1966. Antivivisectionists got less than they wanted though. Driven by fears of pet theft, the overwhelming focus was on how to regulate the sale and transport of dogs and cats headed to labs, for once again, though neither *LIFE* nor *Sports Illustrated* saw a need to spotlight them, cats trotted along with dogs in the new law. A handful of other animals also found some protections in the 1966 Animal Welfare Act, though not from concerns over illegal trafficking. Congress named primates (collectively, ranging from lemurs to chimpanzees), guinea pigs, rabbits, and hamsters as "animals" under the USDA's oversight.

Legislators were emphatic that the 1966 law, and the USDA enforcing it, would keep hands off making any rules on how scientists ran their experiments. The act instead focused on how to identify dogs to verify we weren't trafficking in illegal or stolen animals, right down to the size and shape of identification tags for dogs. Once animals were legally in the vivarium, the act listed minimum cage sizes for all the animals it covered and the obligation to provide clean food and water and adequate veterinary care. I watched my USDA inspector once, on the floor, looking for dust bunnies under the rabbit cages. Clean floors were within her jurisdiction in a way that real bunnies—and dogs, cats, hamsters, and monkeys—seemed not to be. The law covered cleanliness and purchase records and little else; even a requirement to use pain medicines would have to wait until the act's 1970 amendment. Serious oversight of scientists' experiments would one day come, but not for another twenty-five years.

Knowing the USDA's intense scrutiny of dog records, where we got them as well as where they went, I was concerned one day to find that one of our Cornell labs seemed to be euthanizing a lot of healthy beagle puppies. Turned out the kennel manager was sneaking the pups out the back door to his friends and falsifying his records, believing that killing excess animals from the breeding colony was permissible but that giving them away would land him in trouble. "Hold that thought," I told him, as we got buy-in from the administrators to start an above-the-board adoption program. We began finding homes for more of our dogs, as well as cats and the occasional rabbit, horse, or goat. I've been a proponent of lab animal adoptions and retirements since. Some institutions have resisted that. It costs time and effort to run a responsible adoption program, and universities might be legally liable if adopted animals bite a child in their new home. Public relations loom large, as many companies and colleges want to remain under the radar about their animal use, and, yes, it looks bad when someone says they rescued one of our dogs and found them afraid to go up a flight of stairs or walk on grass and slow to learn to keep from soiling the house. Or they take home a lab cat who avoids the litter box in the new home or spends weeks hiding under a couch till they decide to trust their rescuers. We insiders bristle at that word, *rescue*, when we're talking about a dog we willingly gave you, but I yield the point. With a new adoptee's timidness at the novelty of their new world, it certainly looks like we abused the animals in our vivarium, when we could easily have taught them some of the basics of postlab living. Teach a dog to walk calmly on a leash, brave the stairs or the lawn, and, most important, hold their urine and poop until they are out of the house, and prospective owners will swoon over their perfect pooch and thank the lab that found the dog a perfect home, without jumping to Beagle Freedom Project's language of liberating dogs from abuse.

Friends and I wrote up our adoption programs on our Cornell and UCSF campuses, as have other animal-colony managers, encouraging labs that they too can give their animals a chance for a nice retirement after their tenure in a lab.[11] The animal rights group, Beagle Freedom Project, formed in 2004 as a network of volunteers caring for postresearch animals, dogs, primarily, but also cats and other species. It has been successfully promoting legislation in many states requiring labs to adopt out healthy dogs and cats after their tour of service. Beagles are the most numerous dogs in labs, but dogs are still less than 1 percent of the mammals in labs. So why a Beagle Freedom Project and not a Lab Animal

Freedom Project or a Beagles-and-Tabbies Project? And why do the dozen states that have passed the Beagle Freedom Project's proposed legislation mandate adoption for dogs and their tagalong, cats, but no other species?

The Beagle Freedom Project links its direct actions of finding homes for former lab animals (as well as dogs from the meat trade, especially in China) with a chance to educate website visitors to what it describes as cruelty and abuse. The National Animal Interest Alliance is a nonprofit organization formed to promote animal welfare while defending pet ownership and animal use in agriculture and research against the animal rights movement. It too has its own Homes for Animal Heroes adoption program to find postlab homes for dogs, though with the opposite political messaging about what animals in labs experience preretirement and certainly no mention of *rescue*.[12] The words *abuse, cruelty*, and *liberation* are also not to be found. As always, dogs are a sympathetic face for lab animal issues, and cats, though with a lower profile in these campaigns, can certainly pass for sympathetic too. But mice, rats, and even rabbits would be a harder sell in an adoption campaign, and whatever resistance the research community has mounted on the dog-adoption bills would be ramped up exponentially if we were suddenly obligated to find homes for the thousands of mice and rats and rabbits in our labs.

In 1968 the USDA updated veterinarians on how the then-new Animal Welfare Act was coming along. Signs of success included that several breeders had closed up shop, that no allegations of pet theft panned out as actual dognapping, and that the USDA was keeping well out of the scientists' way.[13] The USDA began collecting statistics from the labs it regulated, beginning in 1971. Dogs in labs at the time numbered around two hundred thousand. By 1995 they had dropped to around sixty thousand, and their numbers remained doggedly steady for years, lately dipping down to the mid–forty thousands. I'm not convinced anyone has figured out the reason for the decrease and how much the Animal Welfare Act had to do with that. Certainly, mice seem to have taken up much of the work dogs were doing, and likely pigs too, though our US reporting system is way too spotty for us to ever know. In my years doing this work, I've also seen how we've changed what we report. Until the USDA told us to stop, for instance, we might report the same dog three times in one year, such as breeder Labradors in our hip dysplasia colony who were simultaneously used for vet school teaching and might be on two or more long-running research projects. No matter the cause, dog numbers have dropped through the years, especially by comparison with the teeming and uncounted multitudes of mice and fish. Despite dogs' lower numbers, activists and regulators continue to shine their brightest light on lab dogs, sometimes bringing other animals along but sometimes promoting dogs at the expense of others.

Monkeys later took the spotlight, prompting Congress to revise the Animal Welfare Act in 1985, but dogs got some individual attention then too. Senator Bob Dole added a last-minute provision in the act requiring exercise for lab dogs, a long-running bone of contention in animal-welfare policies dating back to the 1950s. In public comments, as the USDA was rewriting its regulations, animal activists called for large exercise pens or time with personal dog walkers, suggestions that research institutions and their vets saw as unnecessary and expensive. Activists called for exercise requirements for other animals, but the USDA held firm: the amended law called for exercise for dogs alone, not monkeys or rabbits or woodchucks or cats. The final ruling gives dogs the same small cage sizes that USDA mandated back in 1967—a couple of inches longer than the dog (not counting the tail) on each side—with just double that small enclosure sufficient for the exercise mandate.

As for other species—rabbits may want to hop more than their legal cage allotments allow, monkeys to climb, or woodchucks to dig—the law is clear that exercise is for the dogs. Animal ethicists call this differential treatment *speciesism*, giving preferential treatment to some animals based more on people's emotional attachments than to real differences in how much different animals would want the chance to run or hop or climb or dig. But even within a species, pound-seizure debates and vet schools' selection of animals for surgery training reveal how different categories of dogs may get greater or lesser regard, more for accidents of where they were born than for their particular needs.

Institutionalized dogs and cats in breeders' kennels and labs occupy a second tier of concern. Thus vets once practiced first on random-source dogs who'd come to their teaching labs via auctions or pounds and then moved on to commercially reared dogs. Students and professors then worked together to stop buying these second-class dogs for student training, offering spay and neuter services to animal shelters at the price of those animals being a student's first attempt at performing surgery before they are unleashed to care for client's pets. Thus shelter dogs themselves remain a rung below pets who've already found a home. As in human health care, veterinary trainees must learn and gain experience some- where, and vet schools' public clinics may provide that opportunity, at the price of inconvenience, wait times, or simply care in the hands of a less experienced doctor. So too shelter dogs experience longer surgery times and larger incisions, even with the most diligent and well-supervised stu- dents, than a seasoned vet would perform, which means greater risk for postsurgical pain.

Our society certainly stratifies human patients, with the wealthy get- ting access to top-tier care. I spent hours in a student-run optometry clinic getting my first pair of eyeglasses as a boy. I had no surgery scars to nurse, just a long wait, and a perfectly fine and cheap pair of glasses. So too with canine patients: in the research labs, scientists will induce sickness and injury in dogs for the benefit of human health as well as research to benefit the dogs we really care about, the pets who need the best vaccines and antibiotics and medical care. This separation of dogs into various tiers of concern starts with where we get the dogs in the first place.

Pound-seizure activists successfully pressed for restrictions on animal shelters transferring—or being required to transfer—dogs from the shel- ter to a teaching or research lab. According to the Humane Society of the United States, as of 2025 two-thirds of US states still allow pounds to send their dogs to research (or allow municipalities to make a determina- tion) and just under one-third prohibit this; Oklahoma stands alone in requiring shelters to turn over their animals.[14] In pushing labs to avoid shelter animals—who may well have had enough bad luck to last a life- time—in favor of breeding animals specifically for science, we have cre- ated two moral and legal categories of dogs. We spare one cohort, worried at the betrayal of the human-canine bond, and simultaneously bring a parallel cohort to life whose existence (in human eyes) is solely to serve our needs.[15] In the Animal Welfare Act, commercially bred dogs are regu- lated as "Class A" and random-source dogs from shelters, auctions, or other venues are "Class B." This multitier view of the world of dogs has outcomes.

No surprise, the shift to using purpose-bred pups from commercial vendors spawned some megakennels, upward of ten thousand dogs. I used to bring vet students to visit one of these megakennels, and we marveled at how they could keep so many animals clean and in good physical health, how they could train the dogs to sit calmly on a table for a lab exam or a blood sample, and how they could select the dogs with the calmest temperament and the best physical conformation for the next generation of breeders. One of the bigger compromises to animal health and welfare is the seemingly simple question of how to house a dog, what the USDA calls the primary enclosure. I do not like housing dogs on plastic-coated grid floors as the large kennels do, as a mesh wide enough to let their feces drop through will invariably cause some foot problems for some proportion of the dogs. Concrete floors that staff hose down I like even less, as the dogs spend too much time on wet concrete. Solid floors with wood shaving bedding get dust and particles in the dogs' eyes, not to mention how many forests it would take to supply ten thousand dogs with enough wood shavings to keep them clean and comfy. No, if the best dog housing is a life that moves from floors and couches to a grassy park and then back to a cozy bed, even the best staffed megakennel cannot approach that ideal. Pound-seizure restrictions drove the cost of dogs up and likely contributed to the large drop in dog numbers throughout the 1980s. But in a world where our government agencies have either required (Food and Drug Administration and Environmental Protection Agency) or funded (National Institutes of Health) dog labs, something had to fill the void, and that something was kennels full of second-tier beagles.

In 2017 the USDA issued several serious citations—citing three hundred puppy deaths as the worst but not the only welfare violation—for the Envigo company's dog kennel in Cumberland, Virginia. As the USDA saw it, the kennel was seriously understaffed, with a single vet for close to five thousand dogs. That kind of dog-to-vet ratio puts a lot of pressure on support staff to know what they're doing and to keep the vet informed; USDA's reports clearly show staff weren't up to that task, whether for their limited skills and training or for the sheer number of animals. How tragic for many of those dogs, but the legal settlement, even beyond the whopping fine of $35 million, the largest in Animal Welfare Act history, is what strikes me most.[16] A judge allowed Envigo to send dogs to labs that had purchase contracts signed and waiting, but required Envigo to place the remainder, close to four thousand dogs, with the Humane Society of the United States to parcel out to shelters for finding adoptive homes. Prince Harry and Meghan, the Duchess of Sussex, adopted one, and newspapers loved the story. I'm not surprised the court allowed Envigo to meet existing obligations and send purchased beagles to the labs who'd ordered them. Many of the four thousand to leave as adoptees were older breeding animals, like Meghan and Harry's dog, and so, again, no great surprise that these animals with lower value as lab subjects would go out as pets. If Virginia passes a beagle Freedom adoption bill, state law would likely require finding homes for those animals once their breeding life has ended. Cheers to the pups who found their ways to homes instead of labs.

Still, I find myself troubled by this animal-welfare lottery that makes some dogs sudden winners—a home with the erstwhile royals strikes me as a win for one lucky dog—but leaves thousands of others behind. We still use over forty thousand lab dogs per year in the United States. Acting on our behalf, the NIH and other agencies fund dog research, while the Food and Drug Administration expects to see a companies' dog data for many safety evaluations. Congress has recently given the Food and Drug Administration more flexibility in using nonanimal testing alternatives, but dog numbers are so far staying about steady (as are monkey numbers). The United States loves having winners (the annual Thanksgiving turkey whom the president "pardons" every year) and losers (ask anyone without health insurance), and Envigo's four thousand beagles hit the jackpot, going from a megakennel rife with serious problems to loving homes, some of them (formerly) royal. I'm glad for these dogs; the one I met, Hammy, was a real charmer. In the United States, we don't want mega-kennels breeding lab dogs, but our government regulators, acting on our behalf, say that our national health needs still require them. We reconcile that tension in a piecemeal approach: a kennel closed down here, some surreptitious rescues there, a government panel on dogs at the Veterans Administration, a government working group on dogs and other animals in safety testing. I have to admit that if we really and truly *need* these dogs, more of the Envigo pups should have gone to labs. And if we as a society think that would be wrong, we should rethink what we are doing with all the pups who did not hit the jackpot.

When I look at lab dogs, I see the need to improve the Animal Welfare Act standards to give kennel dogs fuller lives, the need to root out operations that will not meet those standards, support for postlab adoption programs, and, most important, reducing our reliance on dogs in every scientific discipline and in testing labs. Rather than simply shifting the burdens of being research subjects to other species, though, I want dogs to carry the others along with them toward the day when animal labs are obsolete.

Here's the vision that I believe dogs can carry, and cats too, though we use far fewer of them in labs. If we can make this work for them, then extend it to other animals. First, we need a national-level review of dog use. Currently, the National Toxicology Program is working on replacing dogs and other animals in safety testing. The National Academies has mounted some reviews of dog research, but they are limited to dogs in NIH-funded research or in the Veterans Administration labs. The USDA counts annual dog use, making it public on a website, but we lack details to even know what kinds of labs—basic science? safety testing?—are using dogs. European animal-use data includes that greater amount of detail; it's certainly possible. My request: the US government should commission the National Academies of Sciences, Engineering, and Medicine to review *all* dog use in the country and set priorities for replacing as many of those dogs as possible, not with other species so much as with nonanimal tests. The review needs experts, not just in dog experimentation but also in in vitro research, and it needs a broad base of representation. The National Academies' 2020 review of dogs in Veterans Administration labs (yes, the VA has animal labs, as does the Department of Defense, the Environmental Protection Agency, the USDA, and other federal agencies) included an animal rights lawyer, and the sky did not fall. Such a panel might decide that studies that presently can succeed only if they keep using dogs or cats or other animals may not actually be important enough to continue; let those wait until we have nonanimal tests that could do the work. That's step one.

Step two: if we must continue to use animals in labs, let's let dogs lead the way in reforming our practices. Let the dog and cat farms continue raising animals for labs, but from day one start training those animals to be well adapted not just for lab life but for life in a home too. Pups need to get over their natural aversion to soiling themselves to live in the small, grid-floor cages at kennels like Envigo (too much information: grid floors stay as clean as they do only because dogs are constantly stepping on their own feces and thereby pushing their dirt through the grid). When labs buy animals for nonterminal experiments, they too should start from day one caring for those dogs and cats in ways that get them ready for life out in the world. These reforms will raise the cost of raising dogs and the cost of using dogs. They will raise the cost of all animal research, if dogs bring along other animals for such improved care. As a society, if we treat the privilege of using dogs in labs as a near-sacred trust, we will pay those costs and develop centers of excellence for dog experiments, such as high-stakes physics research, which costs big money and is limited to central-ized labs that can maintain synchrotrons and telescopes. In the days when scientists developed hepatitis vaccines in chimpanzee labs, chimp science was limited to a handful of institutions that could maintain chimps and to scientists whose projects were sound enough to attract the funding for working with expensive animals in specialty labs. For any scientists who worried that ending chimpanzees' role as lab animals put us on a slippery slope, I say, "You were right." We should indeed see dog research as a last-ditch enterprise reserved for the truly most important projects, and, yes, that slope will be slippery enough that we will follow with similar conver-sations about monkeys, woodchucks, and at some point, carrying the weight of the world's animal experiments on their tiny shoulders, mice.

Lab dogs have played dual roles for a long time. Their size and availa-bility have made them the animals of choice for many experiments, while their place in our affections has guaranteed their starring role in animal-protection campaigns. Just as they were never scientists' only choice as a lab subject, so too they have occasionally shared their public relations role with other animals. We saw how the horses in the French veterinary school stoked British outrage, and British scientists' insistence that they would show themselves more humane than their French counterparts. Cats have made occasional appearances, most notably in their role in an animal-protection campaign in 1976. Henry Spira, whose efforts to spring rabbits from cosmetics-testing labs I describe in the chapter "Rabbit," learned of cat experiments at the American Museum of Natural History. Spira pio-neered a strategy that gave his later rabbit campaign its success, pairing depictions of disturbing experiments (surgeries to injure cat nerves) with questionable usefulness (twenty years of experiments studying sexual performance in injured cats, in the search for important information about sexual performance in injured people), all of it foisted on an animal people could care about.[17]

Frogs too had their moment in the public eye. For decades, only the rare high school student passed a biology class without having dissected a frog, until a lawsuit in the 1980s gave students the chance to opt out. In 1987 Jennifer Graham successfully sued her high school for the option to learn anatomy without killing a frog. The case had legs. Apple Computer developed and marketed its first computer simulation of an actual dissection, an important step in developing alternative nonanimal replacements. Frogs per se were not the stars of this story so much as high school students and their rights, as reflected in the 1988 California Student Rights Bill.[18]

Dogs have been center stage in animal-research battles; now I have one more role for them. In "Chicken" I write about the science of animal welfare. How we treat animals reflects what we think we know about them. Sometimes we anthropomorphize, which is to say, we assume other animals share our human preferences, thinking, "If I were a lab woodchuck, I would want . . ." Sometimes our human intuitions are right; sometimes they're wrong. Anthropomorphism is a starting place for asking the animals, "Is this what you want?" I embrace it, if we pair it with honest reflection and actual evidence. We have no word for the canine version of anthropomorphism (I call it *caninomorphism*), but living with dogs gives us insight into how other minds think and feel. Dogs help us shape the questions we can ask other animals about what really matters to them in their lives.

But before I ask what barking dogs can tell us about understanding animal welfare, I want to visit with a much quieter animal. Rabbits have been in labs for about as long as dogs, but with less public exposure and less public sympathy. In labs even a rabbit in pain will sit quietly hiding it, eyes narrowed but otherwise unreadable, in situations that would set a beagle howling. Most of us find them harder to read than an expressive hound, but they have their needs, so we turn now to a particular campaign in which they figured prominently, the quest to eliminate safety tests that harm and kill rabbits and other animals.

Rabbit

Oryctolagus cuniculus

THE WHISKERED FACE THAT LAUNCHED
ANIMAL-TESTING ALTERNATIVES

Working one holiday season as a department store elf, the writer David Sedaris watched and reported on the crazy that a trip to see Santa can bring out. Lab animals found their way into one child's request, made while he was sitting on Santa's lap. Sedaris watched a mother badgering this kid to tell Santa to get the cosmetics company, Procter and Gamble, to stop animal testing. Pushed, young Jason agreed with his mother that the company tortures animals.[1]

The humor lands, at least for a certain demographic (baby boomers like me) thanks to a New York City activist, Henry Spira, who plastered rabbits' faces on newspaper ads in 1980, calling on Revlon (not yet a Procter and Gamble subsidiary, though it was by the time of Sedaris's "Santaland Diaries") to stop testing eye makeup in rabbits' eyes. The campaign was simple and brilliant. It linked images of fluffy white bunnies, suffering in the hands of the corporations bringing consumers the latest eyeliners or mascaras, asking, "How many rabbits does Revlon blind for beauty's sake?"[2]

The ads created the images people carry of animals in laboratories. I often ask nonscientists—why I have so few friends, perhaps—to tell me what they see when they picture animals in labs. The two images I hear of most are terrified monkeys on the receiving end of electric shocks and agonized rabbits in cosmetics labs. Henry Spira brought rabbits from their long-running, low-profile laboratory lives to the *New York Times* and television. Decades later people still flash to his accusations of the ugly lows that makeup manufacturers will stoop to in making you beautiful.[3]

Rabbits have been lab subjects for well over a century, but they never caused much public ruckus the way dogs did, at least not until Spira came along. Common farm animals, they were a menu item in French cuisine, not loved pets, and so were readily available for their uncontroversial role in Louis Pasteur's rabies experiments. An 1887 *Vanity Fair* article showed Pasteur holding two rabbits in his arms the way a ballplayer might pose with a bat or a musician with a violin. The picture wasn't controversial: rabbits were simply emblems of Pasteur's trade, tools in the development of the rabies vaccine. The *Vanity Fair* article was not an exposé of cruelty perpetrated on this common menu item, though rabies brings a terrible death of pain and paralysis in laboratory rabbits. *Vanity Fair* lauded Pasteur for his efforts taming this dreaded virus that was killing Parisian street dogs and the people they bit.[4] While Pasteur's rabbit experiments met no outcries of cruelty, they contributed to dogs' transition from the dreaded spreaders of rabies on English streets to the loved pets and pitied shelter inmates—all of them vaccinated thanks to Pasteur's rabbits—who became the poster pups of anti–animal lab campaigns in Victorian England.

For years, while activists and scientists fought high-profile battles over dogs in science, lab rabbits stayed behind the scenes, serving in a range of experiments across disciplines. Rabbits are smaller and cheaper than dogs but large enough for experiments for which mice or rats are too small. Their only peek out from behind the curtain came via their role in human-pregnancy diagnosis. In the mid-twentieth century, women gave their doctors urine samples, which lab techs then injected into rabbits. Autopsying the rabbits a day or two later, scientists could see if the ovaries had responded or not to hormones in the patient's urine. "The rabbit died" became an evocative euphemism in various movie and TV scripts for describing a positive human-pregnancy diagnosis (including a man's pregnancy in the 1979 movie *Rabbit Test* and in the 2016 short film *The Rabbit Died*).[5] In truth, whether the women were pregnant or not, the rabbit always died for the examination of her ovaries. People fretted about unwanted pregnancies, but no one fretted about the rabbits.

Few New Yorkers seeing Spira's 1980 ads were rabbit hunters, nor would they see rabbit meat as commonplace on US menus. They certainly would never have encountered the rabbits in pregnancy labs. What most of them would know of rabbits came from Bugs Bunny, the Easter Bunny, and Bambi's sidekick, Thumper. Rabbits were kids' pets, soft and cute and fluffy. Spend some time with them and their appeal is obvious, with their twitching nose and perky ears. Offer them a treat—some like bananas, some peanuts, and some play to the stereotype and go for carrots—and feel their soft lips exploring it as they pull it from your fingers. Pop-culture images of friendly, harmless rabbits were fertile ground for Spira's campaign to publicize the Draize test, which blinded rabbits in eye-safety testing for the sake of human vanity.

The Draize test is a dreadful assault on animals' welfare, performed in pursuit of human safety, not just for beauty aids but for any number of household products as well. In 1933 several women went blind, and at least one died, from using the mascara LashLure. The US Congress soon passed the 1938 Food, Drug, and Cosmetic Act, giving the US Food and Drug Administration (FDA) jurisdiction over cosmetics safety. The FDA promptly hired the pharmacologist John Draize to develop cosmetics tests for skin and eye irritation. In 1944 he published his methods for numerically scoring skin irritation (in dogs, in rabbits, and in human volunteers); allergic skin reactions (in guinea pigs); and mucous-membrane damage (in rabbits' penises). Skin toxicity was his big interest. For eye-irritancy testing, he borrowed from another government scientist who'd worked up methods in guinea pigs and rabbits for scoring eye damage, though John Draize got the credit and notoriety with his name on what became a widely hated animal test.

I am glad never to have seen this test in practice. The FDA used a chart for scoring the amount of redness and swelling Draize rabbits can develop, a chart I studied for my specialty certification in laboratory animal medicine.[6] *Dreadful* is an understatement. My years as a contact-lens wearer, and the periodic flares of pain that entails, make me squeamish about eyes. The Draize test, as refined through several decades, places the animals in stocks, immobilized at the neck in a way that prevents them from rubbing their eyes. Technicians then instill whatever the test substance is—mascara, drain cleaner, shampoo, saline placebo—into one eye, with the other serving as an untreated control, and score the inflammation, injury, and, sometimes, healing, over the course of up to three weeks.

Henry Spira was savvy in framing his campaign with such a tight focus, narrowing all animal toxicology down to rabbits, eyes, and beauty. Rabbits were not then or now the most numerous animals in toxicology labs. At the time Spira launched his Draize campaign, the LD50 test—in which scientists inject or force-feed rats and mice high and low doses of a test substance to find the lethal dose that kills 50 percent of the animals receiving it—was using four to five million animals per year, twenty times the number suffering in Draize testing. In the wake of the thalidomide tragedy in the 1960s, the FDA required labs to supplement rodent safety tests with a nonrodent species, but that burden for various reasons has always fallen more on dogs and monkeys than on rabbits (rabbits are not actually rodents like mice, rats, or hamsters, despite many similarities). Toxicology labs screen for skin irritants, for poisons that act invisibly on the inner organs, and for drugs like thalidomide that harm patients or their developing fetus. Spira chose eyes and beauty for his campaign. Though cosmetics disasters like LashLure were far in the past, Spira focused on beauty products (targeted to women and therefore implicitly presented as unimportant and unnecessary), not cleaning chemicals or industrial solvents or even shampoos, all of which are in more general use and more dangerous to the eye than a wayward mascara brush.[7] "How many rabbits does Revlon blind for beauty?" could instead have been "How many mice does Dow Chemical poison for industrial chemical cleaners?" if Spira really wanted to cover the most animals with the greatest suffering. But I doubt he'd have turned out the thousands who protested in front of Revlon's Manhattan headquarters.

In an interview with *Lab Animal* magazine, a trade journal for animal-lab insiders, Spira explained his strategy. He wanted to expand public concern for lab animals beyond dogs and cats (subjects of two of his earlier efforts) and saw rabbits as "symbols of innocence," unlike easily vilified or ignored rats or mice.[8] Spira said that anyone who has known the sting of soap in the eye could sympathize with animals forced into the experience. But though soap is in wide use for women, men, and children, he chose eye makeup (overwhelmingly a female-use product) for his campaign. He highlighted the jarring contrast of beauty and cruelty and the triviality of companies vying to produce yet another mascara or shampoo.

Spira was a pragmatist. He knew his Coalition to Stop Draize Rabbit Blinding Tests was not, in fact, going to stop Draize testing, at least not in the short term. The cosmetics companies told him the FDA demands Draize testing. The FDA responded that it, in fact, did not absolutely require the test; it would consider whatever alternatives Revlon or Avon proposed to meet the safety goals without harming rabbits. But no one yet had a nonanimal alternative, and the FDA was not likely—nor could it without violating the Food, Drug, and Cosmetic Act—to go back to allowing untested beauty and personal care products and the specter of another LashLure blinding. Spira did not call for women and men to stop using rabbit-tested products altogether, nor did he demand that companies stop marketing the tested products they already had. He did not demand that companies stop developing new products until they had cruelty-free tests that would satisfy federal regulators. His ask was quite modest: that cosmetics firms steer some of their profits—0.1 percent was his suggested tithe—toward funding research to replace the unfortunate rabbits in Draize test labs.

In putting the Draize test in its crosshairs, the animal rights coalition aligned itself with the scientific consensus. Even scientists at cosmetics-testing labs were already complaining about the scientific crudeness and inaccuracies of the Draize test. Just before Spira's campaign, David Smyth of the United Kingdom's Research Defense Society, an organization for the promotion and defense of animal experimentation, not its abolition, wrote that most lab animal uses were not ready for replacement, but a combined effort of pharmaceutical and agrochemical companies could easily launch a "major attempt to find an alternative to the Draize test."[9]

Though no one yet had a replacement for the Draize rabbits, nonanimal tests were replacing rabbits and other animals in all sorts of other labs. Tuberculosis labs once used thousands of guinea pigs, inoculating them with patient samples to diagnose human infections, but in the 1970s scientists developed methods to grow tuberculosis samples in egg-based culture media.[10] Rabbits yielded their role in pregnancy diagnosis to frogs, who by the time of Spira's campaigns were themselves being replaced with nonanimal home-pregnancy tests. For virologists, lab animals gave way to tissue cultures (though many of those cultures required cells the virologists harvested from freshly killed animals), allowing them to develop the polio vaccines of the 1950s. Spira made the issue of animal safety-testing public, inspiring thousands of people to march outside Revlon's New York headquarters. Rabbits were such steadfast standard-bearers for this crusade that they now are the logo for a consortium of animal-protection groups, the Leaping Bunny Program, that aims to direct consumers toward household cleaners, shampoos, and of course cosmetics that meet their certification standards for "cruelty free" products.[11]

Spira did not demand an immediate end to animal safety testing because he knew that, however much scientists questioned the validity of the Draize test, they needed alternative tests that they could trust. And they needed research dollars to develop those tests. Under pressure from the Draize coalition, first Revlon, then Avon, and then the industry group, the Cosmetic, Toiletry, and Fragrance Association, started funding research into scientifically sound in vitro toxicology methods—corneas in a test tube, for instance, made with real human cadaver corneas—that could replace the rabbits. With promises of big grants flowing, not from the NIH but from the cosmetics industry, the effort quickly found a home at the Johns Hopkins School of Public Health, as the new Center for Alternatives to Animal Testing. Thus Spira's fluffy bunnies unleashed the industry dollars that could put the search for alternatives on a fast(er) track.

Dogs were the runaway poster animals for most lab animal protections. The 1966 US Animal Welfare Act came alive after sad stories of a stolen dalmatian dog and shocking pictures of emaciated dogs at a lab dealer's compound moved the public to demand action for lab dogs. Trotting along behind the dogs, some other animals, cats especially, won more protections. The Revlon rabbits similarly gave a tragic face to a lab practice overdue for reform, and they had a reach beyond bunnies in cosmetics labs. As Spira had hoped, the search to replace animals in safety testing extended beyond rabbits to other species and beyond cosmetics to tests for chemicals far more caustic than eye shadow. Cosmetics companies gave the seed money for the Center for Alternatives to Animal Testing labs, and then Exxon and other companies joined in, as they too had safety tests to run.

Scientists meanwhile set to work on alternatives to poisoning mice and rats in LD50 tests and priming guinea pigs for painful, itchy skin allergies.

Scientists with enough research support (i.e., dollars, as well as access to journals that could value alternatives research as publishable science) could work on validating nonanimal tests until the cows came home, but regulatory reform had to keep up. Congress responded to Spira's 1980 campaign with a resolution encouraging regulatory agencies to devote time and funding to developing and validating alternatives to the Draize. The Consumer Product Safety Commission and the Environmental Protection Agency (EPA), two scientific agencies with their own toxicology labs, put moratoria on their use of the Draize while they explored alternatives. The FDA did not go that far; it announced its assessment that Draize was, for the moment, the best available, but they would seek alternatives.

In the jargon of animal labs, "alternatives" does not simply mean substitutes or replacements but also other avenues for using fewer animals in refined, less harmful ways. The FDA and other government scientists knew they could not replace the Draize test immediately, but they could in the interim find ways to reduce rabbit numbers and refine the test. During the decades that the search for a rabbit replacement has been underway, various government and international panels have modified the standard Draize test. Toxicologists decided that three rabbits per test would suffice, rather than the once-standard six to nine. They removed the requirement to run eye tests if they already knew the substance was too acid, too alkaline, or too irritating to the skin. Anything that irritates skin will irritate eyes, so additional eye testing is pointless. Updated standards also allow for the use of topical pain medicines prior to instilling the test substance if it is very likely to be particularly painful, later expanding to allow injected systemic analgesics as well. If rabbits develop severe eye injuries early, labs can stop the test. Toxicologists try to weed out dangerous ocular irritants without touching animals, reserving the Draize as the last gauntlet for screening out compounds that look safe in vitro before releasing them for human use.

Hoping to make the Draize test less inhumane, a 2010 US government panel recommended a dose of topical analgesics, the kind of numbing drops an ophthalmologist uses in human patients, before the testers apply the potentially painful substance.[12] As a vet, I've pushed aggressively, on my campus and in my writing, for expanded pain treatments for lab animals. Scientists should always avoid potentially painful tests and, when they cannot, err on the side of overtreating animal pain rather than skimping on the meds. But they must use pain medicines that are known to be effective in the particular animal species for the procedure in question, at the right dose and the right frequency of administration. Using ineffective treatments may make us feel better, but if it doesn't make the animals feel better, it's one step worse than useless, as it gives tacit permission for painful procedures we should be avoiding, confident that we have conquered all pain with our medicines. I'm skeptical about how much pain drugs can help a Draize rabbit in the face of a seriously harmful chemical and, as a squeamish contact-lens wearer, that worries me. Drugs will help some, but if an ophthalmologist has a patient with a chemical splash to the eye—a splash that never got flushed out but left on the cornea causing further irritation, as in the standard Draize procedure—they'd prescribe more than pain medicines and eye drops (which will actually cause eye damage themselves with prolonged use). Thus I worry these drugs would not seriously improve the experience for the animals. I worry that we inadvertently "welfare wash" experiments when we use pain meds that don't effectively manage the pain.

Refining the Draize to make the experience less harmful for rabbits is just a placeholder for replacing it altogether. John Draize had cast live rabbits as his models from which toxicologists would extrapolate to humans. Few people in his day questioned whether a living animal's eyes were a faithful stand-in for a human eye; Draize's main contribution was not in seeing the similarity between human and nonhuman eyes but, more modestly, in standardizing how a lab would expose the eyes to test substances and score the damage.

Spira's campaign accelerated the search for a test system or systems that could predict a substance's capacity for eye damage without harming a living, blinking, feeling eye in the process. With dedicated grants to fuel the effort, scientists could explore a range of possibilities, from deceased human or nonhuman eyes to an assortment of eye or other cells in culture to chemical tests with no living cells at all.

The in vitro tests (animal cells grown "in glass") of Spira's day now look quite primitive. In 1951 doctors took biopsy specimens from a young patient, Henrietta Lacks, with cervical cancer. Unbeknownst to her, the cancer cells she so desperately wanted gone had a life beyond their role in diagnosing her cancer. HeLa cells were among the first human cells successfully propagated in culture. They spawned an "immortal" line of cells, called HeLa cells, that live on today in labs around the world, with a role, for example, in developing COVID vaccines. They had a shot as a replacement for Draize rabbits, with scientists finding that some substances that harm rabbit eyes killed HeLa cells in culture. As isolated cells in culture, they're a limited-use tool for scientists measuring the types and degrees of injury in a cornea or for monitoring recovery from injury.

A century after Henrietta Lacks's death, in vitro tests are far more sophisticated. HeLa cells entailed the ethical burden that the patient gave these cells with no knowledge or consent and no share of the profits drug companies reaped with their use (though her descendants are now receiving some compensation and she herself some recognition). Other cell lines come from umbilical blood of aborted human fetuses, adding more ethical complexities to nonanimal alternatives.[13] Individual cells of one single type growing in a flask have limited use for modeling even a simple cornea, an organ with five distinct cell layers, each of which protects the eye and any of which could be harmed by chemicals. Scientists can now grow multiple cell types together to more closely model a whole organ, with blood flowing and immune cells and a range of other cells and interactions.

Scientists have not come up with one single test system that could entirely substitute for a living eye, progressing from exposure through injury, response, and healing. Toxicologists use a battery of tests, some of them cultured dog, mouse, or human cells, while others experiment on chicken, pig, rabbit, or cow eyes from slaughterhouses. Regardless of a person's thoughts on meat farming, until we all become vegan, slaughterhouse material gives toxicologists a full thickness cornea to test, a tissue that has no other use. But they are still nonhuman cells. Any test model requires extrapolating, whether across species from rabbit or cow to human or from human cells in culture tubes to human cells in a real person's eye. Human-derived cell constructs cannot replicate eye anatomy the way eyes from an abattoir do, but they at least use human cells, so they can better reveal molecular and genetic changes in injured cells. By combining these various assays, toxicologists can measure ocular dangers about as well as they can with living rabbits.[14] Spira may not have expected that so many years would pass developing a replacement for Draize rabbits, but credit the man for seeing that the test would go away only once scientists and regulators believed they had tests they could stand by, an effort his Draize campaign certainly accelerated.

The dreaded Draize is just about dead as a regulatory safety test, but not entirely. The international regulatory scheme is in great flux, with some countries aggressively limiting animal safety tests, especially for cosmetics. In the United States, government scientists are continuing to collaborate with others to develop the battery of tests that could routinely replace the Draize. In 2023 the US Congress passed the FDA Modernization Act, which essentially breaks with the principle, dating back to the Nuremburg Code, that animal studies should always precede human studies, giving the FDA discretion to rely on nonanimal assays in some circumstances. By no means does this law bar the Draize or other animal tests, but it does let scientists and regulators make decisions based on their interpretation of current science, rather than blind allegiance to the tradition of animal tests. The Draize and other animal safety testing, however, have not yet breathed their last breath, for at least four reasons.

First, mascara is not the only substance that finds its way onto sensitive eyeballs. Patients knowingly apply ocular medicines, while accidental splashes from industrial and agricultural chemicals can send anyone to the emergency room. Pharmaceutical companies will first test new eye medicines in a few human volunteers in clinical trials. Chemical companies do not enlist human volunteers to test what happens should they splash a new pesticide, industrial solvent, detergent, or other substances in their eye. Animal testing will persist longer for some exposures, if only due to regulatory inertia in some countries.

Scientists' resistance to abandoning established practices and regulators' cautious inertia are not the only reasons that nonanimal alternatives will only slowly replace animals. A second reason that animal testing will persist is social skepticism, the likelihood that many potential human volunteers will refuse to test medicines that did not first clear animal tests, however good or bad animal tests are at predicting the human experience. Human subjects committees may balk at allowing a study to recruit volunteers if they do not see convincing animal data that a medicine is safe enough for human clinical trials. Who but desperately ill patients, willing to try any medicines their doctor offered them, would volunteer as guinea pigs for a truly novel drug based solely on computer modeling and outcomes in an organ-culture lab? Presently, nonanimal tests complement animal experiments in bringing new medicines to the point of human testing. But the day is coming, though likely still decades away, when scientists will produce new medicines without using animals in the process. The first animal in the process will then be whatever human animal has decided to trust the AI and cell-culture studies that brought the medicine to the point of human subjects testing. In 2024 the Nobel Prize in Chemistry went to scientists who used AI and machine learning to better understand and predict the three-dimensional structure of proteins, opening doors to designing new proteins of medical significance. Admittedly, their work in these early days of computer-designed medicines included testing in mice some of what their AI developed, but expect this work to increasingly bypass slow, old-school technologies like animal labs, as scientists race to produce vaccines for the next pandemic when it comes.[15]

The race to develop COVID vaccines already condensed clinical testing into a very short time frame but did not succeed in replacing animals. It relied on previous safety and efficacy tests, in animals and volunteers, of a new vaccine technology (called mRNA) developed for other diseases. It relied on tests of the COVID vaccines themselves in animals. The days of rushing insulin treatments to human trials, of testing hepatitis vaccines in children, or other abuses of human subjects are (one hopes) mostly behind us. Will our culture keep up with the science and see volunteers stepping up to safety test drugs that have never passed through an animal's body, without rabbits or monkeys holding our hands along the way? That cultural shift will come, but slowly. A single disaster in which a medicine bypassed animal labs and then killed human volunteers could set progress back for years.

The goals of safety testing are rarely a simple thumbs up or thumbs down. We know that medicines have side effects. We know that many chemical splashes are dangerous. Safety tests give manufacturers, regulators, and doctors knowledge on how to use eye medicines wisely, what to do when workers rush to the emergency room after splashing a weedkiller in their eye, or how to handle and transport hazardous chemicals. The question "safe or not?" yields to more complicated questions that require knowing how a substance might be harmful; whether the harm is to humans or to fish, pollinators, and other species around us; and how to address the harm.

Companies must comply with FDA and EPA requirements for safety testing drugs and chemicals, but they are usually free to go beyond. Thus a third force that could keep companies performing animal tests that regulators do not explicitly require is the threat of liability lawsuits, with animal labs adding to a company's defense if a government-approved product causes, or appears to cause, human harm. Roundup, whose chemical name is glyphosate, is a weedkiller that has been extraordinarily profitable for its manufacturer. Splashed in the eye, it may cause transient irritation at worst, at least in rabbits. Roundup's safety concerns do not stop at the eyes, of course, but it cleared whatever hurdles the EPA erected in the 1970s, deemed sufficiently safe for professionals and householders to spray and pour on whatever weeds they were battling. Its original manufacturer, Monsanto, has developed a range of Roundup-resistant crops, genetically engineered to survive when a farmer broadcasts the weedkiller in a field of desired crops amid unwelcome weeds. Even when it is used as labeled, and with gloves or respirators, farmers nonetheless take in a lot of Roundup in their years of farming, raising worries of cancers or other health effects from chronic exposure.

While the EPA and many scientists find no evidence that Roundup exposure is a cancer risk, others, most notably the World Health Organization's International Agency for Research on Cancer, disagrees, designating it "probably carcinogenic to humans." The different bodies' assessments have led to a regulatory landscape in which Roundup faces more restrictions in some countries than in others, but in the United States Roundup remains on the market with minimal regulations on its use. Scientists may disagree on the data, but juries make the call in the courtroom and have awarded many millions of dollars to cancer patients for Monsanto's failure to warn of a cancer risk it claims its weedkiller does not pose. (In 2018 the pharmaceutical giant Bayer acquired Monsanto along with its thousands of pending lawsuits.)[16]

Animals, in this case rats, have appeared, posthumously, as expert witnesses on the carcinogenic potential of the chemical. Though the EPA allows widespread sale of Roundup with no label warnings that it could be carcinogenic, a jury in the US system can reach its own decisions about the evidence, from animal labs or anyplace else, about the chemical's safety. Every chemical or medicine a company releases for human use is a potential lawsuit, so if protection against such suits goes beyond clearing regulatory hurdles, manufacturers will continue their incentives to guard themselves beyond what the regulators require. They will continue some animal testing beyond the regulatory requirements to hedge their bets against potential lawsuits, and indeed many dozen animal researchers and epidemiologists have continued studying Roundup's safety long after the EPA declared its safety.

Industry sponsorship is a serious source of potential bias in animal and other research, raising the concern that, consciously or not, scientists will produce results favorable to the company funding the project. Thus publication guidelines for animal experiments insist that researchers should always state their funding source in their scientific papers and projects. Evaluations such as the Cochrane reviews, which score the quality of published papers, want to see the source of funding as well as a statement on what role the sponsor took in analyzing data and drawing conclusions. The EPA reviewed Monsanto-sponsored published work as well as unpublished studies in its prospective review of Roundup, but, of course, for any new drug or chemical headed to market, the company with the vested interest in moving the product along will be the major funder of safety studies. Scientists in basic research labs submit their studies, no matter the source of funding, to academic journals with peer reviewers. The goal is that peer review weeds out questionable data before it gets to print, and, after that, others can read and critique the available literature. In contrast, most of the experiments evaluating a medicine or chemical for marketing do not appear in peer-reviewed journals but are submitted directly to the regulatory agencies as part of a company's dossier documenting how safe and effective the product will be. The EPA and FDA will likely never have the funds to sponsor and subsequently be able to rely solely on studies untainted by company sponsorship. Government regulators take on the peer reviewers' burden of evaluating the quality of the data a company submits.[17]

Finally, the Draize and other animal tests have a fourth lifeline. Remember the spectrum of animal labs from basic research to applied safety testing? For all the complexities I've just outlined, regulatory toxicity testing is still relatively simple compared to the basic science of toxicology, in which researchers seek to explore the ways in which chemicals can harm the body. Toxicity tests must meet regulatory demands and serve to inform requirements for protective clothing and to recommend treatments for exposures or to set parameters for human clinical trials of new medicines. Toxicology scientists must instead meet their peers' standards of evidence rather than the regulators' and continue to research the ways in which chemicals can harm the organs, the skin, or the eyes. Though many appear to avoid the word *Draize* in the methods sections of their scientific articles, they continue to use a battery of tests, including eyes from cows from slaughterhouses, human corneal cells in culture, and live rabbits. The name *Draize* is toxic, so authors may say "in vivo testing" instead, challenging a reader to understand how much their test matches the classic Draize assay.

Readers may join me in wondering what to do now, forty-five years after activists decried blinding rabbits for beauty's sake. Should we be looking for the Leaping Bunny symbol and boycotting products without it? Companies can advertise their products as "cruelty free" or "not tested on animals," but those labels have no legal backing. If a company develops a new emollient for a hair conditioner that leaves hair silkier and shinier than the competition and tests that new ingredient but not the conditioner itself in a rabbit lab, will it label the conditioner "not tested on animals"? And how far back should the label reach, as the vast majority of ingredients in shampoos, makeups, and soaps have already run the gauntlet of Draize testing over the decades? When a company claims *it* does not test its products in animals, does it have fingers crossed behind its back, hiding the fact that it subcontracts the animal tests to US or international labs? Some new cosmetics ingredients could also have uses in medicines or in industrial settings. Suppose a chemical could make a makeup stay on the skin three times longer than the most long-lasting cosmetics currently available and that same chemical could be important for topical antibiotics, insect repellents, and sunscreens. Odds are the manufacturers will need to test those ingredients, possibly in animals, for those medical applications, leaving the company to decide to leave that ingredient out of their cosmetics, no matter how alluring the eye shadow it could create or quietly accept that the tested ingredient will keep the "cruelty free" Leaping Bunny logo off the product.

The laser focus on cosmetics that made the 1980 ad campaign so powerful has a downside. A public and regulatory focus on cosmetics testing can carve out that narrow concern without advancing the search for animal alternatives beyond the beauty industry. The European Union was the first jurisdiction to ban the sale of cosmetics (and their separate ingredients) tested on animals (older products are grandfathered in, but new testing is verboten). Other countries and several US states have since followed suit, even as yet others—China was the largest of the holdouts until a policy change in 2021—banned sales of cosmetics *not* tested in animals. Forty-five years since the Revlon rabbits hit the *New York Times*, I'm reasonably confident that Leaping Bunny logo or no, US consumers can mostly use whatever cosmetics, shampoos, or soaps they buy with clean hands—or cleanish hands, given that most ingredients have been animal tested in the past.

The Draize story did not end when China decided to allow imports of cosmetics with no animal-test data. The quest for personal and environmental safety runs up against the desire to limit animal suffering. In the European Union, for instance, where regulators first banned animal testing for cosmetics, a competing rule, the 2006 Registration, Evaluation, Authorization, and Restriction of Chemicals regulation aims to ramp up safety testing of chemicals and may include animal tests in those evaluations, even for chemicals with a "dual use," both cosmetic and noncosmetic.[18] This is true in the United States too, where getting animals out of cosmetics labs—a process that took decades—was the easy part.

Cosmetics testing has always been a tiny part of the impetus for using animals in safety tests. Thanks to Spira and the Revlon rabbits, it's an even tinier part, but tens of thousands of animals of various species still pass through toxicology labs. In the United States, the FDA regulates not just cosmetics safety but the safety and efficacy of medicines as well, whether for use in the eyes (contact-lens solutions, lens implants, injections into the eye to treat macular degeneration), on the skin, or throughout the body. The EPA assesses how various chemicals—weedkillers, birth-control pills, fertilizers, pesticides—will affect field and stream ecosystems when we spray them about, flush them down our toilets, or require workers to handle them. The Consumer Product Safety Commission knows your drain cleaner will be more hazardous to handle than the soap you wash down that drain but makes sure hazards are within limits and that labels contain information—traditionally generated in animal labs—on safety measures when the product gets on your skin or in your eyes. The Department of Transportation also has required Draize and other animal tests through the years; safety data determine red, blue, yellow, and white hazardous materials labels on trucks and packaging, health data that the Department of Transportation until recently mandated come from animal labs.

A challenge in analyzing the FDA's role in animal testing is that it specifies quite a few tests. The thalidomide case had prompted the FDA to require more testing, including tests in pregnant animals for drugs (which is most) that might be taken by pregnant women, as well as testing in a second species (usually dog, rabbit, or monkey) but without dictating specifically which species to choose. Christian Abee, a monkey vet and a panelist for the National Academies' review of the use of primates pointed out that when the FDA requires a second species to complement rodent safety studies they do not come out and clearly state that the animal must be a monkey. They simply disapprove studies that don't use primates. This requirement for a second species adds many animals to the safety-testing process Spira and many scientists have worked so hard to reform, including long-tailed macaque monkeys whose numbers are rapidly dwindling in their jungle homes.[19]

The 2023 FDA Medical Research Modernization Act 2.0 gives the FDA some flexibility in waiving animal tests when validated nonanimal alternatives could do the job, and rabbits are far from the only beneficiaries of this development.[20] Much of the FDA's language is general and aspirational, a promise without specifics that it is indeed moving toward phasing out its animal-testing requirements. In 2025, however, the FDA announced a pilot project to deliver monkeys from their role as the second species in safety testing medicines derived from monoclonal antibodies. Primates are the animals of choice for testing monoclonals or other large, complex molecules that induce harmful immune reactions in people and monkeys, reactions which rabbits, rodents, or dogs may not exhibit. Those immune reactions are harder to model in vitro than screening for liver toxins or eye damage, and thus the FDA has set its sights on an ambitious scientific challenge. The FDA does not explicitly explain its choice of this primate-specific test instead of a goal more easily met, but, given the public concern for monkeys as well as their cost and scarcity, it is likely that species more than feasibility drove the decision.

Could we remove most "second species" testing without sacrificing drug safety? The answer depends on how well a second species rules out unsafe or ineffective drugs before they go from rodent labs to human volunteers. Private drug companies have some answers, hidden in their private databases with no public transparency. I'd initially been skeptical at suggestions that scientists should publicly register animal experiments as they do human clinical trials. So much of animal research in university labs consists of small pilot studies or very basic science experiments in mice or fish, and a public registry would crash under the weight of all this reporting. But once a potential new medicine is far enough into the pipeline that its animal data would serve to authorize human clinical trials, let's start registering those studies and tracking their outcomes. Results will vary discipline by discipline, but publicly available data really could tell us how often dogs or monkeys screen out drugs that mouse labs incorrectly labeled as safe and effective candidates for human use. Of course, in cases where the larger animals provide more reliable data than mouse labs, regulators may want to leave the mice alone and focus solely on the larger animals, until they've got a reliable nonanimal test in hand.

Even without the public campaigns and protests, regulators, animal activists, and scientists know that they have work to do validating nonanimal tests as equal to or better than the animal labs. In the United States, the plethora of agencies with an interest in safety and toxicology data work together under the leadership of the National Toxicology Program, in the Interagency Coordinating Committee on the Validation of Alternative Methods, to promote replacing animals with nonanimal tests that will be at least as good. This committee also coordinates with its international counterparts to recognize one another's safety testing and avoid duplication and also to avoid situations where individual countries' requirements (such as China's requirement up until 2021 that imported cosmetics include animal safety data) drive animal use in any country hoping to sell to them. The United Kingdom and the European Union collect and publish animal-use statistics in a way that should make us all in the United States envious, reporting the numbers of various species in labs (fish and mice included, in distinction to the United States); the severity classification of the experiments (comparable to the United States' pain categories for those animals covered by the Animal Welfare Act); and the types of use, be they basic research, regulatory testing, teaching, or other. In those places, a trend hopefully to be mirrored in the United States despite our current low transparency, animal numbers in toxicology testing are dropping significantly.[21]

Can a single rabbit change the world? Perhaps not, but rabbits do have a way of proliferating. While that one sad rabbit in the Revlon ad may not have liberated all laboratory animals for all time, the Draize campaign effectively used that rabbit to ignite public sympathies and anger and accelerated what many scientists themselves wanted, a reboot of an entrenched system that was well past due for reform.

Chicken

Gallus domesticus

ANIMAL-WELFARE SCIENCE FOR HAPPIER
ANIMALS AND BETTER EXPERIMENTS

My stint as a Cornell University chicken farmer lasted but two months. Laying hens in one barn were crammed together into tiered wire-mesh cages—battery cages—while "uncaged" laying hens carpeted the floor of large pens in another barn, so tightly packed I could not see the manure and litter I was there to shovel. They pecked and pecked and pecked at one another, until I reached my hand in to collect and label their eggs, at which point they paused their squabbling and pecked at me. I hated that job, but at least I got to go home at the end of the day. This chapter, though, is about the chickens' welfare, not mine. It's about the mice, monkeys, woodchucks, and other animals—and how much they might hate, like, tolerate, or adapt to the farms and labs where we confine them. Most people assume chicken welfare is bad on these farms, but what precisely would they change? The air? The lighting? The lack of sun? The cages are crowded: Does that bother the birds? How much less crowded is acceptable to our sensitivities or to the birds' well-being? How can we know and, once we do know, how much more are we willing to pay for eggs to buy the birds a better life?

In this chapter I show how welfare scientists ask animals what they want and need for their maximal well-being. That knowledge can help us design a better woodchuck house, handle mice without stressing them, or give chickens the range of freedoms that they want, not just what we think they should want. Everyone should want better welfare for the animals' sake, but farmers should understand how poor welfare can make farm animals less productive and researchers should know how poor welfare can skew the results in animal labs. Science-based welfare assessments can help us ensure that the improvements we make truly are improvements in the eyes of the animals. Only then can we confront the barriers—money, convenience, inertia, and scientists' fears (often misguided) that happier animals are sometimes actually worse study subjects—and move forward.

Scientists should feel an extra urge to maximize their animals' welfare, as often poor emotional welfare manifests physically as well, even when that's not immediately obvious. In human and nonhuman animals alike, physical health and emotional health affect each other, intimately entwined. The various pains and stresses that animals suffer in labs can affect their biology in subtle ways, skewing the data in scientists' animal experiments. I worked with a physiology researcher who came to me, data in hand, looking for a refund on her animal-care costs. We'd been doing some facility renovation a few months earlier and the distant jackhammering we humans could barely hear had made her rats jittery enough that a solid month of data, measuring blood levels of the stress hormone corticosterone, from a two-month study was useless. A different scientist in her department echoed my admonition that pain and distress can warp data in animal studies. "I know," he nodded, "pain can kill animals!" And since his mice were alive, could he and I, he suggested, not agree that they survived their surgeries precisely because they suffered no pain of any concern? So we want to improve animals' lives but we want to make sure we do it right, in a way that measurably improves how the animals feel. We control almost every aspect of their lives, so it's up to us to know what matters to them, whether in labs, on farms, or in homes alone as we head off to work.

As my first witnesses that there is a science to measuring animal welfare, I call, despite my initially rocky relations with them, on chickens. They played a big role in the development of a new discipline in the 1960s: animal-welfare science. As British farmers were taking up intensive, industrialized agriculture (aka "factory farming"), Ruth Harrison published her 1964 book *Animal Machines*. With her depictions of life down on the veal, chicken, and pig farms of Britain, she stirred enough public concern for farm animals' welfare to push the British government to investigate. The Brambell Committee released its report and recommendations in 1965, including a first version of what animal-welfare specialists now call the "Five Freedoms" of animal welfare.[1]

Alongside its calls for safeguarding farm-animal welfare, the Brambell Report is a manifesto for the scientific study of animal welfare. "Any attempt to evaluate welfare," they wrote, "must take into account the scientific evidence available concerning the feelings of animals that can be derived from their structure and functions and also from their behaviour. . . . We recognize that the scientific information available concerning the behavior of domestic animals is inadequate in many respects for our purposes and that much further work in this field is required before we can be satisfied as to their welfare." They called for funding for animal-welfare research that could both improve animals' lives and increase their economic value for the agricultural industry.

The Brambell Committee's call for more and better animal-welfare science did not stop it from sometimes making pronouncements based on dubious facts and assumptions they did not think to question. The committee decided that fine-gauge chicken wire was an uncomfortable floor for chickens. The floors in battery cages are sloped so eggs will quickly roll down to a collection basket before the hen, or her crowded-together cage mates, step on them and break them. The sloped, springy wire floor looks hard for birds to grasp and is always sagging under their weight, not unlike walking barefoot on a playground climbing net, so they called on chicken farms to avoid fine-wire mesh floors in battery cages. Some British researchers worried that this well-meaning effort to improve caged chickens' comfort lacked evidence and would lead to floors with high rates of broken eggs, losses for farmers, and chickens who did not agree that their lives had improved.

In one of the first-ever scientific investigations of animal welfare, Barry Hughes and A. J. Black asked the chickens what they wanted. They gave thirty-six brown leghorn hens their choice of four different types of floor, and the birds voted with their feet. In over 432 bird hours of surveillance, the birds more often chose to stand on the springy chicken-wire flooring that Prof. Roger Brambell and colleagues had criticized. Hughes and Black may have spared laying hens floors that were not to their liking. They also heralded their experiment as a "new approach to animal welfare . . . in which objective assessment of animals' preferences should ultimately make subjective value judgments superfluous."[2] People knew enough from the book *Animal Machines* about hens' crowded conditions tucked secretly away inside industrialized, ammonia-filled barns to want reform. Chickens were able to lead a new approach to animal-welfare policy, focusing a general call for reform and a general appeal to five freedoms, with their particular input on the reforms that would matter to them.

Poultry biologists built on Hughes and Black's publications on cage floors, fine-tuning questions of what chickens want from a cage floor, how big a cage they choose, whether they want to avoid or seek a cage mate, and even their preferred in-cage location of food and water cups. Other scientists interrogated pigs and cows about their preferences. Nor did animal-welfare science just stay down on the farm; in one of the first efforts at translating farm-animal–welfare science to laboratory animals, in 1981 Marie-Christine Buhot-Averseng referenced Hughes and Black's study of chickens' preferences as her model to ask lab mice what kind of nest boxes—snug and dark, it turns out—they preferred.

Chickens sparked my interest in animal welfare as something more than just hunches of what I might want in a given animal's shoes. This came some years after my undergraduate travails in Cornell's henhouses. The Association for Assessment and Accreditation of Laboratory Animal Care International is the organization that accredits laboratory animal programs, and in my vet tech days Cornell hosted our first accreditation site visit. They assembled a team of animal experts from other campuses, mostly vets, who set to work inspecting our facilities. Like an Animal Welfare Act inspector from the US Department of Agriculture (USDA), they looked at cleanliness and peeling paint. They checked our vet records. But one of the site visitors was different. Dr. Joy Mench is not a vet but instead holds a doctorate in animal-welfare science. We had some young chickens living in converted rabbit cages for an experiment, and she stood watching them, curious at how they walked on the unusual (for a chicken) cage floor. Intrigued, I watched her watching them. She did not need to run an experiment and analyze data; she watched and showed me how their claws kept catching on the metal grid floor. Before this I had not taken the time to really see and score how the cage met the birds' needs. Just an occasional catch on the metal forced them to pull their foot up to take the next step, the way a person does when walking in mud that wants to keep your shoe. Then we saw one whose little foot had slipped down through the hole in the floor, trapping the young bird. Our cages were plenty clean, just like the accreditation rules required, but she taught me welfare is so much bigger and that welfare assessment starts with looking and watching the animals.[3]

It sounds simple enough: watch the animals, ask them what they want, and, if we can reasonably accommodate it, give it to them. I'll get to how well or poorly we accommodate what the animals tell us they want, but let's start with hearing what they are saying. Preference tests are the most basic of animal-welfare science. Give animals one or more choices and see which they choose. A dog at home in the winter may change sleeping locations as the sun moves across the sky, choosing to lie where the sun is shining. One cat may choose canned food over dry; another the opposite (and, being cats, those preferences may change daily). My first animal-welfare experiment was low tech, and nothing I ever published. The Animal Welfare Act requires some sort of elevated resting perch in cat cages, with the stipulation that it be easily sanitized. Typically, that's a wood or hard plastic board a cat can sit on, so it's no surprise to people who know cats that many of them spend their day curled up in their litter box instead.

One day my student and I attached some waterproof plastic foam pads—the kind you might use on a boat or a lawn chair—to one half of the wooden resting boards for some singly caged lab cats. The cushions were soft and easy to clean. My student's job was simple. Enter the room and, before cats reacted to human presence, score each cat for their location: foam pad, wooden board, litter pan, or other. Common sense triumphed, and the foam pads were clear winners. While that would hardly have revolutionized cat care, it could have justified a small, not-too-costly improvement, but it was just inconvenient enough to the animal-care staff, and not a specific legal requirement, and thus the cushions ended in the storeroom once our data collection ended. Like so many scientists, I did not really share my excitement with the animal-care staff, who would likely have come up with their own ideas for welfare improvements to test. I expected them to just keep the animals cleaned and fed and to not interfere with my experimental cushions. In hindsight I wish I'd gathered the cat staff together in a room with some cushioned and some hard chairs, shared the data they helped me generate, and invited them to envision how important those data might be for the cats in their care.

Preference tests are simple and informative if we're planning to give animals only a narrow range of choices in their lives. But they have serious limits. I might offer you ravioli versus okra for dinner and chocolate versus caramel ice cream for dessert. You'll settle for ravioli or okra if you're hungry enough, but you may hate both. You'll pick chocolate or caramel because you really, really like ice cream, even if those two flavors are way down on your list of favorites. If I seriously want to maximize your welfare, I should offer more choices and learn what you really prefer. I should know how strong the preference is and how much it delights you when I get it right or distresses you when I get it wrong—and all the more so if you're not simply my guest but an animal I've confined to a cage.

We can start by asking, "How might I feel in this animal's circumstances? What would I prefer?" and then look for evidence to disprove or confirm that. That may veer too far for some readers from hardcore science toward sentimentalism or, worse, the dreaded sin of *anthropomorphism*. Anthropomorphism means seeing or imposing human attributes in nonhuman animals. Vets and scientists are trained to avoid it. We should instead embrace it, but judiciously. Ants are industrious, and grasshoppers are lazy. Cartoon frogs pluck banjos, while bears wear (admittedly skimpy) human clothes and discuss stealing picnic baskets in human voices. Such fables and stories serve to hold up a mirror to teach us something about human behavior. This sort of anthropomorphism carries little value for improving animal welfare.

Anthropomorphism can be harmful for animals. When we uncritically attribute personality traits to animals, we may then make bad decisions, to the animals' woe. If we judge growling dogs as "vicious," we'll treat them differently than if we recognize and try to help them with their fear. A vicious dog has a bad character and deserves aggressive handling, we may think, but a scared dog needs just the opposite. Scientists describe rabbits or other animals as "stoic." That should mean that if some animals are particularly hard to read, we need creative efforts to understand what they're feeling. But it can slip instead into the mistaken belief that "stoic" animals just don't feel pain all that much, with the result that we under-treat their pain or fail to prevent it.

Animal behavior scientists like David Morton, Gordon Burghardt, and Jane Smith endorse practicing *critical anthropomorphism*, which may start with some empathetic imagining from human experience but then puts it to the test. Scientists run animal experiments because they believe that what's physically or psychologically true of nonhumans, such as that alcohol can exacerbate liver infections or that the right drug can tame anxiety, is often true of humans. But they must put those animal-to-human extrapolations to the test, bringing in supporting evidence and running experiments. And often they do find that what is true in the animal lab is approximately true of the human too.[4]

Turn that around and ask, "How can I critically test my claim that this particular human attribute applies to this animal?" Without testing my assumption, I could force animals into something unpleasant or deny them something important. I may think that if I were running in snow and ice in the Iditarod dogsled race, I'd want shoes. Really? Put it to the test. Examine sled dogs' feet with or without shoes and, ideally, measure their reactions to running barefoot or shod. The animal behaviorist Temple Grandin has relied heavily on her intuitions and experiences, intuitions she attributes to her autism, to envision what cows might feel scary or comforting. But before she sells her ideas to cattle ranchers, she puts her intuitions to the test in welfare experiments with cattle.[5]

Researchers can tap the wisdom of empathizing anthropomorphically with their animals as models. The neuroscientist Garet Lahvis enrolled genetically modified mice in his experiments to model human autism. He standardized behavioral tests of mouse sociability to investigate which genes might make mice, and by implication autistic children, more prone to altered social interactions. He writes how his students report that feeling stressed makes them withdraw socially, an insight that could lead to research questions about his autistic-like mice's social behaviors (and to animal-welfare insights of how poor welfare manifests as distorted social interactions). He rejected the glaringly bright lights common in labs for recording mouse behavior, preferring the dimmer lighting that the mice themselves prefer, as better suited to observing their less stressed, more natural behaviors. Moving on, Lahvis combined anthropomorphic perceptions with objective, quantifiable data to speak out about how the inescapable chronic stress of standard lab mouse care can doom their usefulness in psychiatric and genetic research.[6]

I'm a great believer in the creative power of empathy and anthropomorphic thinking to guide animal-welfare experiments. I'm also a fan of watching the animals we know best, dogs above all, for hints of how other animals might feel and how they might show those feelings. Anyone who has ever taken a walk with a dog knows the excruciatingly slow pace these creatures, though more capable than we are of running and jumping, can maintain. They stop, sniff, and pee at every lamppost until there's no possibility of another drop of urine in their bladder, then stop and sniff, and sniff, and sniff some more. As we slow our walk to accommodate their interests, an insight: if we want to understand what resonates for animals' welfare, we must remember that our human vision–centered appreciation of the world is only one way of living. I could live in a world without much olfactory input, but dogs tell me to be attuned to other animals' different realities.

Watching dogs inspires us to think about other animals and make predictions about their welfare. We can then put our hypotheses to the test and expand our ideas of what we can strive to provide for animals. Trying to be as scientific as possible, we animal professionals sometimes shy from words like *joy*, *fun*, or even *play*. A trip to a dog park should jolt us out of that. No words better capture those dogs' lives. And if dogs can play and have fun, might other animals? If we can observe the signs of joy in a dog, can we look for joy in other animals? And if so, should we not strive beyond eliminating pain and distress and, in addition, shoot for joy, to the extent possible while the animals are in the labs and later in adoptive homes or in monkey-retirement sanctuaries?

Anthropomorphism and what we can observe of dogs suggest the welfare-preference questions we should ask other animals, but simple preference testing turns out to be a weak tool for putting our insights to the test. The cushions that my cats told me they preferred over bare wood went into mothballs. I should have kept going and found ways to ask the cats how much they preferred soft pads. I should have run the next level of test, a weighted preference test, in which I ask the cats, "How strongly do you want a cushion?" Would they tap a lever or push open a heavy door or leave a chamber full of their favorite toys and foods for a chance to lie on their cushion? And I should have run statistics and published my findings and translated my anthropomorphic hunch that cats like soft furniture as much as we do into cold, hard, peer-reviewed science that other cat labs should heed.

Weighted preference tests are a step up from Hughes and Black's chicken floor–preference tests. They're the animal version of economists' consumer-demand studies, where animals are the consumers, not the consumed. In a rodent version of Hughes and Black's inquiry on hens' flooring preferences, Caroline Manser and her colleagues found that rats would lift a heavy door (up to 86 percent of their body weight) to leave a wire grid floor in favor of a solid floor.[7] At the time mesh floors were common in toxicology labs to ensure rats did not confound tests by eating their bedding along with whatever substances toxicologists were feeding them. Studies such as Manser's pushed scientists to review how much a solid floor with bedding derails toxicity tests (turns out, not much) and to move away from the hard, drafty wire-mesh cages rats dislike. Chris Sherwin in 2004 reported that lab mice will work by pressing a lever up to twenty times to leave a cage with nesting material, food, and water to get entry to a chamber with moist Irish peat moss seven inches deep in which they could burrow tunnels.[8]

Animals may tell us what they want, but they won't necessarily get it. Though Sherwin's mice demonstrated their demand for a chance to dig tunnels, no labs that I know of offer their mice seven inches of deep peat moss. No one gives their lab mice seven inches of peat moss in part because a standard mouse cage in the United States is only five inches tall. That's the minimum mouse-cage heights recommended in the National Institutes of Health's rule book, the *Guide for the Care and Use of Laboratory Animals*, and most cage manufacturers in the United States make cages to that standard. Moreover, lab vets would veto this, following the *Guide*'s recommendation of sanitizing the cage every two weeks. How to sanitize all that peat moss? Or should we have staff scrape out mounds of peat moss every two weeks so we can clean the cages and make the mice start all over, digging themselves new tunnels? Cleanliness is next to godliness in the vivarium, frequently overruling animals' behavioral preferences.

Behaviorists ask animals what they want and how strongly they want it, while farm scientists have long spotlighted physical health as the best marker of good welfare. On the farm animal scientists look to production parameters like growth, successful breeding, or milk production. Milk production is particularly useful, as farmers measure it daily. A sudden drop in milk output signals trouble somewhere on the farm, whether that's a facility problem or a rough, new hire whom the cows do not like or trust. Lab scientists have their own metrics of their animals' "production," which they may measure as the number of pups a mouse or woodchuck whelps or the number of times a thirsty monkey responds on a video-game task for a reward of a sip of water. But poor production and performance are crude measures and can be misleading. Welfare scientists will argue that chickens laying eggs and monkeys working a video joystick may nonetheless be suffering, so while we look at these factors, we cannot rely solely on them.

We vets are steeped in our disease-focused training of keeping animals clean and preventing infections. And with good reason: before vets really got a place in labs in the 1950s, serious infections were spreading through lab animal colonies, making animals sick and likely throwing off scientists' data. Sick rabbits with salmonellosis and dogs with heartworm suffered in lab cages, and, as they did, the science projects surely suffered too. Lab vets came to the rescue and cleaned up the animal houses. Credit vets too with tackling laboratory animal pain, as I detail in the chapter "Rat."

Production and health measures are informative because of the close ties between the mind and the body. We know from our experience as humans what our minds do to the body. A sudden fright or injury sends stress hormones—adrenaline, cortisol, and related chemicals—coursing through the body, gearing us up for a fight-or-flight response to danger. When bad situations continue, such as prolonged confinement in circumstances one cannot escape, humans and animals continue pumping out stress hormones, concomitant with a suppressed immune system making them (and us) prone to infections and throwing off other basic life functions such as growth and reproduction.

Scientists measure these stress hormones in blood or urine or saliva samples, but they are a crude measure of suffering. On the one hand, stress hormones may return to normal levels, even when suffering continues, while on the other all sorts of "stressors" are sensations like eating, playing, or sex that humans and animals seek out. Thus stress-hormone measurements join health and physical ones as well-being indicators, useful in conjunction with other welfare markers but rarely sufficient on their own. I would not, running an animal facility, launch a million-dollar upgrade based on what stress-hormone measurements alone told me would lead to better animal welfare.

Animal-welfare scientists pull in physical, health, and hormonal tests and combine those with behavioral tests to build a "welfare-o-gram" for various species, learning how best to house and handle them, how best to keep them happy. Simple and weighted preference assays are straightforward behavioral tests that tell us chickens are fine on a chicken-wire floor and that mice would really and truly prefer a chance to build a respectable tunnel system. We have other useful behavioral tests that can guide us how to best house animals, how to enrich their homes, and how to conduct kinder, gentler experiments.

Scientists can compare captive animals' behavior patterns to how free-ranging animals or animals in substantially larger enclosures spend their time. Do the animals seem to have the opportunity to explore and socialize and forage like their less confined compadres? If they are not performing these activities that occupy so much of their free-living counterparts' time, we need to investigate what's missing or what's thwarting them. Behaviorists score how much time animals spend in normal, healthy behaviors and how much goes to abnormal and unhealthy outlets. Are animals showing compulsively stereotypic behaviors like pointless circling or plucking their own hair? These behaviors usually have roots in some environmental stressors but may persist long after conditions have improved. Kenneled dogs in shelters and in labs may become spinners when spinning is the only movement a small cage allows, but they may then continue turning in circles with the slightest stimulation even when they go to an adopted home full of fun and love.

Welfare specialists have even more sophisticated behavior tests to better probe what animals want as well as what they want to avoid and to explore how they suffer when their preferences are thwarted. A pair of behavioral tests, conditioned place preference and its mirror image, conditioned place aversion, put animals in a two-room test chamber. An animal gets free run of a chamber with two rooms and should be able to tell the rooms apart, through different walls (spotted versus striped) or slightly different floor textures, and should have no strong prior bias for spending time in one room or the other. The researcher then confines the animal briefly to one room, where the animal experiences something the researcher expects the animal either to enjoy or to find "aversive," which is behaviorist-speak for annoying, scary, painful, or the like. Commercial vendors sell conditioned place–test chambers for rodent experiments such as addiction, withdrawal, and anxiety studies. They allow scientists to ask animals if they will avidly seek out or avoid a chamber in which they remember getting a particular drug. The test relies on animals having strong reactions when confined to one chamber, a decent memory, and a strong-enough emotional reaction to an experience to remember it as something to seek out or avoid at the next opportunity. An addicted rat will remember which of two similar chambers, the one with horizontally striped walls or the one with vertically striped walls, gets them a jolt of cocaine. The standardized commercial chambers also find a use in animal-welfare science, but the concept of place preference works with an array of places and chambers customized to the animal and the setting.

The concept is simple. A dog in a car can demonstrate it. For the first visit, a car ride to the beach or to a vet office may be about equivalent for a young dog, full of interesting smells but unfamiliar enough to warrant some vigilant investigation. The dog learns and remembers the differences between these two places and later knows very well whether the car is turning toward the beach (cue all the signs of excitement) or toward the vet office (cue the nervous behaviors). A dog who has had a bad experience at the vet office (as I hope very few do) has the cognitive capacity to remember what befalls a dog in that place and the emotional capacity to score it as "aversive" and to want to avoid it.

An important welfare question in animal labs, unfortunately, is how to kill animals with the least pain and distress, and learned place preferences are informative. Vets know how to do this right, because dogs and cats have helped us learn and refine our practice. With pets in a quiet room in the arms of someone they love, a vet can take one step at a time, to slip them a tranquilizer and then calmly administer the euthanasia medicine. Done right, the dog or cat experiences minimal pain, fear, or distress. This is one way dogs and cats show us the ideal that we want to bring to all our animals in the labs.

In the lab, technicians confront dozens or even hundreds of animals scheduled for killing (I save the word *euthanasia* for situations where we are killing animals in their best interests), whether for tissue collection at the end of the experiment or simply because after the experiment no one needs the cohort of animals or has a home they can retire to. Time constraints alone limit how well lab techs can replicate the patient, gentle, best practices for individual pets in a vet clinic. The American Veterinary Medical Association, starting in the 1960s out of concern for dogs and cats in animal pounds, periodically updates its lengthy tome on how to euthanize or kill animals in a range of scenarios, from feral cats in an animal shelter to whales stranded on a beach. The guidelines do not address slaughter methods for food animals but, even so, run well over a hundred pages for the many situations in which people end animal lives. Laboratory animals command a lion's share of the guidelines.[9]

Carbon dioxide inhalation is the method of choice in thousands of mouse and rat labs around the world, and it is the fate of millions of these small animals. It's quick and bloodless and requires no handling or restraint of the animals and no high-level skill. Just turn on the gas flow, and within minutes the animals have expired. Human volunteers tell us what the animals cannot directly report, that a quick whiff of even a safe dose of carbon dioxide is intensely painful in the nose and throat. Lovers of carbonated drinks sometimes get a minor taste of this pain in the nose. Odds are fair that the animal experience is comparable to the human, and, indeed, simple preference tests tell us most animals avoid carbon dioxide if they can. A pig or rat will quickly leave a chamber full of carbon dioxide, a simple preference we can easily observe. Give them a chance to reenter the chamber for some reward (for the pig, an apple; for the rats, a dark refuge where they can escape glaring lights), even after flushing out the noxious carbon dioxide for them, and their refusal unambiguously tells us of their strong aversion to this gas. Yet it lives on as an agent of death, with the hope that it knocks animals unconscious so fast that their pain and distress last but an instant. Plus, no one has quite come up yet with a satisfactory alternative.[10]

Conditioned place preference and conditioned place aversion chambers scientifically quantify animal behaviors we can see all around us.

Animals do more than move toward the good and away from the bad, as we can see with simple preference tests. Experiments reveal how complex animal minds are, constantly sampling the world around them; reacting with pleasure or dislike (to a behaviorist, negative or positive affective states); and, if the feelings are intense enough and the circumstances memorable, bringing what they have learned to their next encounter or next visit to the site of their experience.

Formal expensive test apparatuses are not essential to show us animals sorting through pleasant and unpleasant experiences, remembering them, and changing their behaviors to get what they want or get away from what bothers them. Slowly wending my way toward vet school, I took a job caring for a small menagerie at a natural science museum, where teacher-naturalists brought animals, mostly orphaned or injured wildlife, to classrooms for their lessons. Every morning I worked from lists they gave me of what animals to pack in carriers for their classes that day. The animals' acquiescence to me holding and lifting them was testament to their sentience. Gone were their innate fight-or-flight reactions to a human being entering their personal space. They had learned, and I did not need any formal test chambers to see this, that human presence was something they could accept.

Wild animals all, our animals had learned to give up their natural aversion to human handling. With experience the distress of human contact lessened, because they all had the cognitive and emotional capacity to process the experience and decide how to face it next time around. A skunk (descented!) should not have let me put a hand under his belly and lift him up for kids to squeal at, but Scooter did. An owl should not sit calmly on my arm nibbling mouse bits from my fingers. A tarantula should not hold still on a human hand, but ours learned to do just that, as I too (sentient and capable of learning) brought my emotional reaction to spiders down to where my arachnophobic aversion lessened. As my relationship with the house tarantula foreshadowed, scientists find that jumping spiders can learn visual cues to help them avoid electrified test chambers, and I'm confident tarantulas too would excel at that task.[11]

These animals are capable of conscious experiences, and their emotional experiences matter enough to them to shape their future behavior. They have welfare interests, or feelings about their feelings. They are sentient. They have what David Mellor calls "welfare-aligned sentience."[12] What happens to them matters to them in a way that damage or injury do not matter to a painting, a wooden table, or (as I read the available evidence) the tree that table came from, no matter how tragic such destruction may be to us. Sentience is not exactly equal to the ability to feel pain, but when animals do show us they feel pain, not just as a physical reflex but as something in their minds, that's pretty strong evidence of sentience. It does not require a high level of self-consciousness, like recognizing that the monkey in the mirror is me.

That animals show an interest in promoting their own welfare tells me I have an obligation to take their welfare seriously. All things being equal, I want to avoid harming sentient animals. I want to promote their welfare in a way that I do not worry about a carrot or a plate of cultured lab cells. In the lab we have so many indisputably sentient mice and monkeys and dogs and others and so much work to promote their welfare, but I also want to know how widespread sentience might be among living creatures.

I want to know what animals are capable of having welfare interests, the potential for pleasures and pains. I need to clarify which animals' welfare I'm concerned about. The monkeys and dogs, definitely. The mice and woodchucks, yes, them too, and probably the fish and frogs as well. A tiny zebra-fish larva sits immobile on the tank bottom for a few days after fertilization. The animal is unquestionably alive. It takes some time before there's even a brain and, even then, for that brain to really start functioning. At what point is it sensible to ask after that creature's emotional well-being? Must I concern myself with one of the most numerous animals in many a mouse house, the tiny fur mites crawling all over the murine residents? I'm not alone in this interest, as many philosophers and cognitive scientists have jumped on the question, and many countries now recognize sentience in their animal laws.

Behaviorists are now bringing learned place preference tests up yet another notch, exploring how thwarted animal desires resonate with the creatures to the point of coloring their outlook on life. Ethologists who study animals' natural behaviors in or out of the lab find that, within a troop of monkeys or a murder of crows, individuals vary. They have personalities that remain more or less stable through their lives. Some are bold, some meek. Some are inquisitive optimists looking for the next best thing in their world. Others are timid pessimists, sticking with the tried-and-true rather than seeking an unlikely bonanza around the next corner. The newest welfare tests measure swings in animals' personalities and moods in response to the pleasures and pains of their lab lives. Animals have a cognitive bias, from being an Eeyore, always pessimistically expecting a low return on effort, to my onetime Boston terrier, Freddie, spending vigilant hours at a chipmunk's burrow, optimistically expecting to bag a trophy.

In one version of cognitive-bias tests, an animal learns to associate a stimulus (perhaps presence in the chamber with vertically striped walls) with some reward. The reward should be absolutely fantastic, such as a succulent mealworm for a marmoset monkey, but only one try in ten delivers the prize. The horizontally striped chamber reliably delivers a boring piece of celery. Given a single shot at entering a chamber, optimists go against the odds, aiming for a jackpot in the chamber with the vertical stripes. Pessimists play it safe, going for a low-value, high-probability outcome in the horizontal. One test of optimism: What to do about a chamber with diagonal stripes? Diagonal stripes are an ambiguous cue the optimist sees as "vertical enough" to invest some effort in exploring them.

Genes and environment influence the cognitive biases of animals' personalities. While personalities are fairly stable over time, they are not set in stone. Chronic pain, social stress, or other suffering can shift an optimistic animal toward pessimism. Some rat strains are genetically prone toward optimism or pessimism, but, when moved from barren cages to cages with tunnels, gnawing sticks, and materials for nest building, both kinds of rats become more optimistic. Rats in unpredictable housing circumstances tend toward the pessimistic, as do rats in chronic pain. A welfare scientist can track these shifts as markers of good or bad welfare by asking the animals how they see the ambiguous cues, be they diagonal stripes or something more relevant to that species's way, visual or otherwise, of perceiving the world.

In a 2012 study of rat optimism, Rafal Rygula and his colleagues taught rats to discriminate between hearing a tone and pushing a lever that delivered a sugar reward versus another tone that warned of an electric shock unless they hustled to press a different lever. If they got it wrong, they got no sugar and instead got an electric shock. (Notice that animal-welfare scientists do sometimes hurt animals in their experiments; not only did they administer electric shocks to their rats, but they made sure the rats found the sugar rewards that much sweeter by keeping them hungry between training sessions.) Presented with an intermediate tone, optimistic rats acted on the belief that a reward was likely. More pessimistic rats pressed lever 2; a sugar treat would have been nice, but they acted as though they assumed they would likely get it wrong and get a shock instead. When tickled, rats make a sound largely outside our range of hearing. It's okay to call it *laughter*; the professional animal-welfare scientists do. Tickled rats tend to come back for more. But not only do they appear to like it: when Rygula switched experiments with the same rats, the tickled rats optimistically pushed the sugar-reward lever when they heard an ambiguous cue rather than act as though they were expecting the worst and must push lever 2 to circumvent it.[13]

The 2024 New York Declaration on Animal Consciousness summarizes some of the data to date on how widespread animal sentience is throughout the animal kingdom.[14] Sentience is just about everywhere we look. Neurologic similarity to mammals is strong evidence that a creature with a similar brain and nervous system likely has at least some rudimentary consciousness. But many animals, with octopuses the best-known standouts, seem remarkably sensitive and intelligent even with a nervous system that evolved quite separately from our own. So scientists look beyond whether a creature has a brain and see if they seem aware enough of their own welfare to weigh choices as they balance their needs for safety, sex, or food. Learned place preference tests and other learned behaviors go a step further, probing how memories of prior good or bad events factor into animals' moment-by-moment evaluation of how to balance their needs and motivations. Just ask a hermit crab, a busy little creature whose tiny brain looks little like ours but yet does the job of guiding the crustacean through the day.

The animal behaviorist Robert Elwood and his students put hermit crabs to the test to probe the extent of sentience, or feelings about their feelings, in the animal kingdom. Crabs rely on a hard shell to protect them from environmental fluctuations and from one another. Most crabs grow their own shell, but not hermit crabs. Instead, they squat in empty sea-snail shells whose owners have died, scurrying naked from one shell to another a size up to accommodate them as they grow. Presented with choices, they have preferences for one type of snail shell over another. Once in a shell that suits them, they leave only reluctantly, for mating, for scurrying to a better shell, or for escaping when bullied by a rival who wants their shell. They also leave their shell when scientists diabolically snake little wires through holes they've drilled into the shell so that they can deliver a small electric shock to the unsuspecting hermit crab. With a tiny shock, the crab reflexively twitches, like a human patient when a doctor taps their knee. That reflex is not necessarily evidence of some higher cognitive or emotional process. It is not strong evidence of sentience. So the scientists up the amperage, watching for behaviors that go beyond reflexive twitches.[15]

One indicator of sentience is watching how animals make motivational trade-offs, weighing the pros and cons of different options as they embark on a behavior. If we are hungry, and the cupboard is bare, we may look out at the pouring rain and decide just how hungry we really are. Are we hungry enough to get wet and cold, or not so much? At a high-enough shock, the crab leaves the shell. But the crab doesn't just flee without evaluating the situation and weighing some options. Crabs have a choice and make a decision; they evacuate less preferred shells quickly, and, reluctant to give up a more preferred shell, they wait for the shocks to subside. Alone in a test chamber or exposed to the odor of a harmless mussel, they scurry out of the electrified shell. But if they smell a predatory crab species in the water, they hunker down in their electrified shells longer, balancing their motivation to escape the electric shocks against their motivation to avoid being eaten.

Hermit crabs' deliberations about shell occupancy show some higher mental functioning that should make us cautious about how we treat them, but there's even more going on in their crustacean brains. After a shock serious enough to launch hermit crabs from their protective shell, they are tentative about going back in, exploring a shell rigorously before trusting it again. This hesitancy to reenter a wired shell lasts at least a day. So something is going on in that crab's brain that's more than reflex. I tread lightly here, not wanting to overinterpret the crabs' experience or to put too human a spin on it. They clearly have a nociceptive response, which is to say that a pain signal gets from the site of the shock to their brain. Their deliberations about leaving or returning suggest an emotionally unpleasant sensation, unpleasant enough that a crab with a fair memory predicts the next day that entering a shell too eagerly could result in more pain. Sentience? I certainly would give them the benefit of any doubt.[16] In truth, I have little doubt that hermit crabs are sentient enough that their welfare matters to them and therefore should matter to me. I do not doubt I have some obligation to promote their best welfare in interactions I have with them.

We mammals have changing moods, but we are not alone in that. Cognitive-bias tests reveal many apparently sentient animals toggling between optimism and pessimism, revealing signs of animal sentience in abundance. In San Francisco the octopus-welfare expert, Robyn Crook, asks octopuses and their kin to tell her about their experiences and their perceptions of pain and stress. She and her colleague, Sarah Giancola-Detmering, trained stumpy-spined cuttlefish, a tiny relative of octopuses, to associate vertical or horizontal stripes with a jackpot reward of a whole shrimp. They then divided trained cuttlefish into a stressed experimental group and an unstressed control group, the stressors being both life in an impoverished bare tank along with chasing the animals in their tanks for three minutes twice daily. Faced with ambiguous diagonal stripes, the unstressed cuttlefish judged that diagonal was close enough to whichever pattern they'd learned would give a shrimp reward and put their energy into testing their hunch. Stressed cuttlefish remained more pessimistic and did not put out the effort to explore the ambiguous cue. Lessons learned: cuttlefish too may have what it takes to suffer or to thrive and to learn from their experiences, and they therefore belong on my list of creatures I should avoid harming. They are sentient. They care about how we treat them. We should spare them unnecessary suffering (I'm holding questions of what constitutes "unnecessary" for the chapter "Flea").[17]

I'd hoped to find a line somewhere between sentient animals whose welfare I must promote and nonsentient animals to whom I owe no more concern than I do to a carrot. I want that bright line so I can feel I'm an ethical vet when I treat a puppy by killing off uncounted (and preferably nonconscious) fleas and ticks and worms. Yet, the more they look, the more scientists find evidence that even honeybees and other insects have enough consciousness to experience better or worse welfare. I live in fear that carrots too will be found to suffer under my knife, though I feel some confidence they will not. Either way, sharp lines are out. Maybe we can quantify sentience and score some animals as more sentient than others and thus more demanding of our concern. Animal-policy scholars talk of "relative replacement" of vertebrates in painful experiments with other, "less sentient" animals, but I am still waiting to see their data on whom they judge to be less sentient and how they make that judgment.

Oysters continue to vex me. In terms of welfare-relevant sentience, are they more like a cuttlefish or like a carrot? In settings where predatory crabs abound, oysters will put energy into growing a thicker protective shell, but it's hard to see that physiologic adaptation as evidence of wel-fare-relevant sentience. They glue themselves to rocks and never budge, and their range of observable behaviors is pretty well limited to opening and closing their shells. Sentience researchers who rely on place prefer-ences or aversions are challenged in confronting a sessile creature who cannot move toward or away from any threat or reward. Perhaps oysters would reap no survival value in developing the nerves that could support sentience, and evolution does tend to weed out unnecessary capacities, such as sight in cave-dwelling fish, that cost metabolic energy to grow and maintain.

Oysters' skeptics may be right to put them in the same moral category as carrots—Peter Singer has discussed them in the three editions of his book *Animal Liberation*, fine-tuning his ethical stance to harmonize with emerging data—but I'm not yet writing them off as nonsentient. For one, they start their lives as motile animals who need to navigate the ocean and find a safe place to settle down, so they have enough neural function, as youngsters, to survive until they've found a nice rock to commit to. No one has yet tested whether their larval neural function rises to the level of sen-tience, but, if it does, do they really lose that when they choose their rock and settle down? Moreover, adult oysters in the shell do have defense behaviors, clamming up if they perceive a threat, which should allow some clever welfare scientist to explore their learned behaviors and the implica-tions for their welfare. And, behold, oysters do respond with a clamped-down shell to loud noises and vibrations or to the odor of predators. Cognitive scientists look at associative learning as another criterion of possible sentience, as when cuttlefish or mice learn that a normal neutral striped wall can be a cue that a reward or punishment awaits. So it's time to look for this capacity in oysters. Couple a normally neutral low-level sound with the smell of a predator, then see if they later clam up just at hearing the sound alone and chart their behaviors to match the unpleasantness of having to hide from predators. For now the welfare lessons from the oyster labs are not that they are sentient or nonsentient enough for us to choose between tofu and oysters on our menu but that we don't yet know enough and should factor that uncertainty into our choice.[18]

The ongoing search for sentient life on our planet redefines which animals have welfare interests we should protect. Already several countries incorporate sentience into their animal-welfare laws. Animal-welfare science can then complement sentience studies to fine-tune what we know about these sentient animals and to chart an ethical course around their interests in their own welfare. Some countries have banned boiling live lobsters or storing them on ice, without banning lobster (or cow or fish or chicken) consumption. Ever the skeptic, I'm convinced lobsters can feel pain and can suffer but not convinced we've yet got good evidence to choose the best alternatives to boiling water or ice. Nevertheless, their sentience credentials make lobsters the objects of some legal and ethical concern and the subjects of experiments to chart how to best respect their welfare.

Animal-welfare science thus can guide better care of lab animals, but reading animals' minds is complicated business. We lab vets are usually the de facto welfare authority in the animal labs, but we receive little training in observing animal behavior or judging welfare. I remember a lab inspection with my ethics committee where the vet in the lab had no idea that the many mice turning somersaults in their cages might be displaying signs of poor welfare, psychological disturbances that could well invalidate the cognitive tests the scientists ran in their brain research—but then neither did the biologists or the nonscientists on the committee. We set that lab and its vet up for a consult with an animal behaviorist just as an MD doctor would send a human patient to a mental health specialist if the patient were obsessively flipping somersaults. Step one in such a situation is recognizing that you have a problem in your mouse or monkey house.

Because welfare science feeds into our practices and affects animals' lives, it's important to do it right. Two cases, one concerning exercise for caged dogs, and the other, cage sizes for breeding harems of guinea pigs, show some of the many ways of getting it wrong. Congress had passed the Improved Standards for Laboratory Animals Act in 1985 to fix deficiencies in the 1966 Animal Welfare Act, obliging the USDA to collect public input, draft regulations, and then redraft them in response to yet more public input, a process that required five years of sorting through tens of thousands of comments. Everyone from the most ardent research proponents to equally vocal animal rights groups had an opinion on what constituted best animal welfare and therefore which updates, some of them potentially quite expensive, the USDA should require. Every issue was on the table, from the new requirement for ethics committees, to a mandate to search for alternatives in painful experiments, to a consideration whether to formally include mice and rats under its aegis, to rules for exercise for dogs and environmental enrichment for primates.

Vets and researchers insisted that any changes in the regulations should be evidence-based, objective, and scientific. Animal advocates countered with their own beliefs, a mix of anecdotes, anthropomorphism, and commonsense claims of what "everybody knows" that animals want. The two camps squared off, with activists claiming to know that dogs need an hour a day of exercise or a play yard of some given size, while researchers presented data to refute this, and vets bemoaned the costs of hiring dogwalkers or building new dog yards. Primate psychological well-being was a new concept that Congress had introduced, and it flummoxed everyone. In lieu of fights about how big a monkey cage should be or how stimulating an environment to give them, the debate centered on who, not what, would prevail. Activists wanted behaviorists such as Jane Goodall to write the rules, mistrusting researchers, lab vets, and even the vet inspectors at the USDA. Science lobbyists wanted the broad-brush rules the USDA ultimately settled on, leaving monkey mental health in the hands of local ethics committees.

Two issues caught my eye as I pored through those thousands of letters the USDA received (the backbone of my PhD studies in vet ethics). Exercise for dogs had been a contentious issue since the early 1960s, with activists clamoring to get lab dogs out of their cages and lab vets saying, "not necessary." Congress forced the issue in 1985, instructing the USDA to write dog-exercise standards. Meanwhile, cage sizes for breeding groups of guinea pigs surprised me as an issue up for debate, and, given that only a handful of commercial breeders dabble much in making baby guinea pigs, the issue elicited much less commentary.

These two issues were the only two in which the USDA explicitly reacted to research papers in its final regulations. In both cases the USDA had started big in its updated regulations, proposing time for dogs out in large exercise pens and larger guinea pig cages that animal activists applauded and that scientists and lab vets saw as needlessly expensive. The research lobby put forth science that trumped the animal protectionists' claims of animals' needs and preferences. In both cases the USDA scaled back its proposals, relying on the data scientists had sent them. Between them, the guinea pig and dog studies illustrate some common mistakes when scientists conduct welfare studies to shape public policy. Guinea pig and dog experiments used video to get the most unbiased data they could collect, free of the distracting effects of human presence on the animals' behaviors. However, they loaded in plenty of bias up front in designing their experiments and choosing what they would consider relevant welfare evidence and, after data collection, interpreting what their animals showed them and what it meant in terms of welfare policies. First, a summary of the studies in question.[19]

The Animal Welfare Act's standards for guinea pig cages defined how many inches of cage size an adult guinea pig should receive, with no added allotment for babies. Thus a sow guinea pig with her litter would get the same space as a pupless female. Three scientists and their video recorders vasectomized males and housed them with a small harem and charted how they used available space in the heretofore regulation-size cages compared with cages a bit larger. Even in the larger enclosure, they found the pigs mostly hugging the walls in a group, spending only around 15 percent of their time out in the middle of the larger enclosures. They concluded and, though it was but a single unreplicated study, the USDA agreed, that there was no cause to mothball existing cages and purchase newer, bigger models.

Activists had been pushing for years to get dogs out of small lab cages. The law requires that a dog cage must be six inches longer than the dog (nose to rump, minus the tail); anyone who enlists their dog to map that out on the floor will agree that that is a very tiny space allotment. The law simultaneously sets what they called a performance standard, that cages must allow a dog "to walk in a normal manner," however impossible that would be for any dog larger than a Chihuahua in a cage only six inches longer than the pooch. Though many labs in fact housed their dogs in larger kennels (four feet by six to twelve feet would be common), vets and scientists pushed back against writing large dog cages, exercise pens, or hired dogwalkers into law as far too expensive a response to dogs' needs.

In contrast to a single experiment asking guinea pigs what they wanted, more effort went into trying to define dogs' exercise needs, with some scientists focused on physical measures, others on stress hormones and medical indices of health, and behaviorists monitoring for signs of stress. The study that most caught my eye video-recorded dogs singly or alone and in small cages or larger pens. That study found that dogs "traveled" more or spent more time moving around than single dogs in larger, bare enclosures. The consensus was that dogs get enough exercise in the cages I described earlier, about six inches longer than the dog (minus the tail). The USDA found itself caught between a congressional mandate to develop exercise standards for dogs in those small cages and a research community showing that what Congress had mandated was not necessary. Ultimately, the USDA finalized a complicated formula through convoluted reasoning, ruling that doubly or group-housed dogs and dogs in twice the mandated minimum enclosure could count as getting enough exercise, with only singly housed dogs in the smallest cages triggering a requirement to get them out of those, in whatever manner and frequency the vet and ethics committee chose.

The guinea pig study's first sin was its low external validity, the technical term for studying the wrong model such that the findings don't apply to the population for which we're running the experiments. In medical research this shows in debates about whether mice are as valid a model of multiple sclerosis as are marmoset monkeys. Scientists in the guinea pig lab had vasectomized the boars to simplify their experiment, eliminating the population fluctuations as sows birth new pups. The assumption is that their feelings about space and crowds will not change during pregnancy, when they swell to twice their width, or while rearing pups. The whole point of proposing a change was that behaviors, and certainly crowding, is in flux when a group of adult animals is not just having sexual relations but producing offspring. Imagine basing human-family policy on surveys of adults without children to get a more uniform sample of responses. Is there good reason to believe a "breeding group" that is not in fact breeding represents the social dynamics and welfare preferences of guinea pig families? I doubt it, but the USDA didn't, and they're the ones who decide.

As Hughes and Black's chickens had voted with their feet on where they wanted to spend their time, so too the guinea pigs voted to spend more time hugging the walls than venturing into the open center of their enclosure, where they would find no cover from whatever predators they might envision. The vets running the experiment were not ethologists and so rediscovered a well-known phenomenon (ethologists call it *thigmotaxis*): most rodents seek cover and hug the wall of their enclosures. This is such a powerful urge that behaviorists measure a rodent's aversion to open fields as a measure of anxiety or of the effectiveness of antianxiety medicines.

The USDA shared the researchers' simplistic assumption that animals' decisions of how to allot their time in different locations or during different activities could straightforwardly lead to welfare practices. They reasoned that if pigs spent only 15–25 percent of their day in the open space at the center of their enclosure, it must not be that important to them. Dogs sometimes stand on their hind legs, but not in the small cages that meet Animal Welfare Act requirements. Pressed during their regulatory update to give dogs a chance to stand up on their hind legs, the USDA wrote, "We do not consider a dog's standing on its hind legs to be a frequent enough postural adjustment to require its inclusion as a minimal standard." For decades regulators and scientists have justified the low headroom of rat, rabbit, dog, and baboon cages by noting how little time these quadrupedal creatures give to rearing up to look or sniff around them. But that quick scanning, though well under 15 percent of their day, could mean the difference between life and death in the maw of a fox or leopard. How might it stress these animals to hear loud or unexpected noises and have no chance to explore to their satisfaction? What amount of standing upright would be frequent enough for the USDA to decide that dogs deserve more headroom than the current six inches above their head of its regulations? Some low-frequency behaviors are quite important for welfare. I spend well under 15 percent of my time (that's three and a half hours a day) in my bathroom, but I would not want an apartment without one.

The guinea pig researchers' biggest errors lay in their experimental design and interpretation, collecting data that did not fully explore what guinea pigs want and need and exploring even less what pregnant, nursing, and growing guinea pigs might need. The contemporaneous dog research that convinced the USDA to back off from requiring more expansive dog-exercise programs erred in recasting the commonsense (and congressional) understanding of "exercise"—that is, a chance to run and play—to a question purely of cardiovascular and muscular fitness. Now no one had ever worried that lab dogs were flabby and out of shape, but in casting exercise as purely physical, scientists could present data that dogs jumping or spinning in small cages got a fantastic workout.

The USDA missed, or dodged, the point of putting dog exercise into the regulations. Pet owners see their dogs' excitement to get out running if they've been cooped up in the house all day. If they knew that the legal size of a dog cage is as tiny as it is, they'd likely expect that getting out would be even more important to lab dogs. That seems to have been the reasoning when Senators Bob Dole and John Melcher slipped the dog-exercise provision into the Animal Welfare Act. Deciding how much exercise dogs need raises the question, "Need for what?" If it's to limit the distress of small cages, welfare scientists might rate how many dogs develop abnormal spinning or self-licking behaviors. They can compare stress-hormone levels in caged and free-running dogs. If they want to measure the richness of canine lives, they can observe dogs' engagement and exploration of their environment and how that increases with social or human interaction or with a more complex enclosure than a bare, concrete-floored exercise pen. If the point of the amendment were to increase dogs' cardiovascular and muscular fitness rather than their happiness, scientists could measure their fitness as easily as can a trainer in a human gym. They can track how much the dogs actually get exercise by jumping or running or moving around.

The guinea pig and dog-exercise experimenters committed the same error Hughes and Black had done in their pioneering foray into asking animals to tell us they want. For behavior-based welfare tests to have any use, they must offer a wide range of choices. Hughes and Black hatched the science of animal welfare in letting their chickens choose among different types of flooring. They presumed that hens would live their lives in cages, so they never got to show how strong a preference they might have for actual soil or grass. The guinea pigs did not get the choice of a recon-figured cage with lots of nooks and crannies for wall hugging or some material over the barren openness of the cage interior that might mimic the safety of leaf cover (in fact, a later study did find just that).[20] Dogs on a beach could easily have given comparative data to putting the caged bea-gles and their "traveling" in perspective. Dogs, guinea pigs, and chickens got the chance to choose between what intuitively look like unappealing choices: a small, barren exercise pen for a dog, a cage with minimal hiding places for skittish guinea pigs, or a choice of artificial floors for caged hens. To understand why welfare scientists would give animals such a narrow range of options, look at the timing and goals of the work.

A sure red-flag warning of likely problems in animal-welfare studies: if labs confronting tighter or more expensive regulations perform or com-mission the studies, even the best-intentioned scientists are at risk of bias in their study design, data collection, and interpretation. In the United Kingdom, the Brambell Committee was suggesting farmers should invest in revamped henhouses without chicken-wire floors. Congress and USDA were moving toward expensive proposals to replace existing cages, build exercise pens, or hire a fleet of dogwalkers. This creates a subtle frameshift from asking, "How can we improve on our current animal treatment?" toward "Does this expensive regulatory proposal make things better for the animals?" Framed as an open-ended exploration of the best of all pos-sible, though nevertheless confined, animal worlds, welfare science will often point to practices far more generous to the animals than the incre-mental changes in regulatory proposals. That's the science that should precede any proposals for larger or different enclosures or changes in how people interact with the animals.

The alternative approach invites biased studies, aiming to show not that the status quo is perfectly fine but that it's no worse than proposed alternatives. Welfare scientists in this situation implicitly reason, "Given that we *will* continue housing dogs or chickens in small cages, what small changes might be in order?" Finding that a small cage is not much better than a tiny cage, such biased studies tip the scales toward sticking with a status quo that fails to meet animals' interests.[21] Too often animal-welfare science starts with a status quo of lab or farm life that researchers know they will not upend, settling instead for small-bore studies to distinguish bad from really bad conditions for the animals. It becomes the practice of trying to squeeze round animals into square holes, having decided from the outset that square holes really are all we are going to give them.

As surely as poor welfare in developmentally stunted animals can skew biomedical studies of animals' physical performance in health and disease, surely it can affect psychological studies, including those in animal-welfare science. Most data in animal-welfare science comes from farm or lab animals raised in confinement and unnatural social groupings and settings and should be considered with a grain of salt. Most studies make little reference to how the studied species, or their closest wild relatives, live their lives.

The USDA's ruling that dogs in tiny cages would merit exercise programs while dogs in small cages would not raised hackles. That dogs in tiny cages maintain muscle tone convinced no one, except the regulation writers at the USDA, that those tiny cages were adequate. In an internal USDA survey, USDA's field veterinarians found plenty of confusion in the new ruling but little improvement for caged dogs in labs they inspected. The Animal Legal Defense Fund filed a lawsuit claiming that flexible performance standards were not standards at all. Judge Charles Richey heard the case and agreed. "'A dog is man's best friend' is an adage the defendants have either forgotten or decided to ignore," though an appeals court threw out the case, saying the ALDF had no standing to sue. John Melcher was a veterinarian who became a US senator and was one of the three main congressional architects of the 1985 Animal Welfare Act amendment that brought us the dog-exercise requirement. Years after the USDA finalized its rules for his and Senator Bob Dole's call for dog-exercise standards, Melcher castigated the agency's and scientists' needless cruelty in an editorial in the *Washington Post*. Whatever scientific data the USDA had found so compelling did not convince him that a pair of dogs sharing a small cage could have enough room for exercise, as Congress had intended.[22]

Animal-welfare science can guide better care of animals, but it requires multiple looks at the same issue to understand how much exercise or what kind of home the animals want, how important those preferences are, how much they may suffer if denied their first-choice preferences, and whether the animals can adapt to their second or third choice and still have a decent quality of life. For a single welfare experiment to tick all those boxes and thereafter determine animal-welfare practices would be a tall order. Imagine acting on findings from single isolated experiments in medical practice without replication and related studies; this is not how science works. Yet, time and again, single studies shape animal-welfare practices because, unfortunately, funding for animal-welfare science is scarcer than hen's teeth, especially for laboratory rodents. Without funding, welfare scientists backlog the many questions they want to tackle. Because of this, much of the welfare science I share in this chapter comes from single, small studies, and all of us arguing for stronger evidence-based standards for animal care confront this challenge to the quality of our evidence.

Animal-welfare scientists face several hurdles in working to improve the lives of animals in lab cages. Money, researchers' skepticism, vets' need to control infections, and general inertia all conspire against progress. These factors all converge in the case of the IVC, the individually ventilated cage, in which millions of lab mice live. Vets' focus through the decades on infection control for animal health aligned with scientists' interest in keeping the unwanted variability of animal infections out of their experiments. These dual interests have resulted in a technology of sterilizable, individually ventilated plastic boxes in which millions and millions of mice are living as you read this paragraph. The mice are isolated from one another, breathing sterile air pumped through their enclosed cage and eating sterilized food. This expensive high-tech containment is essential when housing mice with severely compromised immune systems, common subjects in immunology and cancer labs, who will die if exposed to common bacteria, such as those on a healthy technician's hands. The filtered cage comes with the added advantage of protecting people from mouse allergens that would otherwise be wafting through the air, mitigating a serious worker-safety concern in the mouse house.

These IVCs tick many boxes, and their use has metastasized far beyond the labs working with immune-compromised animals. They are not cheap. Cages may cost close to a hundred dollars each, and the shelving system tricked out with air filters costs in the tens of thousands of dollars, while the cage-wash robots that large institutions favor carry a price tag in the hundreds of thousands. Designed to deliver the efficiency and worker safety that facility managers love, the standardization that researchers seek, and the infection-containment vets prize, these expensive cages and their peripherals have secured their place in mouse and rat labs. The technology first came to life long before anyone was asking British chickens or anyone's lab mice what they might actually want.

A chorus of critics—not solely research outsiders and animal rights activists but behavioral scientists too—speak quite forcefully that IVC systems, especially as they are typically managed, are not mouse-friendly. Joseph Garner and his colleagues decry this industrialization of lab animal care. They write that the high air-flow rates in containment cages, necessary for keeping ammonia levels low in a cage full of mice urinating and defecating, results in mice who are chronically cold-stressed. This kind of constant stress changes mice physiology and may have an impact on the data scientists are trying to collect from their animals. Garner and his colleagues report that the machinery handles bedding made of ground corn cob more easily than wood shavings or shredded paper. Corn cob is gentler on the equipment but abrasive on mice's little bare feet. Vera Baumans and her colleagues asked mice, a la Black and Hughes, their opinions of ventilated cages. They mostly retreated to an adjoining cage without the breeze.

Neuroscientist Garet Lahvis models empathy in psychologically normal mice and its breakdown in mouse models of autism. He watched mice digging in the corners of their IVC cages, whether trying to make a tunnel with the bit of paper bedding in the cage or to excavate an escape hole for themselves. Whatever their goal, he concludes that these mice are chronically frustrated, with no way to live emotionally healthy lives in current mouse houses. Several scientists have determined that their data in IVCs depart from data in more conventionally caged mice—not necessarily better or worse but different enough to caution use of these cages. Immunologists have found that mice in their sheltered IVCs, never seriously challenged with infections to fight off, retain immature immune systems throughout life, making them a poorer model of adult human immunity than a mouse from the neighborhood pet store.[23]

In short, housing mice in small plastic box cages, especially the units with forced-air ventilation, displeases the mice while it disrupts the data they yield. Scientists need to more fully explore just how strong these welfare and data-quality disruptions are to chart a course: Do we continue with these unsatisfactory cages or do a million-dollar overhaul of our vivaria? Large universities and pharma companies, with their budgets of several billion dollars, make these investments in animal housing and should better factor in animal welfare in their planning. They need, and their animals deserve, solid animal-welfare scientific data to inform more animal-friendly facilities and to promote better animal-based science in the process. Some improvements will be difficult without the capital outlay. The system is designed for five-inch-high mouse cages and a few square inches of floor space per mouse. How to give enough depth of bedding for making a proper tunnel in such a short cage? How to let a mouse follow their nature and separate their toilet area from their sleeping area? How to keep the air mice breathe in their tightly sealed boxes free of ammonia and dampness without blowing air they don't like through the cage?

One principle of applied animal-welfare science: since we cannot get into every animal's head, we should give them the space and the tools to have some control over their own lives. Let them decide how much to socialize or to retreat to a corner for some "me time." Let them build an elaborate underground tunnel. Let them build a warm nest and get away from the air blowing in. Let them run in an exercise wheel or forage at night for an occasional treat staff have scattered in their bedding. Meeting these needs and desires in a cage the size of a small shoebox is a challenge. Some of the shortcomings of our present animal-housing systems really do require an overhaul of the infrastructure, but, once an animal facility sets off on this path, change is even harder. Animal cages and all their peripherals map onto cage-size charts in the welfare rules (slightly larger in Europe than the United States, but not by much). Engineers set cage-wash robots to handle those cages, so any new rooms a facility might outfit will follow suit and buy cages that match their robots. Reworking the specialized holding racks to handle larger cages is a challenge, and replacing them a daunting expense. A multimillion-dollar vivarium quickly locks in to certain infrastructure specifications, hence the general hullabaloo raised by vets and scientists when animal activists propose seemingly small changes in the Animal Welfare Act or the *Guide for the Care and Use of Laboratory Animals*.

Short of replacing expensive infrastructure, animal-welfare science can still guide a more modest quest for a happier mouse who will produce better research data. That can include seeking ways to lower the animals' stress when interacting with them, as well as doing what we can within their limited enclosures to let them do what they prioritize. For most of my life, the only way I knew how to get blood samples from monkeys was to restrain them enough for a dose of ketamine, never believing we could get our samples from fully awake animals. That would sometimes complicate my clinical work, wondering if a high white blood-cell count represented an active infection somewhere in the body or was simply the stress of the restraint and the ketamine. Those high blood-cell counts could similarly be influencing the physiological data scientists were seeking in their experiments. Melanie Graham is a diabetes scientist whose experiments sometimes require monkeys. She is working toward better pancreas transplants to manage human diabetes and needs to know that her monkeys are truly diabetic and that they are responding to her therapies. Like human diabetics managing their insulin intake, she needs to monitor blood sugar precisely, and restraint stress can seriously skew sugar levels. She parlays her dual expertise in diabetes research and monkey welfare to refine her experiments so that the monkeys voluntarily present an arm for a blood sample and skip the stress of vets' conventional methods. She and her colleagues emphasize developing a bond between the staff and the animals, so they see humans less as stressors than as friends bringing treats and affection. They break tasks down into small steps—presenting the limb, letting the caregiver touch the limb—with rewards for each step, building up to the monkey holding still for the blood collection. In the process Graham gets better scientific data from less stressed animals.[24]

Brian Hare studies chimp, bonobo, and dog cognition, but only by inviting the animals to play various cognitive games and puzzles he sets for them. The dogs come to his lab with their people for testing and otherwise live in homes. The apes are semifree, ranging in sanctuaries and, like the dogs, are free to engage with tasks or not, as they see fit. He and his students are able to learn a good deal about how their minds, likely healthier than the minds of chronically caged animals, work.[25] This approach to getting data from the least stressed and constrained animals would be a challenge for scientists studying other species (and it's a challenge for Hare and his team too), especially for more medically focused research, but it is an inspiration of how to let animals inform better experimental design.

I am advocating for similar win-win results for the millions of mice whom scientists deploy each year. Mice lack monkeys' smarts, but they too are trainable. For decades vets in vivaria have taught their animal-care staff and researchers how to handle mice. We pick them up by the tail—the most obvious mouse "handle" and farthest from their tiny sharp teeth—to transfer them to a clean cage or to position them for an injection. Welfare scientists since 2010 have explored the ways people stress mice simply by picking them up by the tail. Can people be trainable? I'd have expected that after this frequent encounter—everyone in the lab, from the daily caregivers to the vets to the experimenters, handles mice this way—the mice would acclimate. Clearly, they survive time after time; surely they should learn this is not a big deal. Surely, however, I should see the incorrect anthropomorphism of assuming (I haven't tested this) that being chased and grabbed several times a week by an ogre who outweighs me two thousand–fold is something I should just get used to, assuming I've squirmed free of the ogre every time. Or maybe that's precisely the anthropomorphic insight that I should put to the test and ask the mice how they feel about weekly encounters with ogres and narrow escapes. Do they learn that capture is not so bad, that it never leads to anything worse? Or do they enter each encounter as though it could be the last time they'll escape? Apparently, while mice may come to accept being coaxed into a tube for lifting them or onto a technician's palm, tail restraint is just too stressful. The stress can be great enough to lead to kidney disease, bad for mouse welfare and bad for many experiments. Mice who regularly find themselves hoisted by the tail are more anxious than their counterparts whose humans learn to gently coax them onto their palm, and they are more likely to flee when a gloved hand appears above them.

In short, though welfare scientists differ on how strong the effects are, mice suffer more from being grabbed by the tail and thus make worse research subjects. The animal ethics committee, the scientists, and the vet staff should all be gunning to train all mouse handlers on this "new" (fifteen years and counting) refinement. Two researcher-driven animal-welfare organizations—the National Centre for the Replacement,

Refinement and Reduction of Animals in Research in London and the Three Rs Collaborative in North America—aggressively promote this less aggressive approach to mice and do their best to track the middling uptake of the method around the world. Welfare scientists have surveyed the barriers to facilities moving researchers and carers to keep their hands off the mice's sensitive tails.[26] Even this simple advancement that requires no money or equipment expenditures struggles to overcome the fears that it might slow down the work or, really, fears of something different from what we've always done. The next edition of the *Guide* could change that, with a simple statement that tail restraint requires ethics committees' approval, but why do we need to wait for that? Can vets and scientists simply take the lead without external coercion? Are they trainable?

Chickens hatched a new field of inquiry, animal-welfare science, that shows vets and researchers ways to make life in the animal lab not just less distressful but to go beyond and aim for animals' happiness. Scientists and vets have resisted many of the improvements that welfare science points toward, when better welfare may cost more money, threaten vets' priorities in infection control, or succumb to scientists' fears that they will lose the controls they need in their experiments. And sometimes it's simply inertia that stops research staff from adopting clearly justified improvements in animal welfare.

Animal-welfare science got a boost in 1985. Monkeys and apes, or more specifically, abuses in primate labs, spurred Congress to pass its last serious effort to upgrade lab animal welfare. In the next chapter, I trace some of that history that sprang chimps out of labs, gained better care for the monkeys remaining in labs, and raised the welfare bar for other animals.

Chimpanzee

Pan troglodytes

RICHER LIVES FOR PRIMATES . . . AND
ALL ANIMALS

Working at the zoo in my teens in the 1970s, I helped raise three young chimps. Winston was my favorite. He was playful and inquisitive, and no one could chortle like him, lying on his back for some tickling. Zoo visitors loved to melodramatically hold their noses and laugh at the "smelly monkey," whose outdoor cage let them get as close as four feet from him. Junior volunteers like me rotated sitting by the chimp cage, earnestly telling half-interested visitors about chimp conservation and pleading with others not to toss him their popcorn. We explained that great apes such as chimps are very different creatures from monkeys like rhesus macaques or baboons or marmosets, even though we often lump them all together under the word *primate*. Monkeys have tails; apes don't. And apes like Winston are even smarter than monkeys. They are much more closely related to us humans than they are to any monkey, I diligently explained, while the human apes kept laughing and making faces. And, for the record, he did not smell particularly bad.

If I ever thought about where Winston or the other youngsters in our zoo came from, it was with the fantasy of going on a "collecting mission" myself in Africa someday. Virtually all chimps in the United States in those days had been wild-caught in Africa as babies, forcefully taken from their mothers or anyone else who tried to stop the kidnapping. Baby chimps got diapers and cuddles in zoos. When they outgrew diapers and were getting too rambunctious and strong for a zoo nursery, they went . . . somewhere.

A professor of mine in vet school had one day mentioned with some affection a lab chimp named Winston. Years later I tried to uncover whether my baby Winston in the 1970s was later this professor's lab subject. What I have pieced together is that when Winston outgrew our small zoo, it seems the zoo did indeed sell or loan him to Harvard for a while, to run on a treadmill for studies on the evolution of bipedalism. After that folks who knew him then tell me they believe he went to the circus, though he may have gone to a medical lab. The luckiest chimps of his generation got themselves to some of the better zoos and lived for decades in a family group. Few chimps won that lottery though, and I am now resigned that Winston was not one of them.

Over my decades of animal work, chimpanzees like Winston moved from their caricature as smelly, funny clowns to their current public image as our deeply intelligent next of kin. As chimps gained respect, they brought monkeys along with them. The plight of lab chimps and monkeys then inspired sweeping upgrades in our US animal-protection laws in 1985, upgrades that benefited many other animals in labs, zoos, and farms. With the Animal Welfare Act update in 1985, Congress ruled that all research institutions under the oversight of either the National Institutes of Health (NIH) or the US Department of Agriculture (USDA) would install animal ethics committees to self-regulate their campus researchers' animal labs. And what started out as a primates-only provision—that labs and zoos maintain an enriched environment to promote the animals' psychological well-being—has compelled us to focus on animals' emotional welfare for zebra fish, mice, woodchucks, and virtually all other animals in zoos and labs.

All of us primates, the monkeys with their long tails and we tailless hominids (great apes like chimps and us, their human cousins) share many features, but it's our big brains that most set us apart. Primates' brains are what brings them into many labs, where they model more than most animals can how our own human brains work in health and in sickness. It's not just their smarts—dogs and dolphins rival monkeys in intelligence—but the anatomical similarities to human brains, derived from our common primate ancestors, that make them so valuable to scientists. Primate brains shaped the focus on the mental lives and psychological well-being of captive animals after the mid-1980s. Our laws distinguish nonhuman primates from us humans, but our shared similarities are great. Zoo visitors' laughter betrays discomfort, a recognition that those intelligent eyes peering out at them from the cage belong to animals more similar to us than we want to admit. Laugh at them, put them in cages, lock the lab door: do what it takes to maintain our sense that it is good and natural that we are outside the bars and all the rest of the animal kingdom, even our closest kin, are within.

Television shows in my youth made much of this tension, with chimps in human clothes who were oh-so-close to the human families they lived with but still something comical and easily dismissed. It's like a fun-house mirror. You see yourself reflected but also distorted, and the dissonance makes you laugh. But when you get a chance to look through the bars and into a chimp's eyes, you see the similarity more than the difference. Ham was the celebrity "Astrochimp" whose life spanned the public awakening that primates—monkeys and apes—deserve care and respect. He was about fifteen years older than my buddy Winston, wild-caught in what is now Cameroon in the 1950s. One day in 1961, he was #65, living at the Holloman Air Force Base medical labs. The next day he was stuffed into a cute little spacesuit like a miniature human astronaut, in a diaper, and shot into outer space for a twenty-minute flight up and down. Unlike Laika, the Soviet street dog–turned-cosmonaut who died in her spaceship in 1957, #65 survived, and got the name Ham, aka Ham the Astrochimp, for the deluge of press coverage.

Ham was one of several chimps who trained for space flight, though only two made a voyage. Laika preceded him into outer space, as did the several rhesus and squirrel monkeys whom NASA had launched before him. Dogs and monkeys could show the effects of space flight on mammalian bodies, but Ham's contribution was to model how space flight might affect the human brain. He learned to press levers in response to visual or sound cues, risking an electric shock to his foot for wrong responses, hoping for a banana-chip reward when he got things right. Ham had the right stuff for the mission, and, during his short space ride, he demonstrated that even under weightlessness the mind stayed strong.[1]

I first started working with zoo monkeys and apes in 1971, ten years after Ham's space flight. Zoo visitors saw them as cute babies or funny clowns rather than serious, intelligent animals with lives of their own, animals deserving awe and respect. Zoo chimps in those days wore diapers, a very practical nod to the challenge of rearing baby chimps in human settings once they've been snatched from their jungle families. Though practical, diapers added a layer of anthropomorphism, but the bad kind, where we dress animals up and give them human voices in fables to amuse ourselves or teach us something about ourselves, not the good anthropomorphism that might spark our empathies and imaginations for the animals' sake. Ham the Astrochimp got the anthropomorphic treatment in his toddler-sized spacesuit, lauded as an adorable hero, rather than as the unwitting conscript for the space race.

Then Jane Goodall happened. Millions of people got a look, right there on our televisions during prime time, at chimps as she saw them. Uncaged. They wore no suits or ties or other human clothes. No diapers. They did not rock psychotically in tiny zoo enclosures. They were big, shockingly big, to people who had seen only baby chimps in zoos or on TV. And they had business of their own to tend to beyond childish acrobatics for a laughing crowd. They had grownup things to do, chimp things, not imitation human things. They had social lives, which we learned were not always harmonious. They had kids to raise and food to find. If zoogoers saw a young chimp eating bugs, they'd have squealed in joyful disgust. When Goodall showed them using straws and sticks to "fish" termites out of a hole so they could eat them, I, for one, was in awe. I gave up my *Wild Cargo* fantasies of going to the jungle to catch animals and took Dr. Goodall on as my idol.

Primates, with chimps in the lead, got a public relations makeover through the 1960s and 1970s, as they gradually evolved from *subhuman primates* (the common term in those days) to *nonhuman primates*. Ham's life spanned the years of this makeover. After his space flight, he lived in the US National Zoo, alone, for several years, until he found himself at a zoo in North Carolina, where he at least could be with others of his kind. When he died, young, at age twenty-six (and I wonder if my Winston lived even that long), NASA planned to immortalize him with taxidermy, as the Soviets had done with two of their dog cosmonauts, to be forever in our view in the International Space Hall of Fame. But for a public uproar when the news of that plan got out, you could visit him there today. Send a young chimp to an unknown fate in space in 1961? If that's what it takes to win the space race, yes, and put his picture on the cover of *LIFE* magazine with no worries about backlash or accusations of animal cruelty. Twenty years later, mounting his already-dead body offended our new sensibilities about primates. "Taxidermy is the wrong stuff," wrote the *Washington Post*, warning that it should make any space veteran nervous about their own posthumous treatment.[2]

Despite the evolving public sensitivity about primates, scientists continued pressing monkeys and apes into service behind the scenes for projects for which dogs, mice, and woodchucks were mostly ill-suited. Infectious-disease labs wanted chimps and monkeys for studies of tuberculosis, malaria, polio, and hepatitis. Still today, as many other animals' numbers are dropping in labs, monkeys are still in demand, in fact, so much so that researchers are calling for help addressing a shortage of research monkeys. Toxicologists use them when testing immunology drugs, such as monoclonal cancer treatments, but it's primates' highly developed brains that have set monkeys and apes up for an enduring place in neurobiology research. Since the 1980s primates' braininess has been at the center of debates and protests over their use. For most experiments a wide menagerie of animal models will suffice, but, for immunology and brain research in particular, primates remain in demand.

Monkeys and apes can blame their big brains (their immune systems come in second) for getting them into some of the most invasive animal experiments. In the mid-twentieth century, Harry Harlow ran a now-notorious lab in Wisconsin, studying countless permutations of the ways that separating baby monkeys (and by extension other primates such as ourselves) from their mothers can make them psychological wrecks.[3] Monkeys were the animal of choice for the University of Pennsylvania's head-trauma experiments, given the many functional and anatomical similarities to complex human brains. Monkeys live on in neurology labs where scientists place fine wires into their brains, recording the activity of individual brain cells and examining how the cells communicate among themselves and with body parts outside of the brain. One goal of such work is to provide quadriplegic human patients who've lost use of their limbs a way to use their brain to manipulate robotic arms or operate a computer, work that is now enrolling human volunteers. Monkey brains are a big reason that, while labs are moving away from dogs and cats and even pigs, monkey numbers are holding steady.

Exposés in two monkey neurology labs in the 1980s sparked outrage and legal reform comparable to the dog-trafficking exposés of the 1960s that gave birth to the initial Animal Welfare Act. In 1984 activists with the Animal Liberation Front broke into a head-trauma lab at the University of Pennsylvania, where lab staff had video recorded experiments in which they strapped sedated baboons into car-crash simulators. The activists stole the videotapes and edited them down to a half-hour "worst of" compilation, which shows the researchers occasionally joking about the monkeys and treating them callously. The video, *Unnecessary Fuss*, lives on the People for the Ethical Treatment of Animals website; it's not easy viewing, but it graphically makes the case for welfare reforms. A couple of years earlier, Alex Pacheco, a cofounder of PETA, took a job in a neurology lab in Silver Spring, Maryland, where the head scientist cut the nerves in anesthetized monkeys' arms to study the loss and regain of function. Pacheco got no videotapes, but he took pictures, gathering evidence of infected wounds to bring in Maryland law enforcement to charge the researcher with animal cruelty. The charges were dropped because animal research is excluded under most states' animal-cruelty laws, but the public exposé was potent. These two exposés reminded lab insiders that, even without violence or vandalism, we were wise to keep our monkeys hidden behind locked doors and to scrutinize job applicants. They fueled the growing polarization between animal researchers' critics and defenders. More consequentially, these lab exposés convinced Congress that its faith in scientists' need for freedom from constraints was misplaced. It took action for lab animals as it had not done since dog-trafficking exposés spawned the 1966 Animal Welfare Act.[4]

At the time that stories of dog trafficking for laboratories finally moved Congress to pass the first version of the Animal Welfare Act, science was riding high in the States. Chimps were pitching in to win the space race. Rhesus monkeys from India were conscripted for the successful moon shot of launching a polio vaccine, with the virus grown in African vervet monkeys' kidney cells. Congress came down hard on dog dealing but was reluctant to second-guess how scientists should actually use animals, once they had obtained them, in their laboratories. In debating the new law, congressional sponsors took great pains to clarify their intent to separate animal care from animal use. Senator Warren Magnuson opened hearings on his proposed bill, saying, "I would like to emphasize that the issue before us today is not the merits or demerits of animal research. We are interested in curbing petnapping, catnapping, dognapping. . . . We are not considering curbing medical research."[5]

Under the then new law, USDA's veterinary inspectors would oversee dog dealers and labs' animal-housing vivaria, but the laboratory itself was off-limits. If an animal was "on study"—that is, actively enrolled in an experiment—the inspector had limited jurisdiction. I remember showing my inspector at Cornell some rabbits as I, a vet technician at the time, led her on her rounds. The rabbits had big sores on their backs as a result of the scientists' immunizations. They were ugly, though the animals showed no evidence of pain or itch from them. The inspector did her usual thing, looking for dirt and dust and rust. She measured the cages and had me weigh one rabbit while she watched, concerned that the animal may have gone a few ounces over the legal threshold that moves a rabbit from a four-square-foot cage up to a five square footer. As for what the researchers were doing to the animals (producing antibodies by injecting the rabbits with various proteins), that was off-limits, not to be questioned, even with the ugly skin lesions. Questioning how and why the scientists were running those experiments would have crossed a line into "curbing" medical research. Animal *care* was the USDA vet's jurisdiction; animal *use* was not.

In 1985, spurred on by the monkey-lab exposés, Congress acted, not once but twice. In November it passed a law, the Health Research Extension Act, requiring that all NIH-funded animal studies meet certain welfare standards. In December it passed the Improved Standards for Laboratory Animals Act, which amended and updated the Animal Welfare Act (first passed in 1966). Both laws breached the long-standing divide between animal care and use, set aside the trust it had once expressed in scientists regulating themselves in the important research they were conducting, and required research institutions to establish in-house animal committees to police the researchers. In both laws Congress escalated animal-welfare regulations by requiring ethics committees' oversight of scientists' animal use. As a participant and an observer, I've seen how animal ethics committees, in the United States and around the world, have improved the lot of laboratory animals, as I cover in coming chapters.

Monkeys made it painfully obvious that scientists needed supervision, and the ethics committees created to perform that task were a boon for other lab animals as well. But monkeys' and apes' big brains and obvious smarts got them some special protections in the 1985 act. Congress singled out both dogs and monkeys, our best friends and our closest cousins, with special provisions tailored specifically for them, reflecting the identities we ascribe to them. Both are smart, strong, athletic animals. One sentence in the 1985 act generated a flood of lobbying, as animal advocates saw a chance to greatly expand provisions for dog and primate welfare, and research advocates saw soaring costs for provisions of dubious benefit for the animals.

Congress focused on dogs' fun-loving athleticism in making sure they got some chances to run around and exercise. Despite their obvious smarts, when it comes to housing dogs in labs, the big concern has always been for their physical needs, letting them run and play. Things are different with monkeys and apes. It's not that monkeys aren't athletic and energetic when they're in a place that lets them move. I once watched a rhesus monkey shimmy up a fifty-foot coconut tree in a way I'd never seen zoo or lab monkeys manage—or get the chance to try.

But Congress did not pass language in 1985 giving monkeys tall trees or even just bigger cages for their athleticism. John Melcher, the veterinarian whose second career found him as a US senator from Montana, was one of the lead sponsors of the 1985 update of the Animal Welfare Act. He worried about monkeys' busy minds going stir-crazy, in their lab cages too small to let them do anything to relieve their boredom. He had visited a baboon lab where individually caged animals had barely enough space to walk around and could not stand up on their hind legs for a look around the room. Melcher was similarly distressed at how labs caged monkeys and apes singly in small lab cages. As he visited various chimp and monkey labs, he grew convinced to add language to the act requiring labs and zoos to provide "a physical environment adequate to promote the psychological well-being of primates." As he described it later, his goal was to "create circumstances where life isn't just boring, where they can have

some space, some opportunity to move around and have something interesting to do."[6]

As Senator Bob Dole was adding in exercise for dogs, Melcher added a
dozen words to the 1985 act: "The Secretary [of agriculture at the USDA]
shall promulgate standards . . . for exercise of dogs, as determined by an
attending veterinarian in accordance with general standards promulgated
by the Secretary, and for a physical environment adequate to promote the
psychological well-being of primates."[7] It then fell to the USDA to flesh
out detailed regulations for how to meet Melcher's thirteen-word call for
psychological well-being. Professional primate scientists differed among
themselves, with those using primates as models in human-focused medical research asking the USDA not to take things too far, while primate
behaviorists pushed for expansive rules. The behaviorists pushed in particular for a legal requirement to house primates socially. "A single-housed
chimpanzee is not a chimpanzee," said the chimp and bonobo ethologist
Frans de Waal.[8] No one agreed on a definition of animal psychological
well-being, a way of recognizing or enhancing it, or a way for the USDA to
codify and enforce it.

As with dog-exercise requirements, the USDA came out of the gate
with some proposals that it later scaled back in the face of the research
lobby's criticisms. Dr. Melcher would later complain, "Keeping a primate
alone in a small cage for years degrades psychological well being," he
wrote, asking, "Does a degraded psychological condition also degrade the
research conducted on the animal?"[9] But the congressional law of 1985
specified exercise for dogs, not primates, and psychological well-being for
primates but not dogs (or any other animals). Lobbyists insisted the
USDA keep issues of the mind and body separate, as though playful, energetic dogs needed physical exercise but could never be bored or frustrated
with the other twenty-three hours of the day. They discussed monkeys as
though they were simply armchair intellectuals, with a need for mental
stimulation but no need to get out to run and climb. The USDA
succumbed to pressure, keeping the physical and mental components
separate, and made no increase in the cramped monkey-cage sizes it had
mandated since 1966.

In the process of updating regulations, the USDA called on institutions
to draft their own plans for psychological primate well-being, plans

inspectors could examine during their annual visits. Front and center, the plan must address "the social needs of nonhuman primates." Knowing, however, that lab monkeys would likely still often be singly caged, they borrowed the term *environmental enrichment* from the work of zoo behaviorists and set to listing what a lab could give a monkey to enrich life in the cage. The list included "varied food items" (most monkeys enjoy bananas and raisins, but others go for scallions or mealworms); branches for climbing (if a lab uses cages larger than the legal minimum); "objects to manipulate" (what civilians call *toys*); various puzzles for dispensing foods; and interactions with animal caregivers.[10]

Rabbits inspired a search for nonanimal safety tests that went beyond just rabbits, benefiting rodents and even horseshoe crabs. Dogs' pet-theft protections extended to cats, and provisions for their vet care trickled "down" to other species. The simple requirement for psychological well-being, with the catchphrase *environmental enrichment*, had legs and quickly spread to other lab animals (informally among concerned animal carers and also through the updated edition of the NIH's rule book, the *Guide for the Care and Use of Laboratory Animals*). The zoo industry's accreditation program requires enrichment and welfare programs for all zoo animals, not just primates, but for rats and pythons and all.

Enrichment connotes adding something on top of the barren, solitary cages that many animals endure, be that friends, treats, or amusements. Despite the narrow language of the Animal Welfare Act, welfare specialists expanded enrichment from the strictly psychological to embrace a notion of welfare that includes physical and mental health. Putting environmental enrichment into the regulations—though only for primates and no other animals—forced labs to look at innovations coming out of zoos and adapt them for the confines of the vivarium. Monkeys and apes served as good ambassadors for such improvements, because the same species we keep in labs are also common zoo animals. Institutional administrators had no choice but to go along with the program, freeing up some money and staff time to explore ways to keep the animals active and engaged. Rank-and-file staff were unleashed—well, we at least got longer leashes—to get creative.

"Creative" took many forms, and not all were successful. Not all attempts to enrich animals' lives were truly enriching, and some can be downright dangerous. Like my days catching tadpoles to throw to our zoo turtles, efforts were hit or miss, and we seldom evaluated how whatever "enrichment" we tossed in our animal cages did or did not improve their lives. At times animals injured themselves. Just before I got to San Francisco, a monkey on the UCSF campus decided the appropriate reaction to a rope placed in his cage was to chew it and swallow the fibers rather than to find some creative game to play. Surgery ensued to remove the blocked intestine. Often animals treated their new enrichment like kids with toys on Christmas morning; explore it for a few minutes, then forget it ever existed.

Primates were by far the initial focus of enrichment efforts in labs, but mice and other caged critters were also getting some attention. Word spread that marbles, easy to sterilize and too big for mice to swallow, were a good enrichment for mouse cages. Give a mouse some marbles and, instead of lazing about ignoring them, they set right to work burying them, raising the question with so many of our attempts at enriching animals' lives: What are they thinking? Is this fun? Do they like the marbles so much they want to save them for another day? Are they being neophobic, afraid of new things? They've checked them out, decided they're to be avoided, but in a small cage the only way to avoid something might be to bury it. As long as marbles aren't downright scary, maybe the time and concentration spent in dealing with these unwanted additions to the cage is at least something to do. While marbles got rolled into the informal idea mill of people like lab vets and technicians trying various cage enhancements, in other mouse labs researchers were investigating marble burying as a model of mild human anxiety, a way to screen potential antianxiety drugs. The theory is that the more anxious the mice are, the more diligently they bury the marbles and get them out of sight, so antianxiety drugs that calm the mice and slow their frantic housecleaning have potential as treatments for humans' anxieties. Hanno Würbel and Joseph Garner labeled marbles a "pseudo-enrichment," a cage addition, possibly harmful and possibly not, that is biologically irrelevant to the animals.[11]

Mostly, the enrichment genie was out of the bottle and racking up more successes than failures, to the extent we can measure those. At lab animal science conference "trade fairs," where we shop for the latest in animal cages and vet anesthesia equipment, the companies hawking animal-enrichment paraphernalia, the puzzles and the chew toys and the exercise wheels, are by far the most fun.[12] Online discussion forums (the Laboratory Animal Refinement and Enrichment Forum is the longest running) allow staff, often technicians rather than vets or facility managers, to share their ideas.[13]

Hal Markowitz was an innovator in enrichment efforts for zoo animals, often devising great contraptions to keep all kinds of zoo animals engaged and active. My favorite: he invented a machine that would shoot meatballs through the air of a serval cat's zoo enclosure, so the animal would leap into the air as though chasing a bird on the wing. As the new Animal Welfare Act changes were taking shape, he teamed up with Joe Spinelli, then the head veterinarian at University of California–San Francisco, to start finding ways to work around the limitations of small cages and research protocols to enrich the lab monkeys' lives. His books are fun to read and full of "recipes" for the various contraptions he loved to build, for all captive animals' lives, no matter the setting.[14]

Simply interacting with what we think would be an enriching object we've given them does not measure whether it enhances the animals' welfare. All the animal-welfare assessments of the previous chapter have their place in scoring the various enrichments we might add to animals' lives. As with animal-welfare improvements in general, so too with enrichment: happier animals tend to be healthier animals and better research subjects. The animal-welfare scientist Georgia Mason and her team reviewed 214 varied rodent studies in which scientists compared mice and rats in impoverished or "conventional" housing to animals in "enriched" environments. No matter the disease under study—cancer, stroke, depression, and others—disease severity was greater in the conventional housing, and more mice died. This is serious both because you don't want to kill your lab animals prematurely and because it can muck up the results of the data you're trying to collect. She believes that scientists writing up their studies' methods should label rodents in "conventional" ventilated housing CRAMPED: they are cold, rotund, abnormal, male-biased, poorly surviving, enclosed, and distressed.[15] The early Animal Welfare Act naively drew a sharp line between housing animals and experimenting on them, but, as Mason and others emphasize, they are not so separable. Bad housing does not just harm the animals (even to the point of premature death) but throws experiments out of whack too.

Animal-welfare scientists scouring the literature for the benefits and shortcomings of enrichment efforts face a challenge in that we do not all share a common definition of *enrichment*. More space? More friends? More opportunities to climb or swim or burrow? Nor do we share a common understanding of enrichment's goals. We might measure the success of enrichment efforts by how well they prevent or treat overt pathological behaviors (like mice stereotypically somersaulting in their cages) or evaluate how well enrichments allow animals to perform a wider repertoire of natural behaviors. We can measure how life in a rich or barren environment shifts animals' cognitive biases toward greater optimism or pessimism.

Environmental enrichment embodies the science-driven animal-welfare reforms I describe in the previous chapter. It shares a concern for the inner lives of the animals beyond the mere physical health of their bodies, but it has a catchier name. Who doesn't want to enrich animal lives? In a 2020 survey-based study of burnout, quality of life, and compassion fatigue, lab workers told Megan LaFollette and her colleagues they wished they had more control over the enrichment their animals received and wanted to do more to make the animals with whom they spent their days happier and more fulfilled: happier animals lead to happier staff, who then feel energized and empowered to further work for animal happiness.

Zoos, too, promote enrichment. Mid-twentieth-century zoos started moving away from barred animal cages toward more natural-looking exhibits. Zoo visitors can look across a moat to see the animals in a simulation of their wild habitat or through glass for a closer look. Like a painted theater backdrop, these exhibits look natural enough to humans. To the animals, perhaps less so. Moats and glass still confine the animals as much as cage bars do. Large enclosures do not necessarily make for happier animals. As the 1985 law mandated enrichment for zoo primates and the idea spread to other species as well, zoo welfare directors began questioning the natural-looking (to people) enclosures, questioning how well they met what animals wanted, and they began enrichment programs with cardboard boxes and all manner of conspicuously unnatural cage additions. I've seen signs in zoos explaining that the unnatural artifacts in the animals' enclosures—boxes stuffed with straw and treats, ropes, floating beer kegs—serve a purpose of enriching the animals' lives. Along with the explanations, I've also seen donation boxes. At my old zoo, the box I saw in 2021 announced, "Thank you for your contribution! All donations benefit our animal enrichment program." And there we see the problems with the idea of environmental enrichment and its moniker.

Ethically, *enrichment* implies that we are generously giving the animals a little something extra, but with no serious obligation. It's optional, like tipping for take-out coffee or sending the kids to ballet class. When budgets are tight, and researchers object or staff say it's too much work, we can stop the niceties and the giveaways and keep the animals in their clean boxes with their basic diet, all the water they can drink, and a dusting of sawdust or wood chips to soak up their excrement. In the labs we call this standard or conventional housing. Any slight departure from this we label *enrichment*. It's low priority, something animal techs can play around with when they have the time and the inclination. Instead of enriched-versus-standard caging, what we now call *enriched* should become the standard of care, in contrast with *deprived, bare,* or *barren,* which is what the standard steel or plastic boxes are.

Many universities list enrichment options on their websites for their various species. The intended audience is not the general public, or curious snoops like me, but the scientists whose animals the university maintains. Where I worked, many of the enrichments we listed were optional niceties the researchers could choose for their mice: various food treats, extra nesting material, little huts they could scurry into, or autoclaved toilet-paper rolls they could shred "to reduce boredom." On one website I saw peanut butter available as an enrichment—optional, of course—for breeding mice. We know from living with dogs that they quite like peanut butter. Even with a pill hidden in it, they swallow it down, then get extra mileage working to get it all from their lips and gums. Plenty of other animals love peanut butter too, so why specify it as an optional enrichment just for breeding mice? The reality is that some mice cannibalize their litters unless they have higher fat or protein than a celibate mouse needs. Peanut butter fits the bill. If ballet school for kids and toys in an animal cage are optional *enrichments*, or little niceties, the word does not fit dietary modifications that keep a mother mouse from killing her babies!

Cannibalism is a good indicator we're depriving mice of something vital; filling that void should be an obligation, not an enrichment. Gordon Burghardt, a reptile behaviorist, describes this perspective as an additive model of enrichment, in which we start with barren conventional cages as the barely acceptable stripped-down model and then add various enrichments and see what ones make for happier, healthier animals. Crediting Heini Hediger, a zoo director and early advocate of mental health and welfare for zoo animals, Burghardt speaks of the deprivation that marks all captive animals' lives. He flips the enrichment paradigm and suggests we look at what the animals' best possible lives would be, evaluate what we are depriving them of and how that affects them, and then control that deprivation. What nutrients are nursing mice missing that turns them to cannibalism? What drives are we thwarting that lead monkeys to pluck their fur or dogs to spin in tight circles in small cages? In an additive model, low-ranking technicians throw their love and effort into trying to engage and brighten their animals' lives. And that's good; keep doing that. From the controlled deprivation perspective, empowered welfare scientists, partnering with low-ranking technicians, should be *required* to root out the various deprivations that animals suffer.[16]

In labs, as in forests, rhesus monkeys want to be in a group. They want to groom and be groomed, to watch who's grooming whom, to find safety in numbers, to have a chance of becoming the alpha, and maybe even having a shot at sex. Medical labs frequently house monkeys in separate cages within a room; what part of that deprivation most matters to the monkeys, and what can we supplement to account for the deprivation? An attentive animal caregiver watches how, with just facial expressions and vocalizations, dominant animals establish themselves. To maintain peace in the room, the human caregiver respects the pecking order, giving treats and toys first to the alpha and then on down the line. A monkey alone in a room would be deprived of even that cross-cage social interaction. So animal behaviorists seek ways to ensure not just that they're never housed alone in a room, but if they must leave the room for some sort of lab testing, another animal they know comes along, separately caged perhaps, but nonetheless present.

Grooming and touching are integral to many monkeys' social lives, but less of an option when animals are singly caged. If some dexterous animals somehow get a finger in reach of a neighbor's cage, they could lose it, with a vet finishing the amputation that the neighbor monkey started. In this setting one can easily think that trying to pair monkeys in a shared cage is not worth the hassle and the danger. Yet I remember the first time we paired two of our adult males. The behavior experts watched them for months to be sure they had no question that Tom was the dominant alpha and that Dmitry would not challenge him. They set up a grid between their cages so they could groom without full-body contact should a fight erupt. I was invited for the day of their first full-body contact, invited because I brought the ketamine and sutures in case day one ended in violence. A technician was on standby with a hose to squelch any serious attacks. They slid open the panel between the two cages and stood back. Tom dropped to the cage floor on his side. Dmitry read the signal right and, smacking his lips, rushed over to start grooming him. We did not trust them to live harmoniously 24/7, so we separated them for feeding time and at night, but each time the dividing panel slid open, they got right to their grooming ritual. Each knew his place in their small hierarchy; I don't believe Tom ever reciprocated the grooming. Tom and Dmitry showed me clearly how very important social contact—not just hooting or making faces across a room but actual physical touch—can be for animals.

Zoo behaviorists are the leading thinkers and innovators in enriching caged animals' lives, so I look to them for ideas and inspiration. I visited with Jason Watters when he was overseeing the Wellness and Animal Health Program at the San Francisco Zoo. Looking at a large paddock where some antelope roamed, he explained his concept of the Three Needs that guide his work. Animals need to be able to explore and investigate their environment for food and mating opportunities, to patrol their territory, or simply for executing whatever agenda the animals, not their human caregivers, set. They need to acquire occasional rewards for that effort, if only some freshly sprouted tuft of grass to nibble. Otherwise they learn that exploration is a pointless waste of time and energy. And they need to be able to exert some measure of control in their lives. Without this the world's largest zoo enclosures are boring wastelands, and the animals have no motivation to get up and about. We can bring these insights into better housing for animals in science, though better housing alone cannot fix animals' welfare during experiments.[17]

The mouse neuroscientist Garet Lahvis believes mice are useless for brain research if they do not have normal, fully developed brains and cognitive function. Even without worrying about the animals' welfare, he is skeptical that we can accomplish this in the stultifying environment of the small cages in which we house lab animals. He has calculated the ratio of various species' natural ranges with what we confine them to in labs, claiming a mouse's territory in nature is 280,000 times the cage size in the NIH's *Guide*, while a caged lab rhesus monkey lives in one seven-millionth of its natural range. Lahvis has proposed getting the animals out of cages into environments where they can run and explore and work their minds and grow their brains and just in general be more normal animals when we task them with serving as models of normal humans.[18]

But Lahvis is a bit of an outsider. I helped one of my scientists prepare her application for ethics committee approval, finally agreeing that I would forward it on, noting our disagreement over giving her baby mice enrichment in their cages. She shares Lahvis's interests in developmental neurobiology in mice, as well as his knowledge of the developmental consequences of enriched-versus-impoverished environments. But in her worldview, any sort of enrichment, stimulation, or variety in a young animal's life will derail her window into how brains and neurons work naturally when spared any outside interference. In other words, add a sprinkle of sawdust in the cage but no bedding material for building a cozy nest. Standard lab food blocks were on the menu, but no seeds or nuts to forage for. A running wheel or nest box . . . well, I did not dare to push that far. To me this is equivalent to charting "normal" human development by keeping a child in a small bare room, with no toys or blankets or dietary variety. I see this as an unintentionally successful way to induce psychopathology in the pursuit of studying normal development. Every environment will shape a child or mouse pup's development; there is no such thing as a neutral environment. But that ideology does sometimes thwart welfare staff's efforts to make life a bit better for the animals. As for the ethics committee, they decided to defer to the scientist who they believed knew her field best and allow her to raise her animals in deprivation. A USDA inspector may well have challenged this (especially with me whispering, "Please read this protocol; I'm just making sure this deprivation is acceptable in the Animal Welfare Act"), but, as I always remind everyone, mice are not animals under the act.

Anyone with mice in their home, in a pet's cage or in a garage, knows that mice will avidly tear up whatever paper, cardboard, or insulation is at hand to build themselves a cozy little nest. Welfare scientists use various scoring systems to rate the quality of the nest—a nice round dome gets a top score—and look at poor-quality nests as a sign that the animal is not thriving. Knowing mice's strong motivation for nest building, while also seeing hundreds of mouse cages at my university devoid of any nest-building material, I invited our animal behaviorist to present the case for nesting material to our ethics committee. I was no longer in the vet department but instead managing the welfare-compliance program. I supplied the regulatory language from the *Guide*, and she supplied the science. Mice will use what they can get—about half an ounce of shredded paper is the recommendation—to build a nest. In ventilated cages mice who can't build a nest are cold-stressed, and that stressor affects their biology enough to throw off scientists' data in cancer, immunology, and other fields. But call it an enrichment rather than standard housing, and it becomes an optional nicety rather than a basic part of care. It's a bit like checking into a drafty hotel room and finding that a bed is standard, but pillows and blankets are extras that are inconvenient to provide. The committee listened to her presentation and asked questions. The vets and facility managers pointed out the inconvenience of giving mice some nesting material. It's harder to see the mice during daily health checks. It could interfere with the cage-washing robots. It could plug the floor drains. The head vet emphasized that the *Guide*, with its emphasis on local professional judgment rather than one-size-fits-all rules, did not explicitly mandate nesting material, so, thanking the behaviorist and committee for their work, he made clear mice would not be receiving this extravagance. And they did not.

Ethics committees rarely have detailed knowledge of animal welfare and so rely on vets, who themselves typically lack a behaviorist's training and insight. Let's listen to the welfare scientists and give all our animals more opportunities and reasons to move around as our standard of care, not as some optional enrichment. Give them better diets, larger and warmer cages, and cozy nests. Match them up with compatible cage mates. As Mason, Lahvis, and others call it, "conventional" housing is a known stressor that can bias scientists' data. Researchers should report animals' deprived housing in their peer-reviewed papers' methods section, just as they should report any other factors known to make the subjects abnormal in a way that could skew their data.

The mice in labs do not know how much they owe to monkeys and apes who demonstrated so clearly that caged life is rough on an animal's emotional well-being. Primates and their human champions—legislators, activists, scientists—inspired the efforts to care for their psychological well-being, and, once our eyes were opened, we saw how much this need applied to other animals as well. *Enrichment* is the word we have. To its credit, the word inspires caregivers who want to do their best for their animals. On the downside, it normalizes the impoverished conventional approach to animal care that allows cost, convenience, and scientists' resistance to change to outweigh giving animals the unnecessary luxuries that the word implies.

Enrichment has been a successful meme even when our regulations fail to really drive it. I see progress for animal welfare all around me. Pet lovers can buy enrichments (formerly known as toys and treats) to mitigate the boredom of home life for dogs and cats. For labs, cage manufacturers follow the science, or even seek out behavior specialists to advise them, on how to design and market animal cages that will better meet what animals want. I was delighted at a conference to see zebra fish aquariums on display with optional enrichment inserts. Zebra fish normally do their best to hide from bigger fish, but the clear plastic aquariums in a lab look to me, and perhaps feel to a zebra fish, like being suspended high in a pond of clear water, visible to all but blocked from escape. Given their choice, they prefer to swim above a floor of pebbles rather than above clear plastic, and they are willing to go along with the conceit that even just a picture of a pebbled riverbed is better than clear plastic. And, suddenly, a market for plastic aquarium insets (sanitizable and sterilizable to keep the vets happy) that depict a safe and secure riverbed appeared. If welfare scientists continue to document less stressed zebra fish with these insets, they will become standard fare, as our regs slowly catch up and mandate them as such. The paradigm is shifting.

The research lobby is correct to be wary of slippery slopes or "regulatory creep," in which one new set of requirements for improved welfare grows and grows, adding costs, delays, and bureaucracy to running an animal lab. Give a dog a bone, and soon the activists will be clamoring for all lab animals to get treats and toys. I admit that mostly I like the times when this happens, when our realization that we can improve the lives of one group of animals logically leads to us to expand protections for others, even if they don't have the public sympathies that dogs or rabbits enjoy. One person's "regulatory creep" is another's progress.

I credit chimps with getting monkeys the respect they deserve. Monkeys, along with chimps, brought about new laws in 1985 that benefit other animals, with ethics committees to oversee animal experiments and a concern for animals' emotional well-being that has radiated out to other species. But not every new protection for one kind of animal successfully crosses over to others. A case in point: in their numbers and in the experiments they serve in, mice bear the brunt of animal research, and yet they have been excluded from the protections of the US Animal Welfare Act since day one in 1966.

Research defenders who fear a slippery slope look with trepidation at chimpanzees' liberation from medical research. I started this chapter with chimps, but they have now received protections that are solely their own and are no longer in labs. While the evolutionary line to hominid apes (chimps, gorillas, ourselves) split off from monkeys at least twenty million years ago, their legal divergence is more recent. Government actions starting around the year 2000 sprung chimps from US labs altogether. In 2000 Congress passed the CHIMP Act (Chimpanzee Health Improvement, Maintenance, and Protection Act) to establish and fund retirement sanctuaries for chimps who'd done a tour of duty in federally funded labs and who were no longer needed for experiments. In 2011 Congress directed the NIH to review the continued necessity, if any, of using chimps in NIH-funded research, and two years later the NIH began phasing out its support of chimp experiments. Chimpanzees had been integral to bringing a hepatitis B vaccine to patients worldwide, but that had been before tissue cultures, woodchucks, and genetically modified mice replaced children and chimps for this work. The 2011 report saw little future for chimps-versus-hepatitis battles.[19] Likewise for chimps in HIV research, after NIH had invested heavily in the 1990s to build chimp isolation units, ramped up breeding programs, and set about infecting chimpanzees in a search for cures and vaccines for AIDS. Chimpanzees, the report concluded, had had their day in the labs, and, while some future pandemic might find value in chimp experiments, until then monkeys, mice, and nonanimal tests are replacing them just fine.

By 2015 the Fish and Wildlife Service ended an exception it had carved out under the Endangered Species Act that had allowed chimp experiments for human-focused medical research in a way they would not allow other endangered species to serve in labs. Six months after that, the NIH announced it would send all NIH-owned chimps to retirement.[20] After their years in hepatitis and AIDS research, their stint as Astrochimps, and their service in safety-testing human vaccines, chimps have gotten their reprieve from medical research labs, public or private.

I've heard the occasional grumble that the 2011 chimp report could lead to a slippery slope. I certainly don't expect Congress to expand its retirement program for a few hundred chimps to the tens of thousands of other primates that US labs use. Nor do I see NIH scientists giving monkeys the research exemption chimps got. Indeed, the 2011 review of chimp experimentation articulated a principle that dogs and monkeys must be considered first, before chimps, as experimental subjects. In essence a continued reliance on monkeys punched chimps' ticket out of the labs. The NIH director made this clear, writing, "These decisions are specific to chimpanzees. Research with other non-human primates will continue to be valued, supported, and conducted by the NIH." Rather than setting a chimp precedent that will soon liberate other species from their service in medical labs, chimps left the animal club and got a bit closer to humans, crossing the line of which creatures are acceptable lab subjects and which are not.

Chimps eventually left medical-research labs, but not before their plight in pitifully small cages moved Congress to legislate for richer lives for them. Mice and monkeys are calling out that they too want richer lives. How we house them goes beyond their happiness, however, and can give us better science as well. But even if labs seriously upgrade the daily care they give their animals, they will still need to conduct their experiments. And some of those experiments will inevitably cause the animals pain and suffering. In the next chapter, I show how we can recognize that "inevitable" pain and make progress toward eliminating it.

Rat

Rattus norvegicus

THE PAIN WE DON'T SEE STILL HURTS THEM

I think back on the python Esther's lunchtime rats, animals whose lives and grisly deaths I'd discounted for the longest time. None of them had names, of course, or even identification numbers. The rats came in a cardboard box from a company that breeds animals for science. We opened the box just enough to grab one rat at a time by the tail to toss him or her to the hungry serpent-in-waiting. The snake's strike was fast. The rat's death in her coils, slow.

I was mesmerized. Fifty years later I'm mortified. Those last few minutes of the animals' lives were jam-packed with suffering: the fear, the "air hunger" of their slow suffocation, and the pain. I certainly can't blame Esther for her indifference to her prey's suffering or for how she did nothing to make things easier for them. But I hold humans to a different ethical standard. We need to see and acknowledge the animal pain we cause, even when we believe the pain we cause is inevitable and justified.

Pain is people's biggest concern for animals in labs. People do not want animals to suffer, and, though confining animals to small cages is worrisome, pain stands out as the far greater suffering to worry about. It's been thus for centuries. The vivid image of Victorian-era vivisection (the term both scientists and their enemies used at that time) is of a conscious dog, strapped to a table, under the scientist's scalpel. Animal-defense societies have long decried such lab animal pain, and scientists have responded, long before they had useful drugs for this claim, that pain in animal experiments is rare. "Most vivisection," wrote Samuel Wilks in 1881, "is nothing more than pricking mice with the point of a needle." Defending animal research in a London debate in 1914 (and claiming that it would soon be obsolete, as it would progressively eradicate every disease), Walter Chapple proclaimed that, what, with chloroform and morphine, animal "experimentation in this country is practically painless."[1] Certainly, in the twenty-first century, a researcher embarking on a major operation on a laboratory mammal anesthetizes the animal for the procedure, and scientists working with fish or octopuses or frogs are catching up with that standard of care.[2] If rendering our animals unconscious for short-term painful events such as surgery were the only concern in animal pain management, our work as vets would be largely done.

Pain management for laboratory animals has been my driving interest dating back a couple of decades. Nonetheless, I have faced many challenges advocating for better pain treatments for the animals in my care. I remember all too clearly a time when we had a monkey-research protocol coming up for its three-year renewal review on the university's animal ethics committee. I'd not been happy with how uncomfortable the monkeys appeared in previous months after their surgeries, and the animal-care staff were counting on me (I thank them for their diligence and advocacy) to get the animals better care. How to convince the research team and the ethics committee that the monkeys' pain was both real and unnecessary when improved analgesics were available? The scientist had various reasons to resist my recommendations, mostly fear of the pain meds' side effects on both the animals and the data, and his conviction was that his postsurgical monkeys looked fine to his eyes. Our ethics committee members, most of them research scientists themselves, feared imposing treatments that the scientist resisted, questioning his assertion that his monkeys were perfectly fine after their surgeries, and imposing medicines that might have side effects on the data or on the animals' health. I blamed myself for not having a more persuasive case than being able to say, "I'm the vet who has examined them; they are in pain, and these analgesics are safe and helpful." I'd never been so tempted to leave a job in my life but resolved instead to sharpen my skills and knowledge to advocate for the best-possible pain management for my patients.

In his 2022 book *An Immense World,* Ed Yong celebrates the many extraordinary senses animals use—sight, smell, echo, and more—to gather information about the world around them. A beagle's superpower lies in the nose, following the scent of a rat. An eagle might spot that rat from a mile up in the sky. A bat screams, picks up the echo, and swoops in to capture a moth. Anything we humans can sense, some animal somewhere can sense better, and some animals use senses (like detecting magnetic fields) well outside our sensory powers. Pain, however, stands out as the "unwanted sense." It's the sense that tells us what to avoid. Our sense of touch tells us we have a frying pan in our hands and lets us explore it. But touch a hot frying pan, and pain impulses race past other sensations up their own devoted nerve tracts, called nociceptors, to the spine and brain, triggering an immediate, barely conscious reflex to drop the pan. The brain blocks out other senses for its laser focus on the pain. The conscious brain then springs into action, devising a plan for short-term protection while warehousing memories to keep us away from hot pans in the future. Pain is good for us precisely because it feels so bad. We prioritize its messages and remember them, and that can save our lives.

We all know pain, so it should be easy to discuss, describe, analyze, treat, and in general avoid. It is not. *Pain* and *suffering* are human words we apply to natural phenomena that are difficult to precisely define. Pain and suffering overlap, but they are not synonyms. We may be aware of our many small pains that do not rise to the level of suffering, that may not even rise to the level of going to the medicine chest for a pain pill. On the other hand, we may suffer mightily without bodily pain. Fear, loneliness, and frustration are all nonpain causes of suffering in abundance in lab animal vivaria.

Many animals have nociceptive reflexes that serve to move them from danger. An ant under a magnifying glass (a guilty fascination of mine as a kid) scurries from the focused heat of the sun in a way that a leaf on the sidewalk cannot do. Welfare scientists probe whether animals have the brain capacity to experience the emotional distress of pain. They ask whether animals seem to remember and then seek to repeat or avoid repeating what they'd experienced. They watch them deliberate and make trade-offs, gauging what an animal will choose to tolerate in pursuit of some reward they want. Hermit crabs scoot away from the presumably painful shock of a diabolically electrified periwinkle shell. Laboratory cuttlefish, stressed by being chased around their enclosure, wax pessimistic about whether a chamber will yield a yummy shrimp treat. Chased but untouched, this can't be pain informing the cuttlefish's outlook. Pain and nonpain distress are both forms of animal suffering that we should want to minimize, but to do that we need to recognize the suffering and sort out its cause.

As a vet, I've sworn an oath to "the prevention and relief of animal suffering," so I always need to know if my suffering patients are suffering from pain or from a different challenge.[3] I see pain as something at least potentially treatable with pain medicines. Ibuprofen cannot treat fear or hunger or other nonpain suffering. I'd be equally guilty of malpractice if I treated an animal's fears with a useless ibuprofen dose as I would be leaving another animal's significant pain unmedicated. Worst is simply not seeing their suffering and assuming they're just fine when they desperately need care.

The International Association for the Study of Pain, a consortium of clinicians and laboratory pain researchers, defines pain as "an unpleasant sensory and emotional experience associated with, or resembling that associated with, actual or potential tissue damage."[4] The association emphasizes that preverbal infants and nonhumans can experience pain, even if they cannot say so in precise words. And they distinguish the physical, or nociceptive, sensations from the emotional suffering of true pain. Grab a rose, and the initial prick of a thorn does not cause tissue damage and does not demand a stronger word than "unpleasant." Grasp tighter, and the thorn pierces the skin (that's tissue damage) and ratchets "unpleasant" up several notches. The localized infection that follows will continue the unpleasantness for hours or days.

The definition from the International Association for the Study of Pain makes clear the dual nature of pain, that it is both sensory and emotional. It is physical and mental, to the extent those two are separable. At its simplest, almost a caricature, the sensory part that we call *nociception* is the wiring that brings messages from the site of noxious injury—the stubbed toe, the rose-impaled thumb, the surgical incision—to the spine and brain for action. Reflex actions like dropping an unexpectedly thorny rose do not require much from the brain. Even an anesthetized person may pull away from a firm toe pinch, a sensory stimulus and an unconscious reflex response. At a higher level of consciousness, the brain ruminates on the thorny rose and plots a course of action to prevent immediate or future impalements. The unpleasant pain from the accident leads to action. The emotional response, mild in the face of some noxious stimuli, stronger in response to some others, guides our future actions and keeps us from serious injury.

Human and nonhuman pain can sometimes be obvious to an observer. We see strong responses to sudden, sharp pain in animals. And hear them too, as unambiguously as when we step on a toe and hear the word *ouch* on human lips. Dogs yelp, marmosets screech, and cats yowl. Rabbits can let out a bloodcurdling scream. Rats in a python's coils cry out, but mostly at a high-pitched frequency their mates may hear, but we humans cannot. In the hands of sensitive people, those cries of pain should be sufficient to make us stop whatever we are doing. We cannot undo having stepped on a tail, but we can at least take our foot off of it.

My concern in this chapter is for the less obvious pains lab animals suffer. The Victorian antivivisectionists' revulsion for surgical pain in unanesthetized animals is mostly now a red herring. We now know how to keep most kinds of animals unconscious for surgeries.[5] I don't want to gloss over the possibility of errors, especially in the United States, where training requirements for lab workers are weak, but the principle that all animals must be fully anesthetized for surgical operations is universal. No, where we tend most to fail the animals is after an initial injury or surgery. Whether a deeply embedded thorn or a surgical operation, pain and inflammation continue well after we toss the thorny rose in the rubbish or send the surgical scalpels off to medical waste.

This sort of pain is hard to see in others, human or nonhuman, but that doesn't mean they don't feel it. My sense is that far too many laboratory animals live with unacceptable amounts of pain. But because it is so difficult to see another's pain, vets are challenged to convince scientists to take it seriously. In fact, we vets are challenged to take such pain seriously enough ourselves. If we have to work so hard to decipher how an animal is feeling, maybe the pain's not that serious? Not worth fighting with our scientists about?

Take a vaccine as an illustration of my concerns. A vaccine jab hurts. The injected muscle may experience a quick, reflexive tightening, a minor concern for your health and welfare. It may hurt enough to make you leery (in animals, we label this conditioned place aversion) of going back for a booster. Afterward, the degree of muscle pain will vary person to person. We cannot undo the initial noxious stimulus of the injection, but the ongoing pain from the body's inflammatory reaction may rise to a sufficiently unpleasant level that some people will take a pain medicine like ibuprofen for a few days, spare the sore arm from exercise, or apply cold compresses. That pain is private, unlike the loud *ouch* of the initial shot. It takes a sensitive observer or a sympathetic question for anyone else to know of this moderate, private, but very real pain we are suffering. While we humans can simply head to the medicine chest, caged lab animals have no option for self-medication. Animals get what we decide to give them and no more. Ramp up the painful insult—surgery is the obvious example—and the residual pain after the initial injury goes on longer and stronger. Keeping an animal anesthetized during a surgery is not enough; vets and scientists should prevent or treat their chronic ongoing pains as well.

Our vigilance for postsurgery and postinjury pain begins as the animal wakes up from anesthesia. People will gradually ratchet down the amount of pain medicine they take in the days or weeks following surgery, suggesting that animals' pain treatment needs will similarly subside. We know from experience, though, that injuries and surgeries are not the only pains. Where surgical pain begins with one dramatic bang and then tapers, other pains come on more slowly and build insidiously. In animal labs scientists have a number of tools for modeling painful human arthritis. They inject a drop of some chemical into a rat or guinea pig stifle (which is vet for *knee*), and the pain and inflammation come on slowly. Likewise, many cancers cause exquisite pain as they progress, and, if there's a known human cancer, odds are there's a lab modeling that cancer in lab animals. Separate from experimental manipulations, older lab animals, dogs and monkeys in particular, will develop various aches and pains simply from living to an old age. If we want to help the animals in whom we induce painful conditions and if we take responsibility for animals' lives in our lab cages, we need to be able to recognize and quantify their pain, even if only to euthanize them when their pain and suffering becomes too high.

Various countries' welfare regulations for lab animals enshrine some version of overt, if a bit simplistic, critical anthropomorphism: unless we have evidence to the contrary, we should assume that what is painful to people may be painful to other animals. I endorse that bit of anthropomorphism, and the corollary that pain management for lab animals should match the clinical care of human and veterinary patients when possible. It took a while, but most labs are now on board with assuming postsurgical pain is as real for animals as it is for humans. But the usefulness of the anthropomorphic presumptions of pain faces challenges for the slow onset, insidiously advancing pains that rarely receive adequate care in labs.

Consider the pain of cancer, a dreaded scourge for human patients and the plight of millions of lab animals. Painful? Well, not at first. Cancer is so scary to us not just because of how it wracks a body with pain in its later stages but for how insidiously—painlessly—it creeps up on us. It begins with a painless lump or bump or a silent tumor in our inner organs, and we may not know we've got cancer until it's already spread far and wide in the body. En route to the severe pain of advanced cancer, pain that even potent opiates cannot fully subdue, may be a period where diligent pain medication can help the patient. Apply that to animals in cancer labs. A scientist injects some cancer cells into a rat, causing no more immediate pain than a vaccine would elicit. Even as the cancer cells grow into a large tumor and metastasize to sensitive brain and bone tissue, the cancer may remain painless—for a while. As the disease advances, the animal might find some relief with round-the-clock pain medications. Later, though, the cancer will likely be as refractory to analgesics as advanced cancer in a person or a dog. We know from human experience and from caring for canine patients that advanced cancers are not just painful but often of such a type and intensity of pain that medicines cannot touch it. So I am legally obligated to assume that advanced cancer in a rat is painful, as surely as it is in people. But I also assume that early-stage cancer is painless in both. The general anthropomorphic principle tells the rat's vet nothing about when to start pain treatments, how to know they're working, or how to recognize how severely the animal is suffering. The ethical imperative to minimize animal pain bumps headlong into the veterinary challenge of knowing when the animals are in pain and knowing which pain medicines can help.

Human physicians can ask how much pain a patient is experiencing and how well their medicines are working, and a verbal child or adult can answer in words. They can describe the pain as frequent, dull, sharp, constant, or worsening. They can detail when the pain started and whether it came on suddenly or crept up gradually. They can name the sore spot and report on whether movement exacerbates or alleviates the hurt. The challenges doctors face to adequately treat pain without enabling addictions are serious but very different from the challenges a pet vet or a lab vet has in even just diagnosing chronic pain in their nonverbal animals. Clients with arthritic dogs want effective pain management, as do their dogs. Clients with dogs with advancing cancer want that balm too, plus clarity on when a pet's pain has become unbearable and humanely ending the animal's life is the best option.

Pain diagnosis in even our most familiar of companions, dogs, is a challenge, but understanding how dog vets care for their pet patients can shed light on what lab vets must do for the mice and the monkeys. Remember that what's true of dogs may be true of other animals as well. When it comes to animal pain diagnosis, dogs teach us that, yes, we can get clarity on what they are experiencing, but they remind us that we need to invest time, effort, and attention that few animals in labs receive.

Since dogs won't talk in human words, vets take a medical history from their humans. We ask exhaustive questions of what our clients are seeing. Do their pups still jump up on the bed? Show an interest in food? Perk up when they see the leash and tennis ball in their person's hand? How long has this been going? Getting worse or better? Are pain medicines we've prescribed making an obvious difference? We move on to observing how the dog moves in our clinic. We lay our hands on, probing for areas of pain and reactions to manipulating their joints. A golden retriever may bask in this attention but will wince if I manipulate a painful, dysplastic hip. A Chihuahua may tremble in fear and defensively snap when I touch the sore spot. But the principle holds that competent pain diagnosis requires a thorough history, careful observation, and hands-on evaluation.

A vet can't just send a kennel cleaner to the parking lot to glance in the car at the canine patient, but that's exactly the level of examination most lab animals get. They rarely get the individualized time-consuming attention of the dog in the vet clinic. When they do, the vet faces limitations that are uncommon in the pet practice. Monkeys bite if we try to palpate for painful spots. Rats and mice tense up in our hands, making a physical exam for an arthritic knee fruitless. Not all dogs and cats live with humans who know their animal's normal behaviors through the course of the day, who know how readily they eat, how peacefully they sleep, or how energetically they bound up and down stairs when they're feeling good. Not even our larger lab animals, the pigs or monkeys or even dogs, have human carers who can report on their 24/7 behaviors. Knowing that advanced cancer is painful does not tell us what we need to know about this woodchuck or this one mouse in a cageful of five near-identical creatures at this stage of the disease we have caused them to suffer. Vets and scientists have a long history of underdiagnosing lab animal pain, and, if we don't see it, we don't know to prevent or treat it.

Strange to look back on this now, but when I started my work in laboratory animal medicine, fresh out of vet school, vets normally operated on people's pets without analgesics after the surgery. Vets were slowly letting go of a widely held belief that pain after surgery was a good thing: it kept a dog who'd just had a spay or other abdominal surgery from jumping around and pulling out her stitches or from putting weight on a repaired fracture and rebreaking the bone. Vets also knew that pain was a key to diagnosing some diseases. A cat with a sore mouth would stop eating or might drool excessively; that would indicate it's likely time for some dental work, but the cat would get nothing for the pain. Even if we thought of the possibility, available pain meds for cats circa 1987 were unreliable. Morphine made them manic, and aspirin made them sick. On our farm calls, two people on opposite sides of a cow would lift a broomstick up underneath her belly; if she grunted painfully with the pressure, it was likely (and surprisingly common) that she'd ingested a nail that had perforated her diaphragm and infected the sac, the pericardium, in which the heart beats. If the cow were valuable enough, she may get surgery; otherwise, it's off to the slaughterhouse.[6] As for a treatment for her pain, to keep her comfortable in the interim? Apologies to the cow, we did not learn about that.

Lab vets were actually ahead of most private practitioners on animal pain treatment, at least for dogs (if not for rabbits and the rest), administering analgesics without waiting for definitive proof the patient was hurting. We had to be. The Animal Welfare Act required us to be. The original act of 1966 stopped at the laboratory door, making rules for animal acquisition, transport, and housing but not for running an animal experiment. A combination of animal activist pressure and lessons the US Department of Agriculture (USDA) was learning as it began inspecting labs prompted Congress to amend the law a few years later, in 1970, to impose "adequate veterinary care including the appropriate use of pain-killing drugs."[7]

Unfortunately, our aspirations for obliterating animal pain have always stayed ahead of our capabilities. Despite vets' frustrations at scientists' recalcitrance behind the scenes, research defenders have publicly protested for over a century that we treat our lab animals for any pain. But the available pain meds as I started practice in the 1980s were lacking, in the labs or in the clinic, for man and beast alike. The only opioids (the group that includes morphine, fentanyl, oxycodone, and buprenorphine) in my mouse and rabbit formularies were some older drugs that would deliver less than an hour of pain relief. The logistics and effort of keeping the mice hooked up to a tiny IV through the night or redosing pain meds every hour around the clock were more than any vet I ever knew would have embarked on. It was easier to assure ourselves—to the extent the thought ever bubbled up into our consciousness—that our experiments did not cause enough pain to outweigh the inconvenience, labor costs, or the risks and side effects of these old-school medicines.

The best pain medicines in the world are useless if we don't actually see when our animals are in pain. People assume we vets know when an animal is hurting, but, as I've described, even with our canine patients, we have to work to drag that info out of them. Cats can be even more cryptic, and rabbits are virtually inscrutable. Animal-welfare scientists sometimes lament that most lab animal species are "prey animals" who hide their pain when people, scary predators that we are, walk into the room, the better to avoid being seen as weak and an easy catch for dinner. I'm not convinced that the predator-prey issue matters all that much, and even dogs, who do not see us as predators and who might even benefit from calling on us for help, can be hard to read. Some lab dogs sit in fear at the back of their cage. Others jump in joy that a person has entered the room. The fear and the joy alike hide the pain we need to see.

As I started work in my field, worse than admitting I did not have the knowledge of a true pain specialist was realizing how much we vets and scientists as a group, including the true pain specialists, did not—and still do not yet—know. I had a formulary of rat and rabbit drug doses, including that opioid (meperidine) that might treat pain for forty minutes at most. But quality information, from experiments or from careful clinical observation, on how successfully our available drugs managed animal pain was scarce. One team of vet scientists and their rats helped us lab vets see animal pain in clearer focus. Paul Flecknell and his students at the University of Newcastle in the United Kingdom tackled rodent, dog, rabbit, and other animal pain as something they could measure and, by extension, rodent pain relief as something they could measure, not for modeling for human pain but for improving vet care for animals in labs.

Flecknell built some of his work on what pain scientists do in their animal labs. Pain includes the emotional reaction to physical nociceptive sensations, but pain scientists are able in some experiments to tease the two apart. They manage to study nociception, how the body sends pain messages to the brain, without actually causing true pain. A "hot plate test" sounds like a torture device but, done right, it's not. Scientists can place a mouse on a steel surface, gradually increasing its temperature until the mouse reacts, lifting a foot, and then immediately remove the animal. I've had that experience in a car with heated seats, not at first consciously realizing why I was squirming. In the car it's simple to turn off the stimulus (barely noticed heated seat) well before it could rise to uncomfortable, painful, or what the International Association for the Study of Pain would call "unpleasant." So too, a scientist can measure the animal's barely conscious reaction to the heat, then study how an assortment of drugs or genes changes the animal's nociceptive sensitivity. There's value in discovering how injured body parts send their nociceptive complaints up to the brain and value in finding ways to tamp that down. But heat sensitivity in healthy tissue does not adequately recapitulate the pains of cancer, injury, or inflammation that concern human and veterinary clinicians.

Flecknell, a vet anesthesiologist working as a lab vet, wanted to document the signs of postsurgical pain in lab animals and to measure what pain medicines in what doses would help them. Initially, Flecknell's pain measures were fairly crude: what the animal's activity level, food and water consumption, and weight loss after a surgery were and how much analgesic drugs brought the animals' behaviors closer to a presurgical baseline. They found that, to better measure analgesic effectiveness, they needed better measures of pain. Now, an ethical challenge in pain research, whether humans, dogs, rats, or others are the species of interest: How to study pain without causing pain? Many nociception tests apply a noxious stimulus such as heat, and the animal's response tells the researcher to stop the stimulus. True pain in such tests is of minimal intensity and duration, with little unpleasantness or suffering. But surgical pain is qualitatively different. Measuring surgical pain requires an animal to have gone under the knife, so, in the quest for better pain medicines for animals undergoing operations, these vet researchers subjected their study animals to surgery. And to map what untreated postsurgery pain looks like, they needed (or at least they thought they needed) a control group that received no post-op analgesics. This is standard practice in rodent pain and analgesia experiments, but it's time to change that. In human studies and in testing pain meds for companion animals, the control group either receive the best-available treatment (we do want new medicines to be better than current ones, after all) or are allowed "rescue analgesia" (supplementing the test or placebo treatment with a pain medicine that will not interfere with the experiment) for all the subjects.

In their hallmark study, Flecknell and Roughan anesthetized one lucky group of rats as their no-surgery control group. The other groups underwent abdominal surgery, with one group receiving postsurgical analgesics and the unluckiest group, a placebo. They scoured many hours of video looking for abnormal pain behaviors that animals who underwent surgery displayed which the no-surgery control group did not. They expected, correctly it turned out, that animals who received the right dose of a pain medicine, whether carprofen or the related drug, ketoprofen, would be intermediate, not as painful as the no-analgesia rats but not as painfree (these drugs are potent cousins of ibuprofen, but they cannot work miracles) as animals who'd been spared the knife. Their work was impeccable, but the charts and graphs did not capture what a couple of minutes of video made so vivid.[8]

I had not realized how much I was not seeing until I attended a conference in Seattle in 2003, where Paul Flecknell showed his and Johnny Roughan's rat videos and changed my life.[9] Not pleasant viewing, but he showed us specific signs we should watch for and made clear that this methodical approach could work for other types of animals and other types of pain. His rats twitched their abdomens in spasms of pain, and they lost their balance while grooming. Normal rats do not do that. They arched their backs like cats waking from a nap when their person comes home from work. Normal and cute when a cat does it, but, I learned, cause for concern in my rat patients. More than the specific pain signs I learned to identify was the lesson of how easily even we vets fail to see the "hidden" pain that our animals are showing us. How often had I seen these behaviors before and not realized what I was seeing?

We need to watch our animals for signs of pain, but one common pain indicator is sitting as still as possible, not stretching against a surgical incision or putting weight on a sore leg. Sitting quietly, watching us watching them, the animals' pain is not obvious. So ask them about it. Ask whether animals will stretch and reach for something they want, whether that's monkeys climbing up to grab a raisin at the top of the cage or mice reaching up to pull some nesting material down from where the vet has placed it above their head. It takes a good bit of pain to keep mice from touching up their cozy nest. I like these tests, as I can give the animal some privacy and check ten minutes later on how many raisins monkeys climbed up to eat or how many strips of nesting material mice worked to retrieve over a standardized period.[10]

Pain biologists, with lab animals as models for improving human pain medicine, and animal-welfare scientists, with better animal stewardship as their goal, have much in common and are constantly adding new pain-detection tools to their repertoire. In Newcastle Flecknell's team programmed an automated video system to detect the pain postures they'd first spent hours of human effort identifying. With automated monitoring they could detect mouse cancer pain much earlier than by simple observation, making earlier interventions to alleviate unintended pain in cancer experiments possible.

My next eye-opener came courtesy of Dale Langford, a pain scientist who identified the painful faces mice make and who had come to my university in California to work on pain management for human cancer patients. In the mouse lab, she'd injected mice in the abdomen with diluted acetic acid, known to cause a few hours of moderate-intensity pain. Pain researchers use this as a standard assay in mouse labs, measuring how much a mouse writhes—a behavior that looks to me like the spasms of stomach pain a person occasionally gets—as the pain peaks and then subsides. She trained a camera on the mice's faces and developed a mouse grimace scale, based on studies of preverbal infants' grimaces at the pain of a heel stick for blood collection. Mice in pain, like babies in pain, rats in pain, rabbits, horses, and many others in pain, make characteristic faces when they are hurting. They squint. They puff up their cheeks. The National Centre for the 3Rs in London offers downloadable posters of the mouse, rat, and rabbit pain faces for researcher training. They show the altered cheeks, noses, ears, and whiskers, but the most striking are the squinting eyes (*orbital tightening* is the technical term), which look to me like a hungover person under a bright bathroom mirror light and may similarly serve to decrease the heightened sensitivity we have to bright lights when we are sick or in pain.[11]

Pain faces have largely supplanted the bodily twitching and arching that Flecknell's group identified for pain diagnosis. Revisiting some of Flecknell's videos, I see his rats similarly squinting, along with their cat arches and trembling, but his video technology was not as capable of reading faces as it was whole-body movements. Walk into a mouse or monkey or dog room in a lab, and there's a chart of pain faces to help a person score their animals. Woodchuck and ferret and gerbil scientists could similarly watch for that sad squinting that seems such a universal marker of pain among mammals. Coming soon to a farm or lab near you, welfare scientists can mount video recorders juiced with artificial-intelligence face-recognition software that can tell individual animals in a group apart and alert their caregiver of facial signs of pain or illness.

With these various tools we are better able to see that much of what we do to animals does in fact cause pain. Better, however, is not perfect. We do not yet have totally reliable ways of scoring mouse pain, and even what we do have has not spread widely enough among lab veterinarians. Measuring pain is important, but welfare scientists (as well as pain biologists) need to know more, not just whether the animals feel pain, but how they feel about it. How does it affect their quality of life? Do they, as we humans do, decide when to tolerate or when to nurse their pain? Do they evaluate their own need for pain medicines and weigh that against whatever unpleasant side effects medicines might have? Do they scurry into a chamber they've learned to associate with a dose of pain drug? Does pain turn optimistic animals into pessimists?[12] We can ask the animals, but not with a simple glance in their cages to see if they're grimacing and the pronouncement that they look fine, no squinting and no flattened ears, with no further pain treatments needed.

Poultry producers breed broiler chickens to grow preternaturally fast. Some birds become lame, even in the short sprint to early slaughter. I remember watching my Cornell birds limping or having trouble standing. But is this pain or just the challenge of supporting a heavy body on their young legs? If it's pain, does it bother them enough to do something about it? The UK Ministry of Agriculture along with the Royal Society for the Prevention of Cruelty to Animals sponsored studies in which they asked broiler chickens if they were smart enough to make a connection between analgesic-laced feed and pain relief and motivated enough to want that pain relief. They offered a food containing the veterinary pain medicine carprofen and found the birds with the worst lameness preferentially consuming the medicated food, and walking better when they did.[13]

The US Congress must have had a very Victorian-era image of animal labs when it first added pain treatment to the Animal Welfare Act in 1970. Picture the seedy dognapper at the back door, bringing the howling cur down to a tiny cage in the basement, waiting for the day when the scientist descends, no anesthetic in hand, to launch into an experiment. They did not necessarily grasp the range of animal-research projects, in which Banting's dogs were fully anesthetized for pancreas surgeries but likely received no postsurgical pain management. I doubt that in the 1960s they had in mind the kinds of slow-onset pains of a chimp with experimental hepatitis or a rabbit with a human-induced cancer. They weren't entirely wrong then, and sadly would not be wrong now, to believe that animals suffer severe pain in labs. In imaging the worst, they launched an innovative reporting system that, despite assorted flaws, has been valuable in tracking animal use in the United States.

Congress ruled that every research institution would submit an annual report to the USDA of how many animals it had used, with a vague requirement to somehow show that it was following professionally acceptable standards for animal pain treatment. The USDA compiled these annual reports for Congress and published them for public review, with the aim of tracking the Animal Welfare Act's progress in eradicating painful experiments. These annual reports, along with the USDA's inspection citations, are about the only real transparency we have in the United States for animals in laboratories.

The USDA fleshed out rules for how an institution could, in fact, show it was treating animal pain. It called for labs to list their total animal use, separately listing the animals undergoing painful experiments. The USDA vets had a better grasp than did Congress on how scientists actually perform experiments and had plenty of input from research advocates and animal advocates on what details to require in a report. As the USDA refined its reporting requirements over the next few years, it recognized three realities: (1) many pains are below a threshold of concern, too minor to warrant a pain medicine or to be reported as a painful experiment; (2) distress is a welfare concern along with pain, warranting attention and a place in a facility's report; and (3) scientists do sometimes feel a need to run experiments without treating animal pain or distress.

Under the USDA's current reporting rules, a lab lists its animals in one of three "pain categories."[14] Column C lists animals whose experiments would not cause enough pain or distress to require anesthetics, analgesics, or tranquilizers. A simple injection of a benign saline solution would be listed in column C, as would painlessly killing animals for tissues at the end of an experiment. Column D experiments cross a threshold, with pain or distress of sufficient severity or duration to warrant treatment. Column E lists the animals who underwent painful or distressful experiments but received no pain treatments, because the ethics committee approved the scientist's contention that using drugs would interfere with the experiment. In setting up its column E, the USDA vets writing the rules tacitly allowed what Congress may have thought it was outlawing, to knowingly cause pain in animals and leaving it untreated.

Where Congress called for annual reports on animal "pain," the USDA wrote rules instead for "pain and distress." Since then *painanddistress* may as well be one word, the two concerns conjoined in policies and regulations around the globe. Though always linked with pain on the page, distress is actually a bit of a Cheshire cat, appearing and disappearing as an explicit concern. The USDA and welfare specialists have tried to define animal distress in a usable way, but it keeps slipping out of the frame, overshadowed by pain. For example, the USDA set up its categories for reporting pain and distress but now refers to them simply as "pain categories." It wants reports on anesthetics and analgesics (i.e., pain treatments) and sedatives, which might help with some animal distress. Sedatives, though, are drugs we rarely use in laboratory animals.

The USDA is right in including distress in its pain categories. Pain is far from the only source of lab animal suffering. Scientists keep monkeys hungry or thirsty so they will toggle a joystick on a video test in exchange for a sip of water or a lick of banana smoothie. Scientists induce depression in mice, monkeys, or other animals, modeling psychiatric suffering in search of treatments for human mental health conditions. While this can include painful procedures such as shocks, pain is not the only tool. Various drug treatments; inescapable restraints; unpredictable, scary, loud noises; and isolation from a mother or from social companions can all lead to psychological illness in animals that well exceed a few days of postsurgical pain in the degree of suffering. Congress had called for treating animal pain and for annual reporting on it, and the USDA was right to up the ante and add in distress. For us vets pain and distress differ in that we can treat one with pain meds and the other not, but the animals, whatever the relative amounts of pain and nonpain distress, can nonetheless be suffering.

For our pain classifications (C, D, and E) we have a simple human criterion of the threshold for action and reporting: if it's more than minor or momentary pain about equivalent to an injection, it requires a search for alternative methods and correct placement in the D or E column of an annual report. With nonpain distress, we have less clarity. The USDA has withdrawn the online guidance it used to offer. What's the threshold of fear or hunger or loneliness or depression? Add in the complication that a growing chorus of behavior and welfare experts are calling out conventional lab caging in and of itself, cages that meet the standards of the Animal Welfare Act, as significantly distressful, even without conducting any experiments. If a lab houses a dog in a tiny cage just meeting the USDA standard of six inches longer than the dog, should that dog be listed in column E for the unalleviated distress? Distress, with or without pain, certainly warrants all the caution that animal pain warrants, but don't look to USDA annual reports of animals in pain categories for an accurate measure of animal welfare in laboratories.

In the past the USDA published its guidance on how facilities should categorize and report distress and pain, guidance it has since removed. It once suggested a threshold of reportable distress equivalent to withholding food or water for longer than would be standard, an overnight fast for most human and nonhuman surgery candidates, in preparing an animal for surgery. Try to apply that to a common practice in monkey neurobiology labs, in which scientists keep the monkeys thirsty enough that they will perform tasks day after day, collecting a thousand tiny sips of water while confined to a restraint chair. Our USDA inspector once challenged us to defend why such constantly thirsty animals were not in column E in our annual report for their unalleviated distress. When we updated our peer institutions that we might just report all such water restriction protocols in column E, the pushback was fast and firm. That precedent would require all labs to report their monkeys in column E too, as in significant, unalleviated distress, visible to any curious watchdog groups. The lesson for outsiders looking in on animal labs: absent clear standards on what procedures belong in column E on an annual report, facilities make their own determinations, and skeptics have good reason to read these reports as underreporting the amount of distress lab animals endure.

The USDA posts these facility annual reports (at the time of writing, a little more than one thousand of them) on its website. Most years (the USDA has been inconsistent with this, and reports more recent than 2019 are not easily viewable), it tallies them and lists national totals by species and pain category. Over the past several years, roughly 7 to 9 percent of animals appear in column E, what the USDA calls "pain, no drugs," with over half in column C (nonpainful experiments) and roughly a third in D (painful experiments in which animals receive pain medications). With diligence a person can see every regulated institution's inspection records, as well as their annual reports. If they assign animals to category E, they must include an explanation for why they do not alleviate the animals' pain and distress. A college student can see what their university is reporting on its animal use, and anyone can research the private and university labs in their area, though the information they will find will be disappointingly incomplete.

Anyone who wants to know how many animals suffer pain in US labs is at a loss in a multitude of ways. Recall that only a subset of US lab animals is under regulatory oversight. Of those the vast majority, the fish, mice, and rats, are only under National Institutes of Health regulations, which do not require annual reports of animal numbers and do not recognize the USDA's C-D-E classification system. The NIH's Office of Laboratory Animal Welfare releases individual labs' information in response to Freedom of Information requests, a cumbersome process. It does not compile the reports it receives and certainly does not proactively make information available.

Statistics from the USDA exclude the mice and rats and, even for the animals the Animal Welfare Act does cover, are a complicated read. Category D and E protocols do not separate out painful experiments from those with nonpainful distress. Moreover, the presumption that animals go into column D if any pain or distress they could experience is completely managed by analgesics, anesthetics, or tranquilizers is simply not true. Brain surgery on a monkey, liver surgery on a woodchuck, or pancreas removal from a dog are in column D, so long as the animals received postsurgical pain medicines. Any human who has undergone such major surgeries knows that the following days and even weeks, pain medicines or not, come with plenty of pain. Thresholds of pain and especially distress are vague enough that institutions can decide for themselves how much should count and be reported as suffering. When I look at the regulation-size cages for rabbits, dogs, or monkeys, I judge that any prolonged confinement in cages just barely meeting Animal Welfare Act standards should count as some degree of distress.

Most countries with large animal-research programs similarly require labs to assign a "severity level" to their animal experiments. Ethics committees review projects in the various (typically five) severity bands, and in most countries the experiments likely to cause the most animal pain and distress require special permissions. Labs must then reclassify the experiments if they turn out to have caused significantly less or more animal suffering than anticipated and report them annually. In contrast to the lack of transparency in the United States, the European Union and the United Kingdom both publish their annual statistics and, moreover, include tables of the severity classifications for all the species: the mice, rats, octopuses, and others. In the United States, we cannot divine from what the USDA posts how many animals are in basic science labs, how many are in drug testing, or how many are in safety testing. Want to know how many rabbits undergo the Draize test for eye injury and whether companies class them as untreated pain? So do I, but we do not have a system to give us anything near this level of detail.

In my lectures on lab animal pain, I ask lab workers why their animals don't receive more pain medicines than they presently do. Rarely does anyone suggest that they're already getting the right amount. I certainly do not believe that they do. Barriers may include costs, inconvenience, and the bureaucratic hassles of maintaining controlled narcotics. Do we just not see animals' pain, or do we see it and just not care enough? A culture of care is crucial. I audited a lab once whose work included major surgeries on monkeys and rats. I asked about pain treatments after surgeries. I always do. What looks good in an approved animal protocol does not always match what I find in a records audit. The head of the lab proudly told me how one of his students would bike across town at midnight to check on the monkeys and top up their drugs. "And the rats?" I asked, with a small sermon reminding them that rats' fast metabolisms can make them more in need of frequent remediation than a big rhesus monkey. "Oh, you silly boy," his indulgent smile seemed to say. I expect the rats just suffered unseen, though this is not information you'd find in an annual report to the federal authorities or in their scientific papers.

The hidden nature of animal pain allows us not to see it or to be moved to action, but a great hurdle is scientists' fears that pain medications will wreak havoc on their data. There's enough truth in that concern to justify a look at the evidence. The USDA early on acknowledged a reality that Congress had glossed over and that the *Guide for the Care and Use of Laboratory Animals*, the NIH's animal-welfare rule book, largely sidesteps: researchers do sometimes feel the need to cause animals significant pain or distress, and they do worry that pain medicines could interfere with their experiments in a variety of ways.[15]

Sometimes—and I emphasize *sometimes*—treating animals' pain can in fact make for worse experiments. Let's start with what we know about the side effects of pain medicines. A person takes ibuprofen solely for arthritic knee pain, but a gastroenterologist tells them to stop for a few days before a routine colonoscopy, because ibuprofen can make tissues bleed more easily; it's a complication best avoided in a procedure that screens for bleeding spots in the colon. Opioid analgesics (oxycodone, morphine, and the like) will have effects on alertness and cognitive function tests and are to be avoided. They are also respiratory depressants, so could skew lung function testing at the pulmonologist's office. Xylazine is a sedative that makes blood samples easier on the cats being sampled and on their human phlebotomists, but it sends blood-sugar levels soaring, tricking their vet into thinking the cat is diabetic. The various analgesics and anesthetics can throw off sleep patterns, affect social behaviors, delay wound healing, and suppress immune function.

Despite scientists' frequent reluctance to deploy analgesics for their animal subjects, they, along with vets and ethics committees, often take the opposite approach to administering general anesthesia. It's quite simple to pop a mouse into an anesthesia chamber. Chloroform and ether are out, while isoflurane and sevoflurane are their common modern reincarnations. I've seen nervous researchers, afraid of a mouse biting through their glove, steeling themselves to grab the mouse, place them in the chamber, and turn on the gas to render the animal unconscious for a simple, nearly painless injection. This may help the timid researcher, but anesthetic gases are not a magic vapor; they are unpleasant to the animal and a serious physiological stressor, yet another way to throw off an experiment. These are powerful medicines in human and nonhuman alike.

Scientists rightly worry that powerful analgesics with effects throughout the body can affect data collected from drug-treated research animals. With several animals in the experimental and control cohorts, the side effects of medicines can multiply, and, if some animals are receiving analgesics because they are obviously in pain while others are not, the variability that the drugs and the untreated pain introduce to the experiment only grows. Animal researchers are not irrational to fear that side effects of various pain meds will affect the experiments they are running.

I worked with a lab whose research included brain surgery in mice, with cognitive testing some weeks later. They worried that an opioid pain medicine (drugs like morphine, oxycodone, fentanyl, or buprenorphine) would have long-lasting effects on cognitive development. Another of my labs was studying knee arthritis in guinea pigs and worried that a nonsteroidal anti-inflammatory analgesic like aspirin or ibuprofen would mask or alter the course of the arthritis. A cancer biologist on my campus started her experiments by surgically implanting tumor cells in mice brains, after which she would divide animals into treatment and control groups to test how various cancer treatments might shrink established tumors. She at least granted my request to try analgesics for these brain surgeries and found (though with only five animals tested, I do wonder how valid this is) that tumors did not grow well, presumably suppressed by the pain meds. Would that pain medicines could halt human brain cancers the way this biologist saw them doing in her mice! But for experiments that require mice with brain cancer a medicine that stops the cells from ballooning into tumors is a nonstarter. Results of this trial in only five mice were certainly not something she could publish but were enough that the ethics committee thereafter approved her projects without requiring the pain medicines I'd been hawking.

Scientists investigating new brain-cancer treatments do not want to detour to researching how pain treatments might affect their laboratory animals. If they have a system that is producing data, they want to stick with their apparent success. When scientists do put the effort into comparing results in their experiments when they do or do not treat the animals' pain, they will sometimes find a difference. But are the data from animals who received analgesics more translatable to humans, less translatable, or simply a bit different, with little actual impact on how the animal data might relate to human outcomes?

I found it endlessly frustrating in ethics committee meetings that researchers would make vague statements like, "Analgesics may affect our data," or slightly less vague statements like, "Analgesics can affect immunity and inflammation, so we cannot use them in our experiment." Rare was the scientist who gave the committee some actual data or references backing up their claim that analgesics would not just affect data but that the effect would be detrimental. Yet, for all their concern that pain medicines could be an important variable in their experiments, when it came time to publish, they rarely devoted a sentence to tell their peers what they did about their animals' pain. Jamie Austin and I analyzed several hundred published papers in which scientists conducted a range of surgeries on a variety of animals. The surgeries—bone surgeries, abdominal operations, thoracic surgeries—all required anesthesia and should have received postsurgical analgesics. Overall, in their publications fewer than 30 percent of researchers mentioned pain management for their animals, either as a welfare concern or as a methodological choice that could influence their experiments' outcomes. In the several hundred articles we examined, zero made an explicit statement that they did not administer analgesics, so of course none explained a rationale for withholding pain medicines. Did they all just decide it was not such an important consideration after all, important enough to discuss with their ethics committee but not to communicate with their peers? Did they not treat their animals' pain?[16]

We documented a painful cycle. A scientist plans an experiment and sees that no one publishing in their field uses pain meds for a particular surgery for the species they'll be experimenting on. They secure their ethics committee's approval, based on their summary of the available literature. They publish their results with no mention of pain management for their animals, further contributing to a culture where pain medicines are verboten in some lines of animal experiments. Or, unable to know how other scientists managed the animals in their labs, researchers may use whatever anesthetics and analgesics their vet and ethics committee favor (that is, if they use analgesics at all) but expect reproducibility and replicability, important factors in building on prior knowledge and bringing it to the human clinic, to take a hit.

Scientists should plan carefully how to manage all the potential variability in their experiments, including not just the pain medicines they administer but also the effects of untreated pain itself. And in the interests of greater cross-lab reproducibility, they should fully report their methods in their research articles. Reports from the world of breast-cancer research illustrate the importance of standardizing pain management in rat labs, while challenging the assumption that pain medicines are more disruptive to experiments than untreated pain. The menace of breast cancer does not lie in what it does locally, in the mammary gland, at the site of the initial tumor. If it did, surgery would be a complete cure. Rather, the cancer cells travel through the bloodstream and the lymph vessels and lodge in the brain, the lungs, or other sites. Doctors have long worried that mastectomies or other surgeries could increase tumor cells' aggressive invasion of the lungs. Scientists model metastasis by infusing cancer cells into a vein, then measuring some weeks later how many lodged in the lungs and grew big enough to detect with X-rays. They can then compare how metastatic rates change if animals receive various analgesics without surgery (certainly many cancer patients receive pain medicines, even those who are not undergoing surgery) or if they undergo surgery. The short answer is that surgery's combination of stress, pain, and inflammation depresses immunity to the point that the risk of metastasis is greater—at least in rats. Potent opioid pain medicines like fentanyl and morphine may combat the pain of a cancer surgery, but at the cost of exacerbating how the pain depresses the immune system's ability to fight cancer. The take-home lesson is that both pain medicines and untreated pain can affect experiments, and scientists must design their experiments with this knowledge, erring on the side of their animals' comfort whenever possible.

But not all pain medicines are created equal. Before anyone closes the book on these opioid pain medicines as too disruptive of cancer biology for us to tolerate their use in experiments, it turns out that a different opioid pain med, buprenorphine, the one we use most often in animal labs, actually combats the immune suppression of surgery. How much any of this affects outcomes in other rat-cancer experiments depends on the specific research question. These experiments did not test how well the pain medicines treated actual pain, and for those experiments a scientist should also include ibuprofen-type medicines. Nevertheless, armed with this information, rat-cancer biologists absolutely must understand how painful procedures and the assortment of pain medicines can affect their studies' findings and, at the least, fully report how they ran their study so others can decide whether to accept their findings.[17]

Pain itself messes with the mind and body as much as the pain medicines we take to battle it. Pain and inflammation have their own effects that can skew how an animal's body performs in an experiment. Different painful conditions vary in their consequences. Animals in pain may go off food, whether because their mouth hurts, they can't move around well to get to their food, or they're just plain sick and don't feel like eating. Pain disrupts sleep, and sleep disruption wreaks all kinds of havoc on bodily functions. Mice in pain build poor nests, though that is when they might most want some warmth; a cold mouse is biologically stressed and, again, likely to be a poor research subject. Animals in pain may withdraw from social contact, and we know from human and animal studies how important social support is in health and disease. Yet, for all the ethics committee proposals I've reviewed where scientists explain what pain medicines could do to their experiments if we forced them to treat their animals, I've rarely if ever encountered a serious discussion of how untreated pain itself could affect the data. Again, replicability suffers when every scientist working in the field has animal subjects experiencing various, uncontrolled-for effects of untreated pain.[18] Scientists and vets might find that the expanded use of pain medicines leads to better animal welfare, with better science outcomes, and if they standardize pain management for the various experiments they do, reproducibility and trustworthiness of data might improve.

Experts in their various fields are the ones best situated to compare research outcomes in animals with untreated pain and distress and in animals with pain medicines. In parallel, vets and welfare scientists have been researching how best to recognize and treat animal pain, but virtually every study seems to examine different kinds of pain, using different pain measures. I would love to see a consortium of pain specialists produce a manifesto laying out the standardized pain their subject animals will undergo and the required battery of tests. They should include an ethical standard. Human and companion-animal pain experiments either compare a new treatment against the best available or include a provision for "rescue" analgesia, the amount of extra pain meds required to keep the subject, whether in the experimental or control group, comfortable. The pain lab animals suffer is of our causing, so it's important to get this right. We need to get better at recognizing it, better at finding effective pain medicines, and better at publishing what we do find.

People are right to be concerned at how much pain and distress lab animals suffer. We can do better. Just because dogs are no longer strapped alive to vivisection tables, we should not assume we have conquered animal pain in the labs. Animal rights activists paint lurid pictures of laboratory animals' pain, while research-advocacy groups counter that labs do all they can to manage and prevent pain. The reality I see is that we just don't know enough to support either side's claims. Animal pain is so very difficult to see. Our available pain medicines are getting better and better, but we still do not know enough about what doses to use or how often to administer them. We do not fully know how much, as with human patients, analgesics treat the pain but at the expense of making the patient nauseous, dysphoric, or generally unhappy. We simply do not know enough. I call on scientists, vets, and ethics committees to accept this uncertainty and factor it into any ethical evaluation of a research protocol.

Primum non nocere: first, do no harm. With pain so challenging to recognize, I urge vets and scientists to start with the anthropomorphic presumption that what hurts a person is likely to hurt an animal. With our current state of knowledge, I encourage ethics committees to assume the worst, assume that procedures will cause pain and that pain medicines will only partly alleviate those pains. Even a simple move like replacing injected antibiotics with oral doses can lower the amount of pain animals experience, along with lowering the distress of being chased, restrained, and injected. We will never eliminate all pain and distress as long as we have animals in laboratories, and so I move on to the question of ethical justification and the task of deciding when, if ever, to allow animal suffering in pursuit of the knowledge we humans seek.

Mouse

Mus musculus

LET LAB MICE BECOME ANIMALS

Olive was a lab mouse, one of the rare ones who got a name. She lived in a small lab on the Cornell campus, and every month or two her people collected a couple of drops of her blood, full of antibodies they needed for their experiments. She was a lab pet, as well as occasional blood donor, staying in the lab, well past any need for her antibodies, until old age came to claim her. Most lab mice get an ID number, sometimes a metal tag in the ear, sometimes a tattoo on the tail. Many remain anonymous for however long their mouse life runs, as we tend to see them collectively, a room teeming with five hundred seemingly identical little white rodents. I reckon that in my career, I've been responsible for some ten million mouse lives, as a member of ethics committees, as a lab vet, and as the second-in-command director of university animal care.

I have opinions about mice.

Mice are quite remarkable animals. They've been in our homes since hunter-gatherers in India first took up farming, though they are not domestic in the same way that dogs are. Since then they've traveled with us, uninvited, around the world. Until quite recently the relationship was a one-sided commensalism, in which they got to eat our food, but they did not hunt with us, farm with us, warn us of danger, or keep us warm at night. Impossible to eradicate, mice took and gave nothing in return; or worse, they gave us the various diseases they spread.

Mice are still unwelcome in most settings, except for in their latest niche, where literally uncounted hordes live in medical research labs. Secretive as unwanted "guests" in our homes and farms, mice live hidden in secrecy in US labs. I've explained that the United States has an arcane tripartite system of the welfare oversight of lab animals. The National Institutes of Health (NIH) rules require animal-welfare programs, though only for labs with certain classes of federal funding. The independent non-profit Association for Assessment and Accreditation of Laboratory Animal Care International runs an accreditation program for labs whose scientists think they are good enough to meet its standards. Together they cover some percentage, though no one can say what percentage, of the mice in labs. But, shh, the NIH will release only the mouse numbers that institutions report by a separate Freedom of Information request for each lab it oversees. AAALAC's accreditation reviews are quite rigorous, but you'll need to take my word for that (or not), as it is secretive, publishing no statistics and not subject to Freedom of Information requests. Our signature law, the Animal Welfare Act, sends veterinary inspectors to all US animal labs and publishes compliance inspections as well as facilities' annual reports of the numbers of animals they use. But remember this important quirk of our Animal Welfare Act: the US Congress decided in 2002 that *mice are not animals*.

Mice and rats serve US research labs in obscurity. No one knows how many. In 2021 I published my estimate, based on evidence from the US Department of Agriculture (USDA) website (the easy part), data from the NIH through Freedom of Information requests, supplemented with statistics from state universities subject to their individual states' open records laws. My article got pushback from the research-defense organizations, and the controversy somehow got me my three minutes of fame in the *New York Post*. But it remains the only such estimate based on US evidence. I found that, in the institutions I surveyed, mice are 99 percent of the laboratory mammals, likely adding up to at least a whopping one hundred million mice and rats per year.[1]

I had started my project by asking the thirty largest NIH-funded research universities to voluntarily share their mouse numbers with me, reminding them that the dogs and monkeys in their labs were public knowledge on the USDA's website. As I expected, having run a university animal-welfare program that similarly refused to share information we were not legally required to share, I received either silence or refusal from half the labs I contacted. One university vet was candid: he refused to give me statistics because he feared I would use them to argue that mice should become Animal Welfare Act animals. As I wrote at the time, having confirmed my initial rough guess with actual data, the large numbers are both the argument against expanding welfare protections to better cover the smallest animals (too costly!) and the reason for making that move (too many to ignore!). The knee-jerk resistance to even this small bit of transparency tips the balance for me toward wanting some openness and public accountability.

Mice in federally funded labs fall under the jurisdiction of their institutional ethics committee, as do mice in labs that voluntarily seek accreditation. On the evenings after ethics committee meetings, I would tell my husband of the various debates over approving a particularly painful experiment or sanctioning a researcher for poor or careless animal care. As at most universities, mouse projects, like the mice themselves, greatly outnumbered all the rest. Occasionally he'd stop me to verify, "So this was an issue concerning just three mice in a lab? I'm actually surprised at the effort they invest." Point taken. And on occasional visits to small private companies' labs, I've been happy to see that even with no accreditation to maintain or laws to obey, scientists form ethics committees as though the NIH rules did apply and debate and discuss their company's mouse experiments. At least they did at labs that invited me in for a third-party consult on animal welfare.

Despite the NIH, accreditation, and voluntary oversight some mice receive, it pales by comparison to what a monkey, a woodchuck, a dog, or even a gerbil receives. Because of this congressional decree that mice are not animals in research, no government veterinarian inspects their quarters or looks at how the ethics committee watches out for them. Small institutions with only mice, rats, and zebra fish in-house may be totally exempt from any third-party oversight. We collect national statistics on lab hamsters, in the pain categories C, D, and E, but not on our most numerous lab animals. Informally, even in labs that seek the AAALAC's nongovernmental accreditation or have NIH oversight of their mice, the researchers, vets, and ethics committees know that no USDA inspector will hold their feet to the fire as they do a dog or hamster researcher. Recall the student who would diligently bike across town on a foggy San Francisco night to check on their monkeys but whose boss chuckled at my suggestion that their rats might want similar solicitude. Ethics committees' rigor in self-policing is just not the same as it is for dogs, rabbits, or even hamsters. Research lobbyists claim that mice do just fine with the other two components of third-party oversight, though they have no data to even know how many mice are left outside of these protections. That claim is not credible.

Mice are not fine in our current system, and the vigor with which the lobbyists have fought to keep them in their place reflects the fact that were mice to become legal animals in the eyes of the Animal Welfare Act, labs would need to invest significant money and time into treating them as other animals' equals. I happen to quite like mice, even though monkeys will always be my first love. When we care enough to pay attention, they are fun to watch. They are inquisitive and incredibly agile. Males sing complex mating songs, far out of the range of human hearing, the better to allure females. Females may nestle together to help one another rear their young. I'd always assumed we people look indistinguishable to mice, and we may, but a mouse's world is more olfactory than visual. Our individual human smells distinguish us in ways that mice, even with our efforts to keep the air in their ventilated cages separate from the air we mouse workers breathe, can tell us apart by our scents. On any test of pain and welfare, mice show themselves to be clearly sentient, fully able to experience pain, pessimism, and distress. If we could all care just a bit more about the mice, from whom we take so much, their lives could be at least a bit better.

When the plight of stolen dogs, rabbits in cosmetics testing, and monkeys suffering in neurology brought new protections not just for themselves but for many other lab species, mice reaped some of the benefits. Some of what these other animals set in motion has spread to lab mice as well: the search for nonanimal safety testing, standards of health care and pain management in labs, richer home cage environments, and oversight of institutional ethics committees. In public affections and in regulations, though, they've remained low status. In fact, they've largely followed the opposite trajectory, taking on research roles the larger animals increasingly are spared. As dogs, rabbits, and monkeys raised the standard of care for other lab species, mice—don't blame them—have facilitated the development of new research technologies and new ways to model diseases. Ironically, scientists piloting those new technologies in mice—genetic manipulations in particular—are now bringing them from the mouse lab for use in yet more experiments in larger animals.

Despite the huge costs of high-tech mouse vivaria, mouse labs are still far cheaper to run than monkey, dog, sheep, or other labs. That financial incentive is greater than the regulatory one, which nonetheless still pushes scientists to pick a mouse among the many candidates for an experiment. I've counseled researchers who felt that their successes in their mouse experiments were leading to research questions for which dogs, ferrets, monkeys, or other animals would serve as more informative models. I explained that, though they thought their mouse studies were heavily scrutinized in the slow path to ethics committee approval, the license to use several hundred mice was a much easier prospect than an application to use a handful of dogs. I explained how our prescheduled, semiannual in-house ethics committee inspections paled compared to the unannounced inspections of the USDA veterinarian. A while back my campus received USDA Animal Welfare Act citations for some problems in the vole lab. Voles, remember, are a type of field mouse, but they are not literally a mouse (of the genus *Mus*), and thus they are legally animals. Animal activists watch for USDA violations as it posts them, so activists saw ours and sent out press releases. In the fast news cycle of modern times, these violations in the vole lab, plus a couple in some pig labs, generated no public reaction. Still, compared to a self-report to the NIH of a mouse lab problem that killed hundreds of mice, even if the activists got that one with a Freedom of Information request, would likely have stirred even less interest. In some cases scientists' progress in their mouse experiments was just too exciting not to scale up the studies they needed in other species, and they decided the added scrutiny was worth it. Others decided those other studies could wait and went back to their mouse experiments, because in the Animal Welfare Act and in public sentiment, mice just do not count.

It should surprise you that mice are not legal "animals." I've long been vexed that mice are overwhelmingly the most numerous mammals in labs but are excluded from the protections of our Animal Welfare Act. It's not that mice have not had their day in court or in the halls of Congress. They have; they just keep losing. The 1966 Animal Welfare Act named the limited six types of animals it would cover, but, while hamsters and guinea pigs made the cut, mice and rats did not.[2] Four years later Congress updated the law and expanded the definition of "animal" to include "such other warm-blooded animals as the Secretary [of agriculture] may determine is being used . . . for research." Whatever the secretary of agriculture may have thought about this expanded coverage, the head USDA vet overseeing enforcement was clear that the USDA was under no obligation to uphold the Animal Welfare Act for mice—and would not.

The USDA did not try to claim that mice were not research animals. Citing limitations of labor and money, the USDA reported their internal discussions and their conclusion: "We felt that perhaps with these particular species [mice and rats] there was not any great inhumane care in handling." I have no idea how anyone at the time could say that. Mice were numerous in toxicity-testing labs, cancer labs, and infectious-disease labs. Vendors shipped them around the country in freezing cold and stifling heat, on days when dog or hamster shipments would be grounded under the transportation standards of the Animal Welfare Act. Certainly, adding mice to its docket would stretch the USDA's resources at a time when Congress had just added zoos to their inspection commitments. Only through naïveté or dishonesty, though, could anyone who'd been inspecting labs for four years claim in public that mice and rats were doing just fine in scientists' hands with no need for welfare oversight.[3]

The 1985 amendment to the Animal Welfare Act brought in some sweeping changes such as exercise for dogs and psychological well-being for primates. The most important addition to the law, though, was the creation of animal ethics committees that would oversee scientists' animal projects. Mice and rats remained outside the protections of the updated law, again, not by congressional action (Congress had remained silent about them in updating the act) but by USDA's regulatory decisions. Congress had mandated that the USDA harmonize its regulatory standards with the NIH, which had already been covering mice under its oversight, using standards in the *Guide for the Care and Use of Laboratory Animals*. But the USDA demurred, claiming, "We do not believe it would be in the best interests of animal welfare in general" to devote time to developing standards for what were then and now the most numerous mammals in US labs.[4]

This claim, too, was disingenuous. The USDA oversees hundreds of mammal species without having drafted specific cage sizes or housing temperatures. It covers woodchucks, gerbils, ferrets, and whatever other unusual mammals might be in the lab. It covers wild-caught house mice or mice from pet stores, just not mice "bred for use in research."[5] In a strange manifestation of this mouse-versus-mouse rule, we had a project in San Francisco for a while that started with wild-living house mice (of the genus *Mus*), trapped, brought into the lab, and set up for breeding. The USDA inspector visited them every year, and we made sure the wild-caught mice, legal animals under the Animal Welfare Act, were labeled so our inspector knew which animals were *animals* and which, their off-spring, were just mice. When the initial parent stock eventually died out— mice rarely live past three years—the hundreds of mice they'd produced were "bred for research," and their quarters were no longer an animal room for USDA inspection.

But the USDA remained steadfast. Mice were not animals they were going to go to bat for. Animal rights groups sued the USDA to start seeing mice. The first time the animal rights groups lost for lack of legal standing, as they were not harmed by the mice's predicament in any way. In a second lawsuit, in 2001, the Alternatives Research and Development Fund was more successful. USDA settled the case and promised to start writing up rules for rat and mouse care that it could enforce. The National Association for Biomedical Research and Johns Hopkins University filed a suit to block this, arguing that the USDA did not have resources to expand its coverage, that yet more regulations would be burdensome to research institutions, and (it claimed) that 90 percent of mice and rats were already covered by requirements of other federal agencies or by the nonprofit accrediting group the AAALAC.[6] In quick succession Congress first blocked the funds for the USDA to devote any staff for finalizing animal-care standards for mice and the following year permanently enshrined the USDA's distinction of animals and not-animals in the act, making it the law of the land that the term *animal* "excludes (1) birds, rats of the genus *Rattus*, and mice of the genus *Mus*, bred for use in research." And so mice remain outside the Animal Welfare Act until Congress someday decides that they are, in fact, animals used in research. This is a move it has shown no inclination to make.[7]

Mice are not animals, so, at my institution in San Francisco, the USDA inspector would appear for an unannounced inspection, and I would walk her past rooms with a thousand mice so she could inspect the one room that housed a dozen hamsters. Mice are not animals, so when the inspector read the minutes of our ethics committee meetings, I clarified for her which discussions about researcher noncompliance concerned "animals" under her jurisdiction and which were "just" about mice and rats. Mice are not animals, so our annual Animal Welfare Act report to the USDA did not list them, and no one can see a tally of mice used in the United States in the C, D, and E pain categories in any given year. Mice are not animals, so we had whole buildings off-limits to the USDA inspector. Mice are not animals, so smaller companies and colleges with rats, mice, frogs, and fish are invisible to government inspectors. Mice are not animals, so any whistleblower complaints about their care would not trigger a USDA inspection (though the USDA would have forwarded it to the NIH, which does see mice as animals and would launch its version of an investigation).

This exclusion is wrong and we should change it, by which I mean Congress should change it. Mice must become animals. In the United States, we have a checkerboard of oversight for laboratory animal welfare. First, the USDA enforces the Animal Welfare Act, though only for those species it classifies as animals; for an approximation that's closer to 0 than to 1 percent of the total animals in US labs.[8] Second, the NIH oversees animal welfare, mice and rats included, in institutions that receive federal research dollars such as large universities but also, much less commonly, in some small colleges or private companies without federal grants. Third, rounding this out, many academic and industry research institutes seek accreditation through a nonprofit organization named AAALAC. Like the NIH, the AAALAC includes mice, rats, and fish in its purview and even goes a step further, bringing octopuses and cuttlefish in as well. So, with two of the three overseers of lab animals' welfare unreservedly calling a mouse an animal, I understand the criticism of my call for making mice Animal Welfare Act animals. However, that counterargument is flawed in important ways.

The key is in enforcement and oversight. AAALAC reviews are rigorous and valuable. A team of experts, mostly seasoned lab vets, visits every three years. That gives us three years to prepare for the next inspection. Moreover, AAALAC keeps its assessments confidential, unlike the USDA's inspection reports, which live on its public website. I've never seen much value in NIH oversight, other than that it includes animals the Animal Welfare Act excludes. Enforcement is all through self-reports that institutions make, whether for accidents (a broken thermostat overheated a room full of animals) or noncompliance (the researchers' students did not give the animals their pain medicines), but as long as the ethics committee is tending to the situation, in my experience, NIH takes no further action. In transparency the NIH sits between the AAALAC and USDA; it gives out little information except through Freedom of Information requests. Only with the USDA's surprise annual inspections and transparent dissemination of its inspection reports are labs really kept on their toes.

Everyone working in this business knows this difference between USDA and NIH oversight and how this difference in legal status spills over into differential treatment in the labs. Put three lab vets together in a room, and shop talk quickly shifts from the latest medicines for fur mites to a term we all share: the *problem investigator*.[9] Lab vets inevitably slip into the role of enforcement officer, and the problem investigator, aka the *difficult investigator*, breaks the rules in the animal house. And when vets one-up one another with their war stories, I hear the added disbelief: "And this was not a mouse lab; it was a covered species!" Problem investigators should behave better, but problem investigators with USDA-covered animals *really* should behave better. The ethics committees I've known also look at research noncompliance through this binary lens: Was this a serious issue with a covered species or *just* a mouse or rat, something we will handle in the family without the USDA snooping and without feeling we need to tell the NIH? We let things slide—failure to administer required pain medicines or procedures performed without committee approval—in mouse labs that we would not let slide in a rabbit lab, and the mice suffer.

Making mice legal animals would certainly cost money. How could it not, given how numerous they are? In large institutions with triple oversight, the costs would be minimal, though attitudes would change, knowing inspectors could show up any day. At smaller institutions they would need to establish an ethics committee, if they did not have one already, and bring their animal facilities up to standards. In my ideal world, the government might save some money by phasing out NIH animal oversight and consolidating all the animals under the USDA and the Animal Welfare Act. The big challenge is for the USDA, staffing up inspect all those labs that were previously invisible to it. On large campuses a three-day inspection could triple in time and effort if the USDA still visited every animal room and read every animal protocol that authorized painful experiments. I could see having a system of spot checks. Traffic patrols set up speed checks just enough for speedsters to take safe driving seriously. Our current animal-welfare oversight is more like having near-constant traffic controls but limiting the speed checks to the 1 percent of cars that are yellow or whose license plate ends in 2. So send an inspector to my campus and let them pick a representative sample of animal rooms to inspect and records to read. Give the mice that much. We can do this.

The mouse's second-class status goes beyond regulations and enforcements. Most people in and out of labs would probably agree that scientists who can replace their dogs, monkeys, or cats with mice should do so (and should replace their mice with frogs, fish, worms, or nonanimals if they can). In government reviews of the necessity of chimps, monkeys, or dogs in labs, mice are always lurking just out of sight, with a more-or-less explicit thought that if we cannot replace those animals altogether, can we at least have mice take up more of the burden?[10] For sure, they are less intelligent than dogs or monkeys, so writers slip into labeling them "lower" animals as further justification for placing our research burden on their backs. We should at least be clear about how we are prioritizing some animals over others, especially when a switch from using larger animals to mice ends up inevitably meaning a researcher uses more mice per experiment than they would use dogs.

As much as financial costs drive scientists away from other species, technical advances in animal experimentation have made the shift possible. Woodchucks replaced chimpanzees and children in hepatitis research, until genetically engineered mice replaced the woodchucks. Diabetic dogs brought Frederick Banting his Nobel Prize; now diabetes labs that use animals use mostly mice. Monkeys and even cats who harbor their own immunodeficiency viruses were early conscripts in research on the HIV virus and AIDS. Monkeys continue in that role, but much HIV work has shifted to mice, including curious creatures, "humanized mice," in which mice genetically designed to lack an immune system of their own can host immune cells derived from human fetal tissue. The relative ease of genetically shaping and breeding mice has been crucial; they found a place in human hepatitis research only because, unlike woodchucks, scientists were able to insert human liver genes that could harbor a human-specific hepatitis virus to which a woodchuck or a normal mouse would be immune. Beyond that engineers have miniaturized the technologies, allowing scientists to swap out dogs with mice. In the 1920s Banting's studies would have required several tubes of blood for frequent measures of glucose and insulin levels, far more than any mouse could produce. The only way for him to have replaced his dogs with mice would have been to use dozens of mice in place of each dog, pooling their blood samples for a measurement. Now a pinprick in a mouse vein yields a single drop of blood that is more than plenty for the same glucose test strips that human diabetes patients use.

Lab animals have served to safeguard human health and safety for many years. We do not want to put time and money into human trials of a drug unless animals have told us it's likely to be useful. Animals have been our tasters and safety testers before we risk swallowing a new medicine or even applying a new cosmetic. We make animals hold our hands as we gather confidence to try something new. It falls to mice to serve that role for other animals in the vivarium. Mice have been on the front line for proof-of-concept experiments and new biotechnologies before any other animals get involved. Long before an idea looks sound enough for testing in monkeys or dogs or pigs, it has likely gone through thousands of mice, who may stay hidden in the lab. My monkey researchers who would not mount their bike for their rodent subjects' pain treatments worked out logistics for monkey experiments in their mice and rats but never published the rodent work (and certainly never thanked the mice in their papers' acknowledgments). Logistically, mouse experiments are much cheaper and faster than larger animal experiments. In the short term, this spares the larger animals the risks of serving in certain experiments. In the bigger picture, mice are pioneering some technologies that put those larger animals into lab service in ways never before possible.

I've described how monkeys' brains, so remarkably similar to humans', got them and keep them in laboratories. By contrast, it's their fur that got mice initially into science. In the wild, mice make baby mice in a range of coat colors, tan or brown or black or white. Such "sports" darting in and around grain bins caught the eyes of ancient Egyptians and Chinese writers. No matter the species, if people can get animals to breed in captivity, they'll breed for the unusual and the eye-catcher, whether for various anatomical traits or for variations of fur texture and color. Count mice in on that human obsession to play Creator. In 1895 Walter Maxey and his fellow mouse breeders established the National Mouse Club in England, which is still going strong, setting breed standards and staging competitive mouse shows.[11] As mouse clubs arose across the globe, it was perhaps inevitable that scientists would see value in a small, cheap, easy-to-breed animal whose genetics were on display in their dazzling array of coat colors.

Mice are fast, prolific breeders; a mouse doe may have her first litter when she's barely two months old. Gregor Mendel had established some basic principles of genetics in his study of pea plants in the 1800s, mapping out how pea color and shape were predictably passed down through generations. When scientists wanted to see if what was true of peas was true of mammals, they gathered up mice from various fancy breeders and demonstrated that mouse coat-color genetics appeared to follow Mendelian rules. Abbie Lathrop was a mouse fancier in Massachusetts who noticed mice with lumps and bumps among her ten thousand charges. Sending these off to medical researchers, she and they were able to tie cancer incidence to mouse strain. The connections between genes and health proved much more compelling than the genetic bases of coat color, and scientists at the Jackson Labs in Bar Harbor, Maine, launched the first nonprofit research foundation to study mouse genetics, sending mouse stocks to laboratories throughout the United States.[12] Though dogs and rabbits and others had been lab subjects for years, their genetics were never a big research focus; the fancy mice who spawned lab colonies ushered in a new era of genetics-focused lab animal studies.

In Wales "sin eaters" were the down-and-out people who ate the sins of the dying, taking them and their consequences on as their own. In labs technological advances have cast mice as sin eaters. Charles Banting and his colleagues needed a dog-sized animal for their diabetes work, an animal large enough to allow the surgeries to remove pancreases and to provide enough blood to run frequent checks on blood sugar and insulin levels. Now diabetic mice have taken on that burden, through either chemicals or genes to bring on the diabetes, giving an occasional drop of blood because that's all it takes now to check blood sugar. Woodchucks were the stars of hepatitis B research, until scientists genetically modified mice who can sustain a human hep B infection to take on most of that work. Part of the impetus to forsake dogs for mice is ethics mixed in with public relations; part of it is that mice are smaller and cheaper and all-around easier to use in experiments than bigger animals.

Medical research is not just for human patients. Vets speak of a "one health" perspective, highlighting linkages among human, animal, and environmental health. The intersections of human and nonhuman animal health allow us to try to extrapolate to people what we learn in animals, but, as drugs and devices find their way into human medicine, they often spin off into farm and pet animal medicine too. Human antibiotics, pain medicines, and diabetes treatments all have counterparts in vet medicine, and vets use ECG machines to diagnose dog heart health, even implanting pacemakers in some patients. To be sure, mice are put to service for all these medicines and medical devices that will help people and other animals. But they also serve as subjects and models in research aimed solely at diseases neither mouse nor man naturally acquires. The biggest immunology lab I worked with in my early Cornell days employed mice as animal models for cow mastitis. Up the road another lab exploited mice for studies of canine distemper. Half a world away, mice serve in labs trying to understand and eradicate a very strange disease in an unusual animal, the facial tumors of Tasmanian devils.[13] In perhaps the cruelest twist of fate, researchers have even put mice to work testing the safety and efficacy of vaccines for their nemesis, the cat.

While mice have lagged in welfare protections, they have been far and away on the front lines as scientists develop new avenues of research. Genetics brought mice into the labs, and genetics continues to drive much of the lab use of mice. First, scientists develop highly inbred strains of mice, taking advantage of these fecund animals' speedy reproductive rate. A mouse person starts by crossing brother to sister or parent to child, generation after generation. Weird things can show up with such intense inbreeding, but, past half a dozen generations, with the healthiest brothers and sisters picked for the next round of incest, the breed (or strain) stabilizes. By the twentieth generation, the line of mice is officially inbred, over 98 percent genetically identical. The large breeders such as Charles River (who sold us our rats for the python Esther's lunch) and Jackson Labs have rooms with thousands of such animals, genetically as close as a roomful of identical twins, or even closer. The mice are so inbred that a scientist can transplant some skin or a kidney without any graft-rejection drugs as doctors use in organ-transplant patients, and the recipient mouse accepts it. They are so identical that a male who enters a cage that a stranger male of the same strain has scent-marked—usually a high-stress event for a mouse—settles in as though he had claimed it himself.

We've all heard of the dangers of inbreeding; didn't that lead to hemophilia and madness in an assortment of European royalty (and in *Game of Thrones* characters)?[14] Mouse lineages that survived those initial rough generations of inbreeding gone wrong have won their people numerous Nobel Prizes. Setting up parallel lines of inbred mice with slight differences between the lineages allows scientists to unlock many of the secrets of immunology in the quest to conquer organ rejections in human patients with someone else's heart, kidney, pancreas, or other organ. These successes in mice inspired scientists to want inbred strains of their other species, including rats, guinea pigs, and zebra fish. With twenty generations of inbreeding being the target that produces a line that is 98 percent genetically and immunologically homogeneous, it's not surprising that slower-breeding dogs, pigs, and primates have not yet joined this club. Mice still prevail.

While science observers were thinking in the 1980s that animal labs had crested and were waning, gene jockeying came on the mouse scene, putting inbred-strain research on steroids. Through a complicated process using stem cells, in-vitro fertilization, mother-mouse surrogacy, and harmless viruses into which they packed mouse DNA, researchers (who went on to share the 2007 Nobel Prize) developed genetically modified mice. They began by introducing genes from other species, inserting a gene from a cancer-causing virus into mouse embryos. The mice grew up to become OncoMice, a line of mice genetically programmed to develop mammary cancers at high rates. This was yet another new way to expand the available choices of mouse models in cancer labs, alongside carcino-genic chemicals and the surgical transfer of cancer cells from one mouse to another. The OncoMouse was the first mammal ever patented in the United States.[15]

Scientists then moved on to make thousands of varieties of genetically reengineered mice, some by inserting genes from viruses, sea urchins, or other creatures and others by removing or disabling mice's natural genes. Whereas old-school researchers might remove an animal's organ (as Frederick Banting removed his dogs' pancreases) to divine its function, geneticists moved on to genetically disabling various organs and cells to similarly divine a gene's actions by seeing the consequences of disabling it. Scientists can knock genes out, insert new genes, program inserted genes to turn on or off only in specific organs, and turn the genes on and off at the scientists' will. Thus I might be able to turn on a cancer gene in a genetically engineered middle-aged mouse's brain by feeding them a diet laced with just the right chemical to activate the gene (the antibiotic doxy-cycline is a common choice). Next I engineer immune cells from a near-identical mouse strain, engineered with a sea urchin gene that makes the cells shine green under fluorescent light. I infuse those immune cells and, with the mouse anesthetized under fluorescent lights, watch the glowing green cells infiltrating the cancer. This is a powerful tool, so of course scientists quickly got busy transferring this technology, with greater or lesser success, to zebra fish, rats, ferrets, and others, but mice maintained their dominance as the most common lab mammal to find their genes knocked in or knocked out or otherwise manipulated. This technological advance quickly turned around any decline in mouse numbers that scientists and the engineers scrambling to build vivaria fast enough to accommodate them had envisioned.

Mice are opening technological doors that will increase animal use and animal suffering in other species. As scientists look to develop new experi-mental methods in their animal labs, mice continue to be the gateway animal. Try it in a mouse and, if you can do it in a mouse, try it in a larger animal. Mice have a track record of pioneering technologies that can be used in humans and other animals. Mouse methods are at the heart of genetically modifying pigs to be human organ donors. Wooly mice are an early prototype for inserting extinct mammoths' genes into elephants, to someday generate a herd of genetically reengineered pachyderms who will not bring mammoths back from extinction but will nonetheless populate the tundra as mammoth-elephant chimeras.

Right now gene editing with a new technology called CRISPR is hot in animal labs.[16] Bacteria have immune-defense mechanisms in which they alter their own DNA to ward off bacteria-eating viruses. Scientists explored (another animal-research–driven Nobel Prize) how to use these mechanisms to reprogram bacterial DNA in the lab. Mammalian DNA is enough like bacterial DNA that the same CRISPR mechanism that can edit bacterial genes can edit mammal genes, and of course, in the lab, that starts with mouse genes. CRISPR is more precise than the older ways of knocking genes in or out, and it does not generate nearly the number of mice with unusable genetics that older technologies did. With earlier technologies a complicated breeding scheme in mice with two or more genes of interest might yield only one in sixteen (or even fewer) pups with the right genes for an experiment. This crowds the mouse facilities and the mouse cages and results in killing and culling millions of young adult mice who have no utility for the lab. In that aspect CRISPR could have one mouse-welfare bright spot.

The CRISPR technology has been able to spread from mice to other species more readily than the older methods of genetic modification. It is an exciting tool for a range of applications. It's the basis of Casgevy, a therapy in which blood from patients with sickle cell anemia is collected, genetically edited, and reinfused to them to produce a type of fetal hemoglobin that combats the sickling deformities of the patients' blood that lead to periodic episodes of intense pain.[17] CRISPR is the technology driving the creation of pigs engineered to become organ donors for human patients in need, not because the organs themselves are so different but because of molecular changes that make them less subject to attack from the recipient's immune system. Scientists are fine-tuning CRISPR to genetically alter HIV-positive patients' lymphocyte immune cells, the ones that HIV attacks with such gusto, to remove some of the docking stations the virus uses to enter the cells. With no susceptible T lymphocytes, the virus will have nothing to propagate it and will die out in the body. This is all very exciting, and mice were a crucial step in going from work in bacterial cells to mammalian cells, then whole animals, then human patients.

Genetic modification is not just for mice and fish anymore: CRISPR works well in large animals too, so scientists can now better modify monkey and dog and other larger animals' genes. Scientists—neurobiologists and immunologists in particular—use mice but often want an animal whose brain or immune system more closely models the human. Primates, with their more humanlike brains and immune systems, are alluring. Monkeys are, however, large, expensive, and hard to work with and, until recently, limited in their usefulness because scientists could not control their genes as they can mouse genes. Enter CRISPR and expand the range of diseases scientists are able to program into their monkey colonies. For all of CRISPR technology's exciting potential contributions to human health, it may nonetheless reverse the trend of recent decades in which mice increasingly replaced larger animals.[18]

While the scientific and medical advantages are thrilling, CRISPR threatens to set animal welfare back, for mice and other animals. We have already seen how old-school genetic-modification technologies let scientists create animal diseases they'd been unable to create previously, making cancerous mice, mice with cystic fibrosis, diabetic mice, and any of a number of diseased and distressed animals. Though we have no formal statistics, we insiders saw how transgenics technologies of the 1980s led to an explosion in the numbers of lab mice. I expect to see CRISPR resulting in more monkeys and pigs in labs (if scientists can surmount the present short supply of monkeys; even CRISPR still needs adult monkeys and long gestations to make new monkeys), and I dare not predict mouse numbers. Its efficiency will let scientists make more and more different kinds of mice but with less overproduction of mice they cannot use.

So researchers can now deploy CRISPR to create more diseases in more animals. But more than physically painful and disabling diseases, what worries me most are the efforts to genetically engineer animals with the range of psychiatric diseases—depression, anxiety, and the like—from which humans suffer and from which lab mice and monkeys will increasingly suffer.[19] Harry Harlow was the research psychologist whose monkey experiments in the mid-twentieth century made him a bête noire to animal advocates. He started by depriving baby monkeys contact with their mothers, charting how maternal deprivation ravages a growing primate's mind (he used rhesus monkeys in his lab as surrogates for his big concern, human babies in institutions). He developed increasingly severe ways of raising monkeys to grow into psychologically damaged creatures, with devices such as the "tunnel of terror" and the "pit [of despair]."[20] His admirers called his work the science of love; his detractors to this day brand it torture. People argue whether truly useful knowledge has come from his or his trainees' labs. Harlow certainly never found the cure for depression that he sought, though he may have given enough of a scientific gloss to the argument that babies in orphanages need love and contact more than was then standard. Less up for argument is how Harlow's monkeys suffered from the mental illnesses he generated. Mental illness is some part nature and some part nurture, and Harlow excelled at modeling ways in which bad nurturing leads to disease.

CRISPR gene editing now promises (or, really, threatens) to allow scientists to explore the "nature" part of the mental illness equation, genetically creating monkeys prone to the sorts of psychological suffering Harlow nurtured. Scientists have created and cloned monkeys with sleep disorders, with the range of fear and anxiety to which humans with sleep disorders are prone. Monkeys might blame mice for their role in developing this new technology, though its development relied so much on studies in cells, not whole animals, that perhaps the mice can be absolved—and pitied, as mice remain in the front line for whatever new disease models CRISPR scientists spawn.

Mice might thank monkeys and the 1985 law requiring attention to the psychological well-being of primates, for the decades of research into welfare reforms the mice might enjoy. Despite the patchy legal protections mice get in the United States, many good people around the world are working quite hard to improve the lot of laboratory mice. Animal rights watchdog groups tend to place dogs more prominently in their materials than the little animals so many people shun as vermin. Nonetheless, they promote legislation to improve protections for lab mice, scan publicly accessible NIH compliance reports, and post information about rodents in labs. Advances in mouse welfare include studies to improve their housing, research into murine pain management, and the replacement of live rodents in product safety testing. Legally, the NIH's rule book, the *Guide for the Care and Use of Laboratory Animals*, last updated in 2011, calls for increasing concern for all vertebrate animals' social and behavioral needs and preferences, a mandate that I have seen the AAALAC accreditation process take very seriously

As I describe in the chapter "Chicken," animal-welfare scientists try to give mice their due. They publish their data on how to gently handle mice and rats, how to improve pain treatments, and how to combat the deprivations of the standard plastic shoebox cage and push it toward something a mouse could call home. Their experiments with mice are mirrored in studies for other small lab rodents' welfare. They ask gerbils whether they would stop furiously digging in their cage if they had a little tunnel to curl up in. They scrutinize guinea pigs to see if they grimace in pain the way mice do. Mice hate being grabbed by the tail, to the point that they are higher-stressed research subjects who then produce questionable research data. Building on this knowledge from mouse labs, welfare scientists probe whether hamsters, too, exhibit less stress when gently coaxed into a cardboard tube for carrying them from place to place. I don't know if anyone would have thought to ask the hamsters had they not explored the situation in mice.

Translating welfare findings in mice, however, relies on vets and ethics committees paying attention to what welfare scientists are finding and convincing scientists that better mouse welfare, even without the USDA's Animal Welfare Act inspectors to push for it, make the mice better research subjects most of the time too. Welfare scientists to whom I speak remain upbeat about what they're learning but share their mixed feelings about the slow uptake in labs around the country. Sadly, lab vets themselves can be a barrier, if they don't keep up with developments in animal-welfare science and don't want to push their researchers to take better care of the animals.

I've shown here how mice and rats' second-class status has kept them outside the protections of our most powerful animal law, the Animal Welfare Act. Their expendability, along with their usefulness, has encouraged science to shift the burdens of animal testing from the species most people care most about to lowly mice and rats. Science is full of twists and turns, and we can now watch technologies developed largely in mice, CRISPR gene editing in particular, bleeding over to other species. This is fueling the drive to create more and more monkey models of human psychiatric illnesses and promotes farming gene-edited pigs as organ donors for human patients. As scientists and regulators validate nonanimal tests that can get live animals out of safety testing, other scientists are developing new uses for animals, mice and others, in basic science labs. What irony that mice who received so much of the research burden from other species are paving the way for expanded use of those larger animals in ways that were not possible before.

If caring more about some kinds of animals than others is a sin, I confess my guilt. We all do that. Monkeys are my favorites; others may more enjoy and care about dogs or pigs or cats. Mice are going to be in animal labs long after most other kinds of animals have exited. Their welfare is important to them and should be important to the scientists who extract data from them for improving human health. It's time to upgrade our standard of care for mice and to have our animal-welfare laws reflect those standards.

Flea

Ctenocephalides canis

THE ETHICS OF HARMING ANIMALS FOR HUMAN BENEFIT

A confession: we vets, if we are doing our job right, commit mass murder. I did a stint as the spay-neuter vet in a Society for the Prevention of Cruelty to Animals (SPCA) shelter, where a day in the clinic could include healing and operating on a few dozen animals while killing hundreds or even thousands of others. We were ruthless in our killing, sometimes using chemicals that block respiration or, more commonly, nerve toxins that bring convulsions or paralysis. We didn't much care how the animals died, as long as that flea, tick, worm, or mite in our crosshairs expired completely and left the puppy or kitten or mouse in peace. When a flea abandoned the kitten, hopping to the table in a desperate escape, we pursued and crushed the fugitive under our fingernail with a satisfying crunch. Vets garner profit and praise for this killing.

Our shelter had an open-door policy, taking in all animals regardless of health or temperament, with the result that many animals were simply too dangerous or sick to go to adoptive homes. The vet techs, those underappreciated heroes of every vet practice, spared us vets the unpleasant job of euthanizing the unadoptable. What care they took to keep a frightened feral cat calm or to hold and stroke a parvo-ravaged puppy as the barbiturates in the animal's veins brought unconsciousness and then a peaceful death. The cat and the flea both try to dodge this fate, but in the shelter, in a lab, and in everyday life most people agree that needlessly killing a cat or dog is wrong, unethical, but that killing a healthy flea is obligatory.

The great conundrum in animal research: for experiments we usually want animals whose bodies and diseases most resemble us humans but whose mental and emotional capacities are so different that we commit no sin in harming them. Our closest cousins, chimpanzees, have the physical similarities we want in the lab, right down to a protein on their liver cells that renders them uniquely susceptible to and capable of modeling the human hepatitis B virus. Visit a zoo and watch the chimps for five minutes, and you should be glad they're no longer in cages for medical research. Or find a fruit fly in a banana, and you'll likely be quite untroubled at their use in labs, doubting just how sentient or capable of true suffering their tiny brains may be. Alas, despite their utility for some experiments, in most ways they are just too physically dissimilar to usefully model human cancers, infections, or other ailments. I've yet to meet the animal so like us physically and so dissimilar psychologically to score a perfect ten on my ethical scale.

So many animals have crossed my path since my first day at the zoo: the hungry python, the frightened rats, the itchy puppy, the voracious fleas. Ants gather in my kitchen, minding their own business. Do I have ethical obligations toward these different animals? If so, what are they? It can't be as simple as just leaving them alone or doing unto animals as I would have done unto me; if either is the case, I am a monster chasing a fleeing flea or feeding the hungry python the only food she knows how to eat. Do my motives enter into determining how right or wrong my animal treatment is? Am I justified to kill harmful fleas but not innocent ants, both of them rushing to escape my approach? And what of the animals, in the room but also not in the room, the cows in a can of dog food or the mice who tested the dog's medicines?

As I've told my stories in these pages, I've written as though I naively believe we all agree on what's right and wrong in animal labs. Indeed, some of these cases most of us agree on already; others I'm making my case that I hope will win the day. Scientists should not use animals if the work will not address important subjects, such as hepatitis or diabetes, that affect tens of millions of people. In their labs they should use nonsentient cells or microbes or computers, not dogs or mice or monkeys, whenever those replacements can power the experiments. I've suggested that infecting woodchucks with hepatitis viruses is more ethical than infecting institutionalized children, and I know nearly everyone would agree. Scientists should minimize animal pain and distress in labs, and I am pressing my case that, when in doubt, assume that our experiments will cause pain and either get more generous with pain medicines or avoid performing some procedures altogether. I've made clear that I condemn excluding mice and rats from our Animal Welfare Act, and I reject lobbyists' claims that the animals do not suffer from that. I've decried how people vilify one another over these issues and bemoaned threats of violence toward scientists and vets. I've told my stories as if we all agree that we can be justified in harming animals for human benefit, though constrained by an obligation to set at least some limits to when and how we might harm them.

In truth, we do not all agree about animals. Some believe that human needs and desires always so outweigh animals' interests that people should never hesitate in doing as we please to animals. Others claim that we should never harm animals for even the most pressing human needs. Our lab animal welfare laws and policies reflect what I think most of us believe, that ethical people may sometimes harm animals for human ends but within constraints. People at the opposite poles find some consensus and common ground. But that consensus does not resolve the detailed decisions of which animals we may harm, what might justify the harm, or even what kind of harm we should seek to prevent.

Laboratory animal ethics has been intensely polarized for ages. At one extreme are truly abusive lab workers and scientists and, at the other, activists who threaten violence and vandalism and who harass individual scientists to leave their work. The posturing and name-calling sustain the perception—the illusion, really—that the world is full of misanthropic animal rights "crazies," sadistic white-coats torturing animals, and no one in between. In truth, though, most of us do find ourselves somewhere in between. We want a world free of human illness and suffering and a world without animal suffering too. This includes most people I've known both in animal labs and in animal activist organizations. Whatever the public posturing of "their" side of this polarized arena, many activists know that we cannot yet liberate all the lab animals without jeopardizing human (and animal) medical progress. Likewise, many scientists recognize the animal suffering their experiments cause and want better for their animals, even if they chafe at others, whether vets or activists, pushing them too hard and too fast. Each of us has our own sense of how fast we can and should move to replace animals, how aggressively we should weed out some experiments, how much we believe the animals suffer, and what experiments could justify it. A mentor of mine, the vet ethicist Bernard Rollin, loved to say that, in animal ethics, once we see that we all already agree that there must be limits to what we will do to animals, we can get to work because "now we're just haggling about price."[1] This chapter is my pitch to good, honest people to explore what values they share, examine the facts, respect their differences, and work together toward policies and toward judging individual cases, settling on a "price" we will make animals pay for our medical progress.[2]

Apart from the scientists and the animal activists, most people do not think all that deeply about animal ethics. They eat meat, go fishing, wear leather shoes, and take medicines developed in animal labs. Pushed to think about the animals, the ethic many would espouse is that it's acceptable to kill and use animals for human benefits, as long as the animals do not suffer. Pushed further, informed of how farmed animals, fish on a hook, or animals in laboratories *do* suffer, they modulate their stance, with qualifiers that animals not suffer *too much* or suffer *unnecessarily*.

Our laws reflect this ethical belief, allowing animal experiments but putting some limits on scientists' liberties. Not only do we allow animal testing but our National Institutes of Health (NIH) actively funds it, while the Food and Drug Administration (FDA) and Environmental Protection Agency often require animal data for new products, as do their international counterparts. In allowing some limited use of animals and in setting safeguards to limit unnecessary or extreme suffering, the regulators, ethics committees, and scientists work on this ethical premise: *harming animals requires justification*. These four words—*harming animals requires justification*—underlie virtually all our laws, rules, and practices in how to treat animals, wherever we interact with them. This ethic is permissive: it allows us to exploit animals for human benefit. Yet it is also restrictive: it tells us that our permission to harm animals is not without limits. It guarantees no agreement on specific cases. We need to clarify what animals we hope to protect and what type of harm we ought to spare them. We need to seek ethical consensus and decide how to bring ethics to life in our laws. We need to get serious about weighing justification, currently the weakest cog in the ethical care and use of lab animals.

By "we" I do not mean just insiders like me who work in animal laboratories. Everyone has some stake in animal research, even if they are not scientists running experiments or activists trying to shut down the animal labs. I am looking at you, dear reader. Voters elect a government that allows and supports animal research, with certain restrictions that partially protect animals. Taxpayers pay for most of the basic research in universities. Medicines you take and medicines you give your pets fund pharma companies' animal labs. Laboratory animals belong to all of us, and we share responsibility to divine our ethical duties to them and put those into practice.

Both laws and ethics *require* us to justify any animal harm we cause. In the best of all worlds, legal requirements reflect the ethical, with laws setting some baselines beyond which people may yet strive to perform. I've described how the ethical belief that experimentation was too severe a burden on suffering strays or lost pets eventually closed animal shelters as a source for lab animals, giving rise to the mega kennels breeding beagles for labs. The ethical revulsion at inflicting pain gave rise to laws requiring anesthesia and analgesia. But a general (but far from unanimous) acceptance that some experiments are too important to terminate, even in the face of animal pain and distress, carved out a legal allowance for such studies. Growing public regard for the ethical treatment of animals spurred the legal revisions of the Animal Welfare Act, with its mandates for ethics committees and primate psychological well-being. And an ethical stance that mice are just mice whose welfare is outweighed by cost savings at the US Department of Agriculture (USDA) spawned a misbegotten law casting them as "not animals."

Ethics should usually drive legal reforms, but in the animal world legal requirements have sometimes preceded the ethical. While preparing the legislation for its congressional vote, senator (and erstwhile country vet) John Melcher toured a number of animal labs, with a particular focus on lab monkeys and chimps. His ethical revulsion in seeing their lives prompted him to insert the provision for the psychological well-being of primates into the law. That legal advance required better care of monkeys, though what I saw around me in those days was how it unleashed animal-care staff to get creative about monkey care, confident the new laws would compel their bosses to go along. With this new approach to monkey care written into law, workers quickly saw the ethical imperative to support welfare programs for all their animals, and that too became law, most recently in the 2011 *Guide for the Care and Use of Laboratory Animals*, the NIH's rule book. If the leaders lead, the ethicists will follow, as will the welfare scientists and vets, seeking new knowledge on the most ethical route to animals' well-being.

Harming animals requires justification, but does this apply to all animals? How we treat animals depends not just on our commitments to do the right thing but on what we think we know about them. As a specifically *veterinary* ethicist, my perspective comes from hands-on work with actual animals, not theoretical constructs. Thus my philosophical musings always include wanting to know the best available facts about how animals think and feel and about how well the data we derive from our experiments on them produce useful and necessary information about human health and welfare. And so I watch how technical and veterinary advances have fed into this coevolution of legal and ethical standards for animals. We are now better able to recognize the subtle signs of pain in a range of species and have better drugs to treat the pain, shifting how scientists and others judge the ethics of withholding pain meds. The legal mandate to safeguard monkeys' mental health prompted behaviorists and welfare scientists to refine their measurements of animal welfare, which again has raised the ethical and legal standard of care for caged animals. The one-time ethical choice between testing cosmetics and cleaners in rabbits' eyes or risking danger to human consumers has tipped, because of shifting ethics and awareness but also because scientists have succeeded in developing many nonanimal safety tests. Law is catching up with the ethics and the science, if slowly, with the FDA Modernization Act allowing the FDA more flexibility to waive animal tests.

The coevolution of legal and ethical requirements starts with animals, and three questions loom. We need to consider whether we have ethical obligations to all animals, fleas included, or can we legitimately give them no more mind than we would a chair? Among those animals to whom we grant a place in our moral sphere, do we see an ethical justification for prioritizing some over others? Finally, we need to consider how to balance human wants, needs, and desires against animals' interests, as substantial efforts to improve animals' lives will cost money, change our eating habits, and potentially slow some medical progress.

People offer many reasons for why we should care about animals or try to treat them well. I have to assume most readers believe that we should try to treat people well (despite what we see of how humans around us actually treat other humans), and for many ethicists that's a starting point for asking what it is about ethical treatment of other humans that should cross over to the rest of the animal kingdom. Chimpanzees share close to 99 percent of our genes; if that implies we should treat chimps better, what does it say about mice (90 percent genetic similarity), chickens (65 percent), or fleas (60 percent)?[3] Dogs clock in with 94 percent genetic similarity, but should they get extra ethical points as our best friends, with thousands of years over which we have domesticated each other? If that's why we should treat them well, what of less constant friends? Are cats domestic enough to get the same perks dogs get? How about pigs?

Activists and scholars work to expand our moral vision beyond our own species by spotlighting the species most people regard most highly, with the myriad reasons we hold them dear. We gape in awe at whales, moving many people to want to protect them (from humans who don't share that awe, evidently). We look with affection at dogs and cats and feel good that we have anticruelty laws to protect them, again from ne'er-do-wells who do not share our regard for them. The Great Ape Project works to put chimps and their (and our) kin, the bonobos, gorillas, and orangutans, on our side of that moral line, with all the rights that humans enjoy, leaving (for now) all other animals on the outside.[4] In explaining her desire to spring Kaavan, the elephant, from a zoo in Pakistan and bring him to a Cambodian elephant sanctuary, Cher said, "He was suffering. Elephants are just like we are. They're so family-oriented and so emotional."[5] Animal activists cheer and animal-using industries fear that these efforts to move some animals across the invisible moral and ethical line to the human side will create gaps in that barrier through which others will slip. Chimps today; rhesus monkeys tomorrow? If rights for orcas, why not rights for sea lions? Or octopuses? Or sea anemones?

For me, sentience buys the entry ticket for welfare protections. It's not the only criterion for protections more broadly; I really do want to save the redwoods and preserve ecosystems. But for welfare specifically, I start with sentience, the capacity for pleasure and pain, for suffering and happiness. The eighteenth-century English utilitarian philosopher Jeremy Bentham said it first. In his most famous passage, quoted in virtually every book on animal ethics ever written anywhere, he gave us a tool for scoring which animals to include in our moral sphere and which we might exclude: "The question is not, Can they *reason*? nor, Can they *talk*? but, Can they *suffer*?"[6] And if animals can suffer, which is to say, if they are sentient, we have a moral duty to take that suffering into account in whatever we would do them.

In "Chicken" I describe some of the tools scientists use to probe which animals have sufficient cognitive and emotional capacity to qualify in our eyes as sentient. They find animals somehow making choices to maximize their own welfare and to avoid suffering. And if these animals' suffering matters enough to them to shape their actions, I say it should matter to us too. As researchers look around, they keep finding animals they score as "probably sentient." Hermit crabs scurry away from electric shocks and have the smarts to stay away from wired shells. Cuttlefish turn to pessimism after a few minutes a day of being chased around in their tank. Scientists, vets, and gourmet chefs should take these sentient animals' suffering into account as they plan experiments or the night's menu. But having sentience gets an animal only so far. The question is not quite so simple as "can they suffer?" though that gets an animal's interests in the discussion. All three may be sentient, but no cuttlefish can expect the treatment that a dog or their person receives.

Fair or unfair, we do treat different species differently, including those far closer in capacities than a dog and a cuttlefish, saving the very best for ourselves. The English animal rights activist Richard Ryder coined the term *speciesism* in 1970 to challenge our assumption that simply being human or nonhuman should determine what rights and protections a creature merits. He likened this to racism or any other -*ism* in which a dominant group looks in the mirror and decides that membership in their superior group should confer special privileges. Ryder, whose ideas Peter Singer amplified in *Animal Liberation*, did not deny that we animals differ among ourselves; rather, he argued that it was indefensible to treat others differently solely on the basis that they belong to a different species. Singer urges us to treat similar interests—human or nonhuman—with similar consideration. If a dog, a pig, and I (and an ant, for that matter) are all equally hungry, it is speciesist and therefore prima facie wrong to hog the available food solely for myself or to feed the dog till their stomach is bursting while leaving the pig famished.[7]

The US exclusion of mice and rats from our Animal Welfare Act stands as the most naked example of speciesism, with lab hamsters getting their annual USDA inspection that mice and rats do not. In their intelligence and emotional complexity, mice and hamsters are about equal. Their differential treatment is rooted in economics, as our government does not want to devote resources to welfare protections for our most numerous laboratory animals. The exclusion lives on, however, because of speciesism, because not enough people care about these sentient creatures simply on the basis of the species they belong to. This is unjust.

I've known cardiology researchers to switch from using dogs for their heart research if a pig, with a roughly similar heart, will do. Dog labs are usually a bit more expensive to run, and lab-reared dogs a bit pricier than lab pigs, but not often enough to tip the scales. Researchers I've known have been happy to make that switch, in part for their own and their families' sentimental preferences for dogs. Scientists know that, though the laws mostly treat dogs and pigs similarly (both are covered by the Animal Welfare Act, though only dogs figure into state-by-state laws requiring postlab adoptions), they work under a glaring spotlight if they stick with dogs as their research model. The Animal Welfare Act does single out dogs as participants in exercise programs, a provision likely rooted again in sentimental feelings of what we owe our best friends if we cage them for lab life. Maybe this privilege is warranted, as dogs do seem born to run, run, run in a way that pigs and other animals do not. Or maybe it's just unfair favoritism, and pigs would want to get out to root around as much as the dogs want to get out and run, even though they'll cover less ground in their foraging than a dog might. Their interests may differ, but Singer's call to reject speciesism would have us do our best to respect both.

Speciesism is a complicated charge to prosecute, as our differential treatment of dogs and pigs shows. In many experiments dogs and pigs make interchangeable research subjects, producing data of roughly equivalent value. The two are also about equal in intelligence and sentience. Their welfare interests in avoiding pain, in pursuing a social and sexual life, in living in an environment where they can make their own choices to explore and enjoy are all about equal. Where the two could serve equally well for a science lab, even at a vet school that trains students who will become mostly dog vets, most folks will choose the pig over the dog as a practice animal. Researchers know that dog labs are under greater scrutiny from regulators and from activist watchdog groups than pig labs, so even if students would learn more using dogs in their classes, the professors will buy a pig as a stand-in. Some scientists have personal, emotional preferences to avoid using our canine best friends in labs if pigs will suffice. This does not mean dogs do, or should, outrank pigs in a moral hierarchy. But to the extent the two share an interest in a richer life than lab cages provide, dogs receive preferential treatment in the Animal Welfare Act's mandate to enroll them in an exercise program.

Dogs and pigs present different animal-welfare challenges in the lab. A focus group debated whether toxicology labs should use more minipigs and fewer dogs and monkeys, deciding that, of the three, pigs are the easiest to maintain in groups, don't need trees to climb, and don't want exercise yards to run in and so end up adapting best and suffering least in labs. I counter that dogs often live better lab lives than pigs, with lots more human socialization and playtime. In San Francisco our vet techs would sometimes bring lab dogs to hang out in their office (we used very few dogs, and they would not have been able to do this if we had more); pigs never got this treatment. Just the fact that lab dogs are more likely to find a loving postlab adoptive home might push people to choose a dog for their nonterminal experiments.[8]

Equal consideration of equal interests has little to do with the various preferences we give one animal over another in practice, including when scientists decide on the animals they will choose for their experiments. Do some animals somehow warrant lesser care than chimps or dogs or even hamsters because humans typically see them as food (pigs), vermin (mice), or just too distantly related (octopuses) to merit equal concern?[9] Sentiment, fellow feeling, kinship, and self-defense have their place in animal ethics, but we should remain mindful of how much our biases and preferences affect animals and lead to their suffering. When we cast mice as a "lower species," as so many do, we put a huge burden on them.

I'm not entirely comfortable with using sentience as the only measure of moral concern. It is, after all, one more way in which we look in the mirror, decide what we like about ourselves, and use that criterion to expand our ethical universe to those others who share this trait we value in ourselves and to shut the others out. Sentientism gives us permission to do what we want, whether that's eating nonsentient plants and questionably sentient oysters, driving the highway at dusk with no regard for the bugs dying on the windshield, or ruthlessly killing dog fleas and mouse mites. I do all these things, but look how convenient this ethical position lets my group, the sentient ones, trample over the others, the outsiders. We should be cautious about ethical theories that neatly fit the preordained moral outcomes we want them to deliver. Still, I stick with my focus on sentience because it pushes us to take seriously animals it would be convenient to dismiss, be they mice or fish or octopuses. It serves as a brake in how readily we dismiss the interests of animals who may not have won our hearts but whose suffering is as real to them as the suffering of our sentimental favorites.

Sentient octopuses long ago convinced Europeans to grant them protections in animal labs. A handful of countries now explicitly include animal sentience in their laws, not just in labs but pretty much anywhere animals exist. Details vary. Landlocked Switzerland, for example, took the lead in saving live lobsters from being plunged into boiling water, with Norway, New Zealand, and other countries following suit. No country gives sentient animals equivalent rights to people, but they converge on moving (some) animals from the status of owned objects to the status of sensitive creatures with interests that we must accommodate.[10] "Good mice" in labs are threatened by "bad mice," their feral cousins who find their way through nooks and crannies into labs' feed rooms, contaminating the mouse chow and spreading their parasites. The tropical rat mite is a common parasite that feral mice bring into lab vivaria, and, as they spread out through a mouse room, they even breach the filtered cage lids that keep out viruses and bacteria. Glue traps are efficient at catching the murine invaders, but so very cruel, so the ethical and legal presumption is that we avoid them. Wild mice, after all, are sentient creatures. As for their mites, well, we kill them with abandon, with the same merciless chemicals for killing a puppy's fleas. Now what will we do should someday behaviorists report evidence of mite sentience, that they feel pain in their small mite brains? Will we ban the crueler pesticides? I won't try to predict.

The United States' animal-welfare laws do not presently mention sentience, and they do not include all sentient animals under their aegis. For the USDA to see that mice are animals would require an act of Congress. Surprisingly, the NIH's Health Research Extension Act (Public Law 99-158) refers only to "animals," giving the NIH the flexibility to embrace octopuses, a move it has in fact made, and lobsters, who deserve to follow. Ethics committees need not wait for the law to catch up with the science and morality: all sentient animals have welfare needs that researchers should consider. Scientists factor in several considerations in choosing animals for their labs, but to the extent they shift burdens from monkeys or dogs to animals they class as "lower," they need to assess whether that's in fact an ethical move. How many hundreds of mice, possibly with greater suffering per animal, will a scientist use in swearing off dogs for their experiments?

As we debate the justice of privileging some animals over their equally sentient relatives, the real elephant in the room is our speciesist assumption of human superiority. As Peter Singer notes, our physical and mental differences from other animals challenge our ability to recognize how similar their and our interests are, even when we focus on something as universal (at least among mammals) as pain. A thick-skinned horse, he writes, may feel a slap differently than a young child does. But it's our emotional processing of pain that really complicates things. Do we humans have heightened suffering when we layer on our anxieties over a trip to the dentist with whatever pain the dentist has in store? Do we listen to friends' accounts of their own pain in the chair and ratchet our fears even higher? Yes, but we also go there knowing the pain will have limits, that the dentist will use pain medicines, that we have chosen to endure this pain to limit other suffering down the line, and that, despite our anxieties, we have little to fear. But all of us (and by "us" here I mean our great community of clearly sentient creatures) process pain through our fears and anxieties and occasional experiences that the pain was not after all so bad. Even fish can amplify the distress of a procedure if its occurrence is random and inescapable and beyond their control, just as they can adapt to something once they've experienced it and found it not so bad. Equal consideration of interests requires us to consider this totality of pain and its associated emotion and to do our best to mitigate it. If we jump on treating our human pains and ignore those of other animals and especially if we cause those animals pain for whatever we may want from them for our needs, that is the speciesism that Ryder, Singer, and animal liberationists struggle against.

Singer puts this challenge to scientists and animal ethics committees. Consider a hypothetical, severely brain-damaged infant whose parents have just died. Let's say that no one in the world knows or loves or cares about this child, whose mental capacities and therefore range of interests are far lower than even a mouse's. As a research model, this child has a human body and so in most ways should be a better research subject than a mouse, with no need to extrapolate the data across species lines. Presently, no human subjects or animal ethics committee would consider using such minimally conscious children to replace fully sentient mice or even dogs or monkeys in experiments. No law would allow this. Singer points out that the only plausible basis for the distinction is our own speciesist biases that favor our species over others.

Singer is right: this is speciesism. And for now it's the decision the vast majority of people would endorse. Not only do people decline to treat animals' and humans' similar interests similarly, scientists in the lab, most of the time, treat fairly trivial human interests as far more important than animals' strong interests in avoiding harm such as pain and suffering. Harming animals requires justification, and, as we expand the range of animals whose welfare we must consider, so too are we expanding what we see as harm. England in the 1870s pioneered laws to spare animals (dogs, primarily, but horses, cats, and others too) from the pain of unanesthetized surgeries; now we ponder how to promote the psychological well-being of octopuses. As we add more animals to our moral sphere, we are still expanding the list of kinds of harm we should try to spare them.

Animal pain has hogged the spotlight as the ultimate harm in animal labs, but the 1985 laws for dog exercise and monkey psychological well-being reflect the growing recognition that pain is far from the only harm. Fear, cohabitation with an aggressive cage mate, confinement, cold, hunger, loneliness, boredom, bright lights: all these stresses we induce in our confined animals and have an ethical obligation to recognize. This is quite a catalog of bad situations we want to spare animals, complicated by how each species has its own priorities and welfare needs. A monkey wants to climb if scared, and a woodchuck wants to burrow. Put them in cages that deny them these needs, and that includes cages that USDA and NIH list as acceptable, and their fear and distress will mount. An aspirin tablet does nothing for these painless but distressing animal types of harm. An ethics committee that scrubbed all animal protocols free of pain and suffering should be very satisfied at a job well done. But their job is not completed, as I urge scientists and their supporters to expand their horizons. With a clear mandate to focus on pain and distress, ethics committees barely discuss some other important concerns, uncertain whether to even count them as harm to the animals. Death, confinement, and deprivation deserve a closer look.

Consider death. Few research animals leave the lab alive. Scientists kill animals for tissues they need to study. And they kill animals—they are, in fact, obliged to kill animals—as early as possible if those animals are on track to develop serious diseases, cancers, psychiatric conditions, or other incurable conditions. But should we score death itself, however humanely executed, however better than some other fates, as a harm we should avoid? Certainly the prevailing ethic of meat eating, for those who take animal welfare seriously and yet continue eating meat, is that death is not a harm to animals if they have lived a good life and died a painless, stress-free death. This ethic prevails in the laboratory too. Working within the requirements of animal-welfare laws that cast the killing of healthy young animals as ethically unproblematic, ethics committees rarely discuss killing as a harm to be minimized.

Animal researchers see killing their animals as a moral imperative rather than a harm to avoid, at least for experiments that cause illnesses and injuries. Mice and rats are the workhorses of any cancer labs that still use live animals. Remember that, as in humans and dogs, most early-stage lab animal cancers are painless swellings that do not diminish an animal's quality of life. Most rodent cancer experiments I've witnessed involve killing the animals long before their tumors get large enough to interfere with moving around or become evidently painful or otherwise symptomatic (such as seizures in a mouse with brain cancer). Only rarely does an experiment focus on late-stage cancer in which pain and distress mount. Otherwise, in cancer and other labs, the prevailing ethic is that killing animals is crucial to limit animal suffering.

Two situations prompt ethics committees to question killing healthy animals, one driven by animal species and the other by animal numbers. Well before the Beagle Freedom Project began racking up legislative state-level successes requiring labs to offer healthy lab dogs and cats to adoptive homes when an experiment has ended, many research institutions were allowing or encouraging researchers to stay the killing and give animals to the vet team to place in homes. Arnold Arluke and Lesley Sharp are anthropologists who've spent time in animal labs and witnessed the impulse many lab techs have to save favorite animals from death, keeping them in the lab as unofficial mascots or finding them homes and sanctuaries.[11] Ending animals' lives just feels wrong when it's a species—dogs, cats, and primates in particular but not just them—that resonates with a person.

Tiny numbers of lab mice find their way to homes as pets. Instead, millions of young, healthy mice meet their end just as they wean from their mother, as scientists test their DNA and cull out animals with the wrong genes for their experiments. Labs and commercial vendors overproduce animals so as always to have a cohort of the right age, sex, and genetics ready at a moment's notice for scientists' experiments. Concerned about the sheer numbers of such mice, European laws now require labs to track and report the numbers of animals bred and then immediately culled as unusable. This required accounting reflects a concern that killing per se is a harm to an animal, even if a minor harm noteworthy only when the scale of the killing climbs to great heights. Otherwise, if an animal's death truly is not a harm, why care? For larger animals adoption and retirement sanctuaries are the appealing and practical alternative to killing animals who've outlived their lab usefulness. Small animals too find their way to homes—I lived with my pet rat, Hazel, and, before her, two meadow voles, Jack and Earnest. But the sheer numbers of small rodents in labs overwhelm what adoption programs can handle, so limiting overproduction remains the key to limiting how many small animals labs kill.

Adoption programs do not serve to alleviate suffering, but they do point to another truth. Animals are capable of living fuller lives than they typically experience in the confines of the lab. Even within the lab, they deserve greater opportunities to do the things that matter to them. Depending on the animal, that could include basking in a sunny window (or under a sun lamp in our hermetically sealed vivaria), exploring play areas, or socializing. When Congress called for enrichment and exercise programs, they were likely thinking of the suffering dogs and monkeys experience, even in cages the USDA lists as acceptable, and looking for ways to mitigate the suffering. But they opened the door to the idea that, even if animals are not outright suffering, even if they don't have a clear image of what they're missing, this deprivation is a harm. The USDA received pushback on its attempts to legislate welfare beyond the absence of suffering. As one lab vet groused in a letter to the USDA, "We were told that we were supposed to exercise dogs. We are not supposed to make them happy."[12] But I say that we are supposed to strive to make them happy. We should shift our frame from suffering versus not suffering to suffering versus happiness.

We harm animals when we choose to deny them the good things in life, even if they have no knowledge that those good things—a romp in a dog park or a cozy bed in a loving home—exist. Our obligations do not end with simply preventing bad things in animals' lives; we should commit to giving lab animals the brightest lives, however short those lives may be. When lab workers push for environmental enrichment, they implicitly claim that minimizing distress does not qualify as adequate compensation for the animals from whom we take so much. We should think of the best possible lives for the animals and rethink the idea of harm to include any type of animal care or experimental use that denies their flourishing. "You have no idea what you're missing," we could say to them, and that reflects poorly on us. We should strive to provide for animal flourishing and to challenge any justification for giving the animals anything less.

I review animal protocols by listing a birth-to-death inventory of all the welfare challenges and opportunities, the procedures that cause pain or other distress, and the times when the animal could grab a chance for some extra play or exploration or enjoyment. Will the animals be born healthy or genetically destined right from the start for some illness we are researching?

Will the scientists take samples to analyze while the babies are still with their mom? How will we house them? Has the researcher listed every step of the experiment, including how they treat placebo and control animals? Will I have a free hand to treat sick animals, or does the project forbid some medicines? I need that kind of inventory to ensure we've covered everything before the experiment starts. When the time comes, how will the scientists kill their animals, or might they find them homes?

These questions fit well with the rubric that dominates laboratory animal care: the "Three Rs" of alternatives in animal research: replace, reduce, and refine, from the 1959 book by microbiologist Rex Burch and psychotherapist William Russell, *The Principles of Humane Experimental Technique*. Laboratory animal people now revere it with near-biblical status for its articulation of the Three Rs framework, which helps scientists and ethics committees organize their quest for ways to minimize harm to lab animals (or, in their words in 1959, "reducing inhumanity" to animals). Scientists should *replace* sentient animals when they can, *reduce* the number of animals to the minimum needed for valid results, and *refine* their experiments with pain medicines, gentle handling, and any of a range of interventions.[13] Don't use lab animals if you don't need to; use as few as possible; treat them as well as you can: it's a perfectly fine framework for reducing harm to animals, starting with the most zero-harm act of replacing the animals entirely, not with "less sentient" animals (if someone someday develops a credible hierarchy of sentience) but with nonsentient cells, microorganisms, and computer simulations.

Just about everyone in animal research learns their Three Rs. Any university or company website that admits to using animals in labs professes their allegiance. Most countries, the United States included, base their lab animal–welfare laws on the application of the Three Rs. Lab workers frequently refer to the Three Rs as an ethical framework, but, as a stand-alone ethical framework, it is severely limited. It's aspirational: we should replace animals when we can, and, when we can't, we should use as few as we can in as gentle a manner as possible. The Three Rs give us little ethical guidance on how we should judge cases, how much we should inconvenience ourselves, or how many dollars we should spend on animal care.

Replacement shines as the most important of the Rs. Scientists experienced in running animal experiments may not have expertise in cell-culture methods or in computer modeling. They may not have access to human volunteers or expensive hospital equipment, like a CAT scan or other human-size devices. They need to pull in collaborators with complementary expertise. This is an advantage of a large research campus, where one professor's busy lab has computation experts sharing a bench with cell experts, while the animal researchers are in the mouse or monkey house running the experiments. For all the labs I've visited as a campus vet, I recall no one turning to cells in a flask or numbers in a computer simulation simply to follow the Three Rs. Rather, scientists avoid the complications of animal experiments—the variability among subjects, the challenges of getting samples, the expense of animal experiments—when they can get crisper, clearer data with other types of experiments. That's good for science and good for animals, but it still requires scientists making the effort (and being pushed to make the effort) to explore these nonanimal methods.

In the aftermath of the Nuremburg trials and various scandals around the use and abuse of human subjects, ethicists and regulators developed the norm that, in general, scientists should not experiment on people without firm animal data in hand. There are times, however, when data from human volunteers (as well as autopsies and reviews of hospital records) should precede animal experiments. As scientists now have better-developed imaging technologies and genetic screens, they can run relatively noninvasive experiments with people, reserving animal experiments for further exploring what the human experiments tell them. If a gene seems a likely cause of a disease in the human experiment, they may have a reason to manipulate that gene in animals and track the effects. If an MRI experiment suggests a certain part of the brain carries a particular function, they may see a need in manipulating (typically, by destroying) that part of the brain in animals to document resulting functional deficits. Thus nonsentient and inanimate objects may replace lab animals, but so may highly sentient human volunteers.

We also speak of *relative replacement*, a term that requires some caution, as it runs on a hierarchy of ethical concerns that may not map well onto animals' biology. A monkey scientist will want to consider dogs or, more likely, pigs as a replacement for their monkeys and use rats or mice as a substitute for pigs. Cat researchers see ferrets as a somehow more acceptable subject, not for better data and not because ferrets are any less

sentient than cats but because public (and the researcher's own) sentiment sees cats as animals that civilized people do not harm. Often swap-
ping out one sentient mammal for another, especially a smaller one, leads
to much larger numbers of animals in use, first to reengineer all the pig
experiments to work on a rat-size body and then because scientists typically use more small animals per experiment to get better numbers for
statistical analyses. And smaller animals may suffer more, especially if scientists do not train and acclimate them as they would a pig or monkey,
using brute force instead of positive reinforcement.

Occasionally reduction and refinement compete. As vet students, my
trio received four beagles in succession for learning our skills. Students a
few years our senior had one dog per semester. With the same number of
practice surgeries, our dogs had fewer unnecessary surgeries per animal (a
refinement), and more of those were nonsurvival procedures in which we
killed the dog on the table and spared them yet another round of postsurgical pain. But our training took four times the number of lives, the opposite of reduction. We don't have measuring units of suffering, so it's a challenge saying which approach has the least welfare impact, and much
hinges on whether we score animals' deaths as a harm or a relief.

A full-throttled deployment of the Three Rs requires homework before
scientists jump into a project or start shopping, for example, for marmosets or mice for their multiple sclerosis studies. First, they need to know
what nonanimal replacement tests could answer at least some of their
experimental needs, such as what they can accomplish in a petri dish with
antibodies and cells. They need to avoid unnecessarily duplicating the
work of others, including somehow knowing that another lab has pursued
the same experiments and gotten no positive findings (which makes the
case for publishing negative results as well as positive). They need to know
how to refine monkeys' daily care and housing, how to recognize and manage pain, how to nurse a weak or paralyzed marmoset, and how to collect
samples without distressing the animals. Congress created an Animal
Welfare Information Center when it updated the Animal Welfare Act, a
national library resource to support the search for alternatives. The generated various bibliographies on these welfare topics could barely capture
the available literature and information, even in 1986. The staff there run
workshops where scientists and librarians learn alternatives-search strat-

egies. In practice I never found the center useful, especially for scientists embarking on a new line of studies with which neither they, the vets, nor the ethics committee have much experience.[14]

AI-powered literature searches should help researchers locate more of what they need. I've often found that a good search on a topic may yield several dozen articles that could be worth reading for a planned experiment but then, sadly, find the most promising ones locked behind journals' paywalls. If the campus does not have a paid subscription (and even big-money campuses will not have a paid subscription to every journal), scientists can take the time to write authors for reprints or spend the money, fifty dollars perhaps, for access to an article. It's so much easier, like Aesop's fox eyeing the grapes just out of reach, to decide the article is too sour, too expensive, to bother with.

Scientists are both collegial and competitive and hold their secrets until they're ready to publish, yet one more setup for unnecessary duplication. Increasingly, I am warming to proposals I'd first heard at meetings on lab monkey welfare that scientists should preregister their animal experiments, with their study design and hypotheses, just as companies preregister their human clinical trials. We would finally have a way of knowing what percentage of lab animals actually produce data in peer-reviewed journals or product-licensing applications, plus scientists could tap into what their competitors, I mean, their colleagues, of course, are working on and decide to join forces or step aside. This departs from the current standard scientific practice, which rewards priority in crossing the finish line first with a discovery.

The COVID pandemic was a horse of a different color from science as usual. It offered a glimpse of a different future, in which the urgent global search for diagnostic, treatment, and vaccination made scientific research briefly more communitarian, more focused on the group achievement than on success in this or that individual lab. COVID was a high-enough priority to promote this different approach to balancing competition and collegiality. If people saw lab animal testing as a near-sacred trust to be acted on only in severe circumstances, we could rework how scientists share their data to collectively minimize and reduce reliance on animals.

Russell and Burch focused on the possible harm of animal experiments, taking care that pursuit of the Three Rs should "be promoted without prejudice to scientific and medical aims." They presumed, or at least they wrote as though they believed, that scientists are doing high-quality, important experiments and that measures to improve animal welfare should defer to the scientists' needs. They did not want to constrain scientists. No countries had animal ethics committees when they published their book in 1959, and thus they put the ethical burden on scientists themselves to reduce harm to animals.[15] But Burch and Russell did not say what to do when a scientist has thoroughly pursued the Three Rs to the extent possible and still comes up with a research project that will cause animal suffering.

Ethics committees nowadays oversee scientists' use of the Three Rs to lower the welfare costs to animals in the labs. They seek out alternatives *to* the use of animals (i.e., the various nonanimal replacements) and alternatives *in* the use of animals (fewer animals, with refined experimental techniques that entail less harm and greater chances at happiness). They focus quite earnestly on reducing the costs the animals might suffer. And there, from what I have seen of current practice, they mostly stop.

Harming animals requires justification, ethically, but, lacking a legal mandate, we currently do not grapple with justification as we should. Simply put, justification of animal harm means that any animal suffering, confinement, or death is worth the sacrifices animals involuntarily make. We know the individual subjects virtually never gain from their involvement in an experiment, so if we will allow their suffering for others' benefit, their use must be (currently) irreplaceable, and the potential benefit to others must be huge.[16] An experiment must have a high probability of producing quality, reproducible data in service of unimpeachably important goals in advancing human, animal, or ecosystem health. Proposals that do not meet a high standard must not receive approval. Other experiments, those with high-welfare costs for less than the most compelling goals, must proceed slowly and incrementally, verifying with each step just how much the animals suffered and just how much the experiment has produced.

An ethical system like the Three Rs that looks only at the harm and harm reduction, without probing what benefits could justify animal harm, is weak. It lacks balance. In their 2020 book, *Principles of Animal Research Ethics*, bioethicists Tom Beauchamp and David DeGrazia call out the flaw of casting the Three Rs as the totality of laboratory animal ethics. They

propose six principles, organized as principles of animal welfare and principles of social benefit, which they believe a spectrum of people, from
research-skeptical animal activists to strong proponents of animal-based
research, can embrace and that an ethical scientist should apply.[17]

Beauchamp and DeGrazia's principles of animal welfare are essentially
an update of Russell and Burch's, searching for harm reduction and for
maximizing animals' freedoms and pleasures. It's in their principles of
"Expected Net Benefit" and "Sufficient Value to Justify Harm" that
Beauchamp and DeGrazia go beyond the Three Rs. They reflect the trend
in animal-welfare regulations in many countries, advocating that scientists' efforts at harm reduction alone (i.e., the Three Rs) are insufficient to
justify animal experiments. A scientist can ethically use animals only if the
expected net benefit of the experiment is great. The experiment must
tackle important goals with a high likelihood of success (a cure for cancer
would qualify; a new mascara or cure for baldness would not). The experiment must lead to clear, accurate data, and those data must (in medical
research) be relevant to human beings. It must translate.

The animal behaviorist Patrick Bateson proposed his "Bateson's Cube,"
a graph showing three independent axes to score for any given animal
experiment after the scientist has explored potential nonanimal replacements.[18] It's a visual reminder that animal welfare, the quality of the
experiment, and the benefits it seeks can all vary independently. I'd add a
fourth dimension (though I wouldn't know how to draw the resulting
shape) of translatability. The four scales do not have units; each ranges
from low to high. A high-welfare, high-quality, high-translatability experiment in service of a high-benefit outcome gets good marks. Scoring low on
all four sends it to the trash bin. A mixed score calls for serious deliberation. Each scale requires its own separate analysis, and the wisdom and
expertise for each are different.

Consider the marmoset monkeys and mice in studies of paralyzing
experimental autoimmune encephalomyelitis, the common animal model
of human multiple sclerosis. Both are sentient creatures who will suffer
from paralysis and whatever nerve pain they hide. Marmosets, as climbing
animals, may suffer more from a loss of limb use than mice, who can pull
themselves along with front limbs. On the other hand, the marmosets, if
only because we care more about them, are more likely to receive nursing

care during the illness. It's likely a lab would use far more mice than marmosets for such studies, because they can: mice are cheaper and more available. Either way welfare costs are high for both. The experiments may be of similar quality, but a mouse experiment is more likely to achieve statistical significance, enrolling as it will more subjects. But mice don't yet "win" this horse race, as the neurologist who shepherded ocrelizumab to market found that experimental autoimmune encephalomyelitis targets a different immune cell in mice than what MS hits in humans, so the mouse data will not translate well to human medicine. To the extent that scientists trust the wrong animal model, they increase the risk of bringing a useless or even dangerous medicine to human clinical trials. Multiple sclerosis is a serious disease afflicting millions of patients. That gives a high score for benefit, but, if the drug under development is a "me too," no better than existing drugs and invented solely to give its developers a cut of the MS market, its benefit score is shakier.

The prospective and ongoing evaluation of medical-research projects calls for varying experts' evaluations as the project moves along. Notice how the relative suffering of monkey versus mouse and the ways to recognize and mitigate it require expertise from animal-welfare scientists, daily caregivers who spend time with animals, and vets. The neuroimmunologist masterminding the experiment can contribute knowledge of what humans with MS experience. Scientists and statisticians are best poised to evaluate the quality of the experiments and the data they yield. Add in clinicians and clinical scientists to assess whether a drug from the marmoset or mouse lab proves safe and effective in humans. Epidemiologists can provide data on how well the new drug is affecting the incidence, longevity, and quality of life of patients. But who scores the benefit of the new treatment other than the drug company's accountants? These questions move the review from the technical to the ethical and require more voices for a thorough evaluation. For me the prospect of a new treatment for MS is a worthy goal, if patients rate the improvement to their health and quality of life as substantial and any side effects manageable. But many diseases scientists research are less debilitating or affect fewer people or produce side effects the scientists believe to be negligible but which patients score as unacceptable. A treatment may add only a very short period to patients' lives, possibly just extending a painful, terminal disease. So let's

seek patients' voices on scientists' animal experiments. Are they researching what patients need them to study?

When I first arrived at the University of California–San Francisco in
1999, the Human Subjects Committee reported to the same vice-chancellor who oversaw other ethics committees (a conflict of interest, for example). The animal committee lived with technical committees like radiation
safety and environmental health. They reorganized us, with one adminis-
trator directing both human ethics and animal-subjects reviews, but the
animal committee's work remained highly technical, still the domain of
researchers and vets and safety specialists, without adding in the important questions an ethical review must ask.

When the Three Rs and harm reduction are the sole consideration, a
committee composed mainly of scientists can review animal protocols on
its own. The "justification" in our welfare laws is a technical justification,
not an ethical one. Scientists and vets can judge if an experiment requires
individual animals to submit to multiple surgeries. They can review the
likelihood of animal pain and the plans to prevent or treat it. They can
examine a scientist's claims that available technology does not yet allow a
full replacement of animals. Our university committee, primarily a group
of researchers, spent hours reviewing scientists' explanation of how many
animals they planned to use. Scientists would dive into the explanation,
checking the researcher's math or demanding a better explanation of the
statistical tests they would use. Back the protocol went to the researcher,
who would fiddle a bit, possibly changing the numbers but frequently just
rewriting the explanation. "You know," one scientist explained to me during a lab audit, "we all just make up those numbers with huge overesti-
mates so we won't reach the limits of our approval and need to pause while
we apply for permission to buy more mice." Yes, I'd suspected that, but
number crunching was in our technically focused committee's comfort
zone and where it invested its effort. It could set aside questions like, "Why
would anyone want you to do this research?" But to finalize the approval
of a protocol using 24,328 mice when the numbers on paper added up to
only 22,000 was unthinkable to them.

The right technical questions are essential, though a myopic obsession
with the numbers of animals is misplaced. Committees I've worked with
focus on two Rs, reduction of numbers and refinements of experimental

procedures but are weak on the most important R, replacement with non-animal alternatives. In the chapter "Marmoset," I describe a consortium of experts who ran a huge multicenter set of mouse and human studies to determine how well the genetic changes that inflammatory diseases such as traumas and burns induce in mice correlated with humans. The experts' verdict was "close to random."[19] It's higgledy-piggledy, a coin toss. It's useless. That particular program's findings may apply to a small subset of mouse experiments in the fields of immunology and genetics, but no scientists in that subspecialty should expect ethics committee approval without clearly explaining why their work with this discredited model could produce benefits. Since the Three Rs debuted in 1959, scientists have run countless experiments on a couple of billion (my estimate) animals. The data are growing to allow this sort of retrospective analysis of hundreds of models scientists use. Are mice living (and dying) with debilitating experimental encephalomyelitis producing useful information for human MS? Do woodchucks yield relevant data in human hepatology? Do animals suffering as psychiatric models of post-traumatic stress disorder, anxiety, and depression produce clear, reproducible knowledge that translates to the human mental health patient? And in all of this, are scientists staying abreast of the growing number of nonanimal tests in their field, which may make their animal use obsolete?

Even in the narrow role US animal committees currently play, scoring harm reduction without judging benefits, I would not defer solely to scientists. We scientists do like our facts, but we also know that most of our facts are open to interpretation and therefore to our biases. We deal in probabilities. We may see a 30 percent likelihood an experiment will succeed as a reasonable gamble. Others will see that as a 70 percent failure and say no. We assess a relative likelihood that pain or pain medicines could skew data and, if the two conflict, then make a judgment, not a technical but an ethical judgment, to err on the side of protecting animal welfare or data. What scientists see as facts often are really interpretations of facts. I describe in "Chicken" how veterinarian scientists discovered the fact that guinea pigs do not use the added floor space of a larger enclosure, supporting the policy proposal that they should not get larger cages. But that "fact" was really an interpretation of the finding that the pigs spent less time in the center of the enclosure than hugging the walls. How scientists frame questions, collect data, and interpret it are all subject to perfectly innocent bias and groupthink. A broader range of voices in all areas of science, especially when the "facts" will determine how we treat humans or other animals, can unveil the biases that keep scientists from seeing that their technical assessment has a big component of ethical judgment.

Nonscientists and unaffiliated members on animal ethics committees are not there just to witness how diligently scientists judge the technical details of their peers' projects. They have a voice in scoring those technical issues too, less encumbered by the scientists' worldview that science is good and must always get the benefit of the doubt. Their involvement is all that much more crucial if the Institutional Animal Care and Use Committees in the United States truly become ethics committees, weighing the harm to animals against the hoped-for benefits of the work. This is not easy, especially given that the harm all falls on the research subjects who do not stand to benefit from the work.

Weighing the benefits of animal experiments presents a challenge, one that ethics committees are only now starting to take on, by law in much of Europe and by choice in scattered US labs. Whose benefit are we talking about? We mean human patients, certainly, and those of us who will be patients one day even if healthy now, along with future generations. Much of human medical progress crosses over to veterinary medicine, but that's not a requirement for a justified experiment. MS does not affect nonhumans, but it's nonetheless a disease most of us would harm animals to cure. Should we harm animals for research that will benefit other animal species, domestic or wild? How do we decide that mice should suffer to make dog or cat vaccines, or, for that matter, how do we justify making puppies deathly ill to produce a better distemper vaccine? Is it simply a numbers game, a few hundred lab puppies to make a vaccine for millions of pet dogs? Or is it an allowable prejudice that home dogs are somehow more deserving of our care than lab dogs? Should childhood diseases get higher priority than adult health issues? Than geriatric adult issues? I've seen proposals to spare animals a research role in conditions we bring on ourselves, through having sex or driving recklessly or waging wars or smoking or eating poorly. I steer clear of this sort of blame-the-victim approach to limiting health care or health research to those few upright people who've never done anything risky, those people with the resources to live healthy.

The types of benefits are just as numerous as the beneficiaries. We use animals to search for cures for cancers and MS and puppy distemper. We are only recently getting away from safety testing nonessential products such as mascara or, to avoid the misogyny of focusing on women's cosmetics, tattoo inks. Should a committee approve animal testing if a new tattoo ink were to require it? Perhaps some "diseases" are too trivial to warrant animal testing. Male baldness may not deserve experiments in mice or in stump-tailed macaques, long the monkey of choice for baldness experiments. As a bald man sitting on a committee with no shortage of other bald men, I might have rolled my eyes at the thought that a scientist at our prestigious university would trifle with finding me a "cure" I saw no need for. Even baldness, however, can be less trivial than it looks. Female baldness, until we fix our society's cruel judgment of women's appearance, merits a different ethical review. That's one more argument for including more voices, including patients and clinicians, on an animal ethics committee.

Curing diseases is noble. And profitable. Economists put financial price tags on diseases, such as lost worker productivity, that somehow factor in the numbers of affected patients and the severity of their disease. A company greenlights continued research projects with an eye to the financial bottom line. Will an MS drug pay for itself and generate profits? These financial estimates are proxies for what really matters, patients' quality and length of life. A single company's profits may or may not reflect real societal value. Jeffrey Aronson and Richard Green write that over 60 percent of medicines are "me-too" drugs that echo the first-in-class drugs initially released for a treatment.[20] Sometimes they are a genuine improvement, and the world is fortunate to have them; at other times they offer no real advantage for patients, just a chance for their manufacturer to get their hands on a lucrative market for statins or headache medicines or baldness cures. Should these "me-too" drugs cost lab animals pain and suffering?

The harm-benefit analysis shifts when a potential medicine is in the research pipeline. As scientists initiate the preclinical testing in animals that will show the FDA their medicine is safe and effective enough in lab animals to allow clinical trials in human volunteers, the odds of successes at that stage should be high. Batteries of nonanimal tests and computer screening should have already weeded out anything likely to harm animals, so the odds of them experiencing harmful unpredicted toxicities should be low. People are right to argue that a 10 percent success rate this late in the process is too low. Contrast this late-stage preclinical testing with the more basic research bench scientists perform, where guarantees of success are far lower. Any harm-benefit analysis should reflect these different levels of certainty.

Harm-benefit analysis is tricky in basic research, where the potential applications to a particular affliction are remote or totally unknown at the time of the experiments. I've reviewed animal projects from scientists obsessed about the role of some cell or gene or chemical in the body, spending years exploring all the nuances of how that cell functions in health and in an assortment of diseases. They need to explain the value of pursuing this cell so hotly until it's given up all its secrets; otherwise, the activists' accusations that scientists just keep pulling in grant dollars for doing the same thing over and over with minimal variation sticks. As a medical scientist, I recognize my own biases here, in believing that any cell or gene can go haywire and trigger some disease or other. I am predisposed to believe that everything we learn about cells and genes holds some promise of usefulness, and that brand of bias alone is reason for wanting a wider range of people in the ethics committee meeting. We vets, whether we have a lab of our own or not, are just too close to the ideology of science to provide the balance a harm-benefit review requires.

Science defenders point out that a scientist may not have specific applications in mind when pursuing some basic research, and time and again old knowledge from one field of science pops up as unexpectedly useful in the face of a new epidemic or even just in retooling some information in pursuit of some new cure of an old disease. Jerrold Tannenbaum prefers the term *foundational research*, to better grasp how studying biological functions may not have a direct application in its sights but can nonetheless unearth some new knowledge that will one day possibly—no guarantees—show some usefulness. In 2012 US representative Jim Cooper launched the Golden Goose Awards. The award riffed off the Golden Fleece Awards, in which Senator William Proxmire lambasted what he saw as dumb science projects that wasted government funds. The Golden Goose Awards by contrast celebrate research projects that might look useless but then later reap great rewards. The animal-research advocates at the Speaking of Research group tout these awards in the animal world. One example is the 2019 Golden Goose for Jack Levin and Fred Bang, whose studies of bacteria, horseshoe crabs, and blood clotting led to the limulus amebocyte assay, which let scientists use blood from horseshoe crabs to replace live rabbits for screening for toxins that contaminate medical supplies. Their work grew from their medical interests in clotting and infection, but they had no idea at the time that it would lead to a vital test for keeping medical products free of toxins and replace rabbits in the process.[21]

How could an ethics committee tell a scientist not to use animals in studying some obscure cell whose role in health and illness has not come to light . . . yet? Or deny animals to a brilliant scientist out to buck conventional wisdom and revolutionize how we see a certain disease and its treatment? Who knows what prizes could await if we unleashed the iconoclasts or supported years of focus on that obscure cell? I will tell you how. First, claiming that we should allow all basic or foundational animal experiments because we never know what results may flow does not work. That a scientist cannot name a specific disease (or yet-to-be-encountered disease) their basic studies may one day cure does not necessarily make the work useless, unworthy of animals' sacrifices. But if basic science is really such a crapshoot, with no idea of what will turn out to be important, then factors other than scientific merit loom large, particularly financial and animal-welfare costs, as the basis for green-lighting some experiments and parking others.

Second, we already have systems, not perfect but functional, for deciding what projects to support and allow. We have systems because science costs money and ties up equipment, and both are finite resources. Scientists do not get green lights or greenbacks for every hypothesis they want to test, no matter how a burning question hounds them personally. They compete mightily for research resources, whether dollars or lab space, to show that their questions are important and their plans for answering eclipse any competitors. The NIH calls on Scientific Review Groups to help it assess and rank grant applications, so it can distribute limited research dollars to the most promising experiments. Innovation is good, and some out-of-the-box ideas deserve consideration. Reporters love to write about that quirky idea that couldn't get funding but turned out to be a game changer. They love those stories because they are rare. Wild, new ideas sometimes are wild because they're foolish. They look new because, with so many scientists working on the same problems around the world, truly new mold-breaking findings are, unfortunately, usually false.[22] Do they justify funding or animal-use privileges? If in service of a truly important need such as a presently untreatable disease, perhaps they are. Committee reviews need not choose a binary aye or nay but something more nuanced, a small pilot study with frequent progress reports before approval to expand the work.

Like government research-funding agencies, private pharmaceutical-company executives decide on what research projects will receive limited research and development dollars. Those fiscal resources are finite. By contrast, an animal ethics committee does no such thing, treating the approval to use animals as an unlimited, infinite resource. We have no system, either nationally or locally, for deciding how much animal suffering to allow. We do not have a system for ranking animal experiments and allowing only the best to proceed. Yet we need only look at funding agencies and corporate boards for a model of how to think of animal approvals as a precious finite resource to distribute, just as precious research dollars are apportioned.

Experts alone cannot judge whether a project's goals merit the privilege of using animals. Scientists own the technical question of whether animals *are* necessary for a worthy goal. They should not have a monopoly on the normative "should we" question of what animal projects they *should* pursue. In current practice science experts advise on what projects to fund. Ethics committees, especially those dominated by scientist members, determine what animal-harm reductions are sufficient and may express some opinions about the technical merit of a project. But they cannot represent a pluralistic society or the animals' interests in sorting important from unimportant research. Scientists alone should not determine how the animal-welfare costs of a project balance against the benefits of the knowledge it yields.

The route to rounding out ethical evaluations of animal projects includes bringing in more human stakeholders, from animal protectionists to patients with diseases under study to people without a committed agenda of either permissiveness or restrictiveness. It could include, borrowing from pediatric human subjects' ethical reviews, appointing someone specifically in the role of the animals' advocate. In the next chapter on

laws and policies, I share my dream ethics committee, drawing on practices in the United States and other countries.

I'm happy to have seen and, in my way, participated in an evolution of
thinking about which animals merit our ethical (and someday legal) consideration. I'm glad to see the notion of harm expanded so that we move
from solely asking how we can limit the badness of pain and distress and
move toward seeing any deprivation from animals' maximal flourishing as
a harm to eliminate. I applaud every effort to proactively develop standards that exceed our current legal requirements, not to dodge new regulations but to recognize that ethical obligations will also exceed legal ones,
for people of good intent. Justification? Can we move beyond the Three Rs
of harm reduction to a serious scrutiny of the reasons for running an
experiment? Can we find a better way of deciding that, just as some experiments simply do not merit funding, some experiments simply do not
merit the privilege of using animals. Yes, I think we can, even in a world
where we admit that human interests will likely continue to outrank animal interests, especially when humans alone are in the driver's seat.

Rhesus Monkey

Macaca mulatta

THE PEOPLE AND POLITICS IN THE ANIMAL HOUSE

Eddie was a lab rhesus monkey, confined to a custom-built restraint chair a few hours a day for his role in a neurology experiment. He could shift his body around a bit to get comfortable, but the restrainer held his head rigidly still. Scientists could track how his eye followed a blip on a computer screen, simultaneously recording from nerve cells via fine wires they placed into his brain. If he followed the blip correctly, he got a measured sip of water. Hundreds of successful tries in a session got him hundreds of small sips. Some days, if the blips were too confusing to follow, he got less water. Animals' thoughts are often hidden from us, but anyone could see when he was thirsty; at times, as the caregiver opened the door to let Eddie back into his freshly washed cage, he scoured for whatever drops of rinse water might still cling to the cage walls.

Between sessions Eddie sat in his cage, in a room with other singly caged monkeys. Empty water bottles hung on every cage, as the monkeys quickly drank their day's allotment, building up a thirst they could try to slake in the next day's testing session. They ate their monkey biscuits and got dry treats—raisins sometimes, but never juicy grapes—that would not interfere with keeping them thirsty. They watched one another constantly, even developing a social hierarchy based on calls and facial expressions with animals they could not touch, seeing who was getting treats they thought they might deserve.

Scientists call this way of studying animals' brains the "awake behaving model." They can watch the brain's cells firing away as the subject manipulates a video joystick. Scientists defend it as basic research, conducted with no explicit or imminent new therapy in mind, research that will unlock the secrets of how the brain works, which decades later may find a useful medical application. These studies do actually feature in the successful development of deep-brain stimulation treatments for parkinsonism. They have led to clinical trials of brain-computer interface implants that are giving paraplegic patients, immobilized in their arms and legs, the ability to operate computers. Implants in clinical trials restore speech in patients paralyzed by amyotrophic lateral sclerosis (ALS). As some of these devices bring info from the brain to drive a computer or to power speech, scientists are also busily bringing information in the other direction, with an implant in the brain's visual centers that can read data from a video recorder mounted in a blind patient's pair of eyeglasses. The first scientists mapping the living brain fifty years ago may not have had these specific applications in mind, but their work in monkeys and other animals—not every single experiment, surely, but the accumulated knowledge—paved the way for these remarkable advances.

It's another example, like marmoset monkeys and multiple sclerosis, in which decades of animal suffering preceded medical breakthroughs. It's a practice that has sat squarely in animal rights groups' crosshairs for decades—and not without cause. Eddie's scientists worked to study brain-eye integration in normal monkeys (to the extent a monkey in such conditions has a normal brain). Other labs study brain-cell function after some sort of injury or deprivation, often during early development, such as closing one eye with sutures, deafening a part of the brain through exposure to loud sounds, or injecting a neurotoxin to mimic parkinsonism.

Day to day in the monkey room, progress in refining and replacing awake-behaving monkey research seemed glacial. I might meet with the researchers to discuss Eddie's weight loss and negotiate an increase in his water ration for a few weeks. Turn the water flow up, and the research could slow; turn it down, and the animal stays thirsty. We used Eddie's weight—a crude welfare measure to be sure, but we already knew his body was physically healthy while his mind constantly dwelled on his thirst—as our main guide for titrating how much water he could receive, so he would stay just thirsty enough for the researchers' requirements. Some of the monkey scientists welcomed us vets' prescriptions for stronger pain medicines after surgeries; others actually refused. Eddie's lab decided that some animals, Eddie included, could go to a retirement sanctuary after their time as a research subject ended, though the grant that covered the studies would not pay for our work in shipping them to new homes. As for the benefits of the work, they remained mostly abstract to anyone, the ethics committee and us vets included, not working in this specialized field. Small details of how certain cells responded to sensory input from the eyes or the fingers seemed no more interesting to us nonspecialists, if we even read our scientists' papers, than hearing that a coin collector acquired an uncommon 1954 penny in good condition. On weekends the monkeys got extra water while the students crunched the week's data. Then on Sunday night a lab member would come to campus to remove the monkeys' water bottles and get them started on the week's "fluid regulation," our euphemism for water deprivation, and the week's data collection.

Taking the long view, I can report on refinements scientists, behaviorists, and vets have made to improve animal welfare in awake-behaving science. Wireless technologies allow data collection without rigid restraint in a primate chair. While Ham the Astrochimp suffered electric shocks to make him perform his computer tasks, thirst had replaced electric zaps in primate labs by the time I'd come to San Francisco and learned about this research. Now scientists are learning they can train smart monkeys with lots of treats and positive reinforcement to more willingly show their prowess in video tasks. Too many monkeys still live in single cages, but there is nothing about the awake-behaving model that requires this. And, as for justification, progress for paralyzed and parkinsonism patients makes the work an easier pitch to an ethics committee. Peter Singer has surprised me in his review and qualified acceptance of monkey tests to further develop deep-brain stimulation therapies for current and future patients.[1] You'd still be a fool to trade places with a monkey like Eddie, but, even so, you might score this as incremental progress toward a better balance of animal-welfare costs and human benefits.

No single scientist can claim credit for the medical advances brain implants may deliver. And no single person can claim credit for animal-welfare improvements for Eddie or the monkeys who have followed him. Let me tell you about the people who've pushed for reforms in our practice and in our rule books, who've developed the knowledge to allow those reforms, how we make sure labs are following the rules, and where we go from here for that better balance of costs and benefits.

Plenty of people have had their say in what kind of treatment monkeys like Eddie receive. Sitting in their cages, Eddie and the other monkeys in the room watched us humans come and go with rapt attention. Even behind the masks and goggles we donned to shield them from our infections and us from theirs, they certainly knew who was familiar and who was not. They knew who brought treats and who would put them in their restrainers. They remained on guard to distinguish friend from foe from stranger who could turn out to be either. Their daily lives included the caregivers feeding and cleaning their cages, the trainee scientists running the experiments, and the vet nurses making their rounds. Less frequently, I or another vet popped in, sometimes with our animal behavior specialist (a role not every campus has) or sometimes to meet the head scientist and discuss the various monkeys. When technicians leaked their concerns about how the lab was treating the animals or had a sick monkey who needed care, we vets were the first responders. More than once frustrated staff members upped the ante and sent anonymous complaints to the US Department of Agriculture (USDA) when they found the vets unwilling or unable to stand up strong for the animals. Twice a year a couple of ethics committee members visited the monkey room with one of us vets touring them around, explaining to them what we saw as problems or successes of how we cared for the scientists' monkeys. Once a year the USDA inspector showed up unannounced to inspect the monkey room and the monkeys' records. Every three years the Association for Assessment and Accreditation of Laboratory Animal Care International (AAALAC) accreditation organization sent a team of vets and scientists for a similar inspection. Once in my twenty years on the campus, the National Institutes of Health's Office of Laboratory Animal Welfare came around for a general check on our animal-welfare program, an effort they launched to get around to the largest NIH grant recipients. Add in the occasional plumber or electrician, and that's quite a parade of people looking at Eddie as he looked at them.

Some people Eddie never saw, and they never saw him, no matter how much they wanted to. The animal activists, imagining the worst, waited for their chance to expose and condemn what they were forbidden from seeing. We in the labs felt their presence. We, the vets, scientists, ethics committees, and staff, tried to balance what the scientists' experiments seemed to need against our best estimates of what the monkeys and other animals needed. Our version of that balance was far from the animal activists' version, and they did their best to publicize that, mixing what facts they could glean from our USDA inspection reports and our NIH grants along with old pictures from who knows what labs to build their case against us.[2] Privately, we complained to one another that their information was at best partial and misleading and that their pictures were not even of animals from our labs. Had they asked us for a visit or for pictures of our monkeys so they could more accurately portray (and campaign against) our animal labs, most assuredly we would have said no. Eddie likewise never met the wealthy benefactors or the NIH administrators who funded the research. He never saw the legislators who wrote an Animal Welfare Act that gave him some protections while denying him others, including the most crucial permission to cage him for experiments in the first place. He did not meet the vets and scientists who wrote the NIH rule book, the *Guide for the Care and Use of Laboratory Animals*.

Skeptical outsiders question how well our various standards and rules are enforced, so I offer here my insider's view on that process and how we can improve on it. That story is extra complicated in the United States, as I've described in the chapter "Python." In the United States, two different federal laws cover some lab animals while still leaving others without federal oversight, and a voluntary accreditation system plus state and local laws in some locations fill in some of the gaps.[3] But the most vigorous enforcement of welfare rules fails the animals if the rules don't live up to the ethical requirement that every harm to animals be justified.

Our rule books reflect competing beliefs about how much scientists can be trusted to do right by their animals. The earliest regulations assumed we all share an understanding of what's best for animals and so mapped out ways to catch bad people, dognappers, and cruel scientists and put them out of business. Dognappers are a rare breed, nowadays, but USDA inspection findings are fairly easy to see on its website. Facilities' self-reports to the NIH Office of Laboratory Animal Welfare, though better hidden, are accessible through public records requests. Both sources show scientists, breeding facilities, and animal labs still breaking the rules. The USDA issues citations when scientists run unapproved experiments in which animals fall ill or die. Its inspections can shut down a dog-breeding kennel for multiple problems, from insufficient vets for the thousands of dogs to multiple illnesses and puppies' deaths.[4] On the other hand, inspectors will cite for silly things, such as researchers not signing their written application for an ethics committee review. We do still need welfare compliance oversight and inspections, and, as a lab insider, I want that to be animal-focused, not simply another form of government bureaucracy.

Research lobbyists believe our current laws and their enforcement are sufficient as they are. I frequently encounter variations on this statement, first (but no longer) posted by the Foundation for Biomedical Research (FBR): "The United States Department of Agriculture (USDA) has set forth federal regulations governing the care and use of animals in biomedical research that are considered more extensive than those covering human research subjects."[5] Considered by whom, they don't say, but it's clearly someone either uninformed or dishonest. Even ignoring the point I've hammered that the USDA does not cover 99 percent of the animals in labs, the claim that animal regulations are more extensive than human subjects' protections is nonsense. And the FBR knows this. Animals are in labs precisely because we plan experiments no one could possibly perform with human volunteers.

Consider this tale of two subjects in tests for new brain-cancer treatments. The mouse conscript and the consenting human volunteer will both undergo surgery to receive a cancer drug directly into a brain tumor. Both projects require ethics committee approval. The animal committee reviews the surgeon's—a student or technician most typically—qualifications for injecting drugs into the brain. The human committee does not, as only a licensed, specialty-certified, practicing brain surgeon would perform the operation. The mouse's cage must meet certain specifications and is inspected twice a year by the animal ethics committee. No one provides the human patient with a cage or inspects it. And, while the human subject is a cancer patient, the mouse experiment starts with a first surgery to implant tumor cells into the previously healthy brain, so that first surgery too requires review, along with a review of the later surgery to deliver the drug once the cancer has developed. The "extensive" regulations even cover how to kill the mouse at the end of the experiment, but none for how to kill the human patient. The human regulations center on the patients' informed consent; they or their guardian must fully understand the experiment's goals and the risks they take in participating. Their involvement ends the minute they say no. Animals may grunt, bark, or squeak their nonconsent, but that gets them nowhere. Even when scientists take the time to gently train animals to cooperate or assent to research, we remain incapable of getting their informed consent. They may cooperate with an injection but cannot know what viruses, cancers, or other types of harm are swirling in the syringe. So technically, FBR is correct: the mouse regulations are more extensive. But we can dispense with the insinuations that the mouse *protections* are in some way more extensive.

And, of course, animal-protection groups want stronger laws and stronger enforcement, and some are willing to bend the truth to that end. Balancing out FBR's deceptive claim is People for the Ethical Treatment of Animals (PETA) with its own distortions, wanting you to believe that there are "no regulations that govern the conduct of an experiment."[6] Animal advocates have complaints about the laxity they see in the USDA's Animal Welfare Act rules but, despite complaints, see USDA inspections as an important tool, even filing lawsuits against USDA plans to reduce its inspections of some facilities.[7] So, yes, we have laws that have greatly improved the situation of animals in lab cages, and, contrary to FBR's claims, those laws most certainly do still allow great animal suffering. The reality is that regulations in the United States and around the world start from the premise that we may ethically harm animals for medical progress and then leave much to the discretion of the scientists, ethics committees, and vets in the animal rooms. My job here is to walk readers through the strengths and gaps of our welfare standards and their enforcement and chart my vision for a more humane future.

The USDA sends Animal Welfare Act inspectors, most of them veterinarians, to every lab with "covered species" at least once a year. These snapshot inspections can find facility problems and, rarely, a sick animal that staff failed to report (and if the USDA inspector is first to notice an animal is sick when staff have been in and out of the animal room for days, you know there's a serious problem in the facility). The inspectors rely mostly on reading ethics committee minutes and animals' medical records. As the compliance manager for a large university's animal program, I certainly chafed at some of the paperwork violations the inspectors noted, knowing that, in outsiders' scrutiny and publication of our inspection reports, they would highlight the number of citations we received, glossing over the ones for failing to sign a document or file it by an arbitrary deadline. The inspectors knew how to burrow into a case. If the ethics committee had discussed researcher noncompliance in a pig lab, I knew to have the pig's medical records on hand as the inspectors read our committee minutes or to bring the inspectors to the animal room for a look-see. I also knew how to negotiate an inspection write-up with the inspectors, who might trust me to fix some problem they were debating over whether to cite, both of us knowing that next time they came to call, that uncited problem had better be fixed. To outsiders that collegiality, that coziness, which I so appreciated as a compliance manager, keeps too many problems hidden from public scrutiny. How many, of course, outsiders cannot say, given the limited transparency in these hidden interactions.

Our welfare protections in US animal labs rely on a system of "regulated self-regulation," in which institutions must empower an animal ethics committee, whose duties include inspecting the institution's animal facilities twice a year and investigating any concerns of welfare noncompliance. The USDA reviews this committee's records (at labs that it covers) and labs must self-report their compliance efforts to the NIH and to the accreditors at AAALAC. By law the head veterinarian has a seat on this committee, and the committee must include a nonscientist and someone otherwise unaffiliated with the institution (and one individual may serve in both capacities, setting themselves up to be a very lone wolf in a committee room full of scientists and vets). The committees have the authority to halt animal-use protocols if they've gone awry or to bar individuals from animal-research privileges.

The animal-care program at the University of California–San Francisco was under fire, and with good reason, when I got there in 1999. Though the largest of the University of California's animal-research programs and perennially one of the top national recipients of federal research money, it alone in the UC had never pursued third-party AAALAC accreditation. The campus kept racking up USDA animal-welfare citations, which animal rights groups made sure to publicize. News of an experiment in a squirrel monkey lab prompted San Francisco City Hall hearings in 1998, as the board of supervisors debated their jurisdiction of a state university experimenting with federal funding. The city never went further than a resolution urging UCSF to get itself AAALAC accredited, but we were on notice that we were under their magnifying glass.[8]

The university hired me to spearhead a training and compliance unit, our animal-welfare assurance program. In my new role, I would walk the beat and root out problems before the USDA inspectors would find them, reporting on my lab audits and inspections to the animal ethics committee. Still new to the campus, I visited labs, introducing our new compliance program and gently putting scientists on notice that even the mouse labs would now be subject to oversight. I reviewed the basic federal regulations and campus rules as if they were new, and many scientists reacted as though they indeed were. One sighed his understanding that this was necessary; he also had a lab across the bay, in Berkeley, "where they really take this stuff seriously." The ethics committee expected me to audit labs and report on the occasional bad researcher whom they could nudge toward compliance with another round of lab training or, if their misbehavior reached unacceptable levels, hound them out of animal research. They expected 1 percent, at most 5 percent, of labs to require committee discipline. Instead, I told them my estimate that 90 percent or more were deficient in some way, including our committee chair's own lab.

The most common failure, which vet friends around the country report to me as well, is to somehow drop the ball on administering and recording doses of pain medicines and animal exams after surgeries. They might give just one of two drugs that their approval requires or completely fail to give the pain meds at all. On paper this can look trivial, but for the animals in the cage, with no option to ring a bell for a nurse, it may result in them languishing in unseen and unnecessary pain through the night. The committee could suspend every lab or lab member guilty of this failing, report all this to the NIH, and hope that the NIH doesn't stop the research dollars coming to such a thoroughly noncompliant university and that watchdog groups don't file information requests for our NIH reports. Or we could paper it over as a records-keeping issue, with a tsk-tsk or a warning letter from the committee, and hope for improvement (which we sometimes saw and sometimes did not) on the next round of audits. The committee members, mostly insiders, are usually the institution's own research scientists, like the vet and like the compliance staff. We do not want to be overly harsh on our colleagues, threaten our university's good standing with the regulators, or derail experiments in progress. Conflict of interest? Yes, as much as I have good news to report on how diligently the committees work and how professors serving on the committee can be incensed at their peers' callousness, the conflicts are real, as can be the outcomes for the animals. And the nail I continually hammer, when self-regulation is the main compliance oversight: the mice and rats will continue as second-class animals in this system.

The USDA has several tools in its enforcement chest. It can confiscate animals or revoke dealers' licenses to sell animals or research labs' registration to conduct research. One big stick is the fines the USDA can levy, historically a wrist slap, thousands of dollars at a company or college with billions of dollars in operating budgets. In 2024, however, the USDA with the US Department of Justice shut down a large breeding facility, forced it to send four thousand dogs to homes, and collected a $35 million fine. That hit me as extreme, until I read the inspection reports online, including several instances where staff never reported sick dogs to the vet. This could be heralding a change in USDA practices, but a tough federal stance during one government administration can collapse in the deregulatory fervor of another. The USDA offers some transparency of its enforcement, and anyone can go to the USDA website to see inspection reports for their local college or biotech company (if they have animals that the USDA covers and not just mice or fish or other animals the Animal Welfare Act excludes). Drastic measures for drastic welfare violations are appropriate, but tough legal enforcement is only part of the work in upgrading standards for animals.

The NIH enforces the Health Research Extension Act of 1985, the law that requires animal-welfare programs at labs that receive federal funds. It excludes many private labs, from biotech startups to big pharma companies. In its favor it covers the animals that the Animal Welfare Act excludes and may even soon take octopuses under its protections. I've never been impressed with its enforcement program, though I've used it as a cudgel with errant researchers: "Take better care of your animals, or we will need to file a self-report with the NIH." Otherwise I've found NIH collegial to a fault, even letting me wiggle out of filing a written self-report (available to animal rights watchdogs through Freedom of Information requests) if I promised that the animal was not actually used on a government-funded grant. Its big stick is that researchers or whole institutions can lose their government grants. Institution-wide noncompliance is not common, but the NIH does enact smaller holdups on individuals' grants.

Accreditation through the AAALAC combines elements of the two laws. At large campuses where I've worked, a team of half a dozen site visitors spends close to a week on campus, after they've prepared by reading a thousand-page write-up of our animal-care program. Unlike the USDA's inspections, it is not remotely a surprise inspection but planned for months in advance. The intense months of preparation campuses invest in the run-up to these preannounced triennial visits reflect their willingness to let things slide in between times. The AAALAC uses the same rule book that the NIH does and covers the same range of species (vertebrates plus some of the brainier invertebrates like octopuses). It is even less transparent than the NIH, though public universities must release their AAALAC correspondence after Freedom of Information requests. I've served as a site visitor a few times, and have been visited for inspections a few dozen times and can tell you that the process is really thorough, with site visitors, our peers from other research campuses, pushing us to a level of performance that may be higher than at their home institutions. Animal activists look askance at the extreme lack of transparency and the potentially overly chummy you-inspect-my-animals-I'll-inspect-yours nature of the system. My insider report is that the accreditation system works well, but, wrapped as it is in secrecy, my testimonial is hard to verify, and I cannot blame their skepticism.[9]

A good auditing program and better government oversight can ferret out incompetence and bad behavior and correct them. We can make progress without increasing costs or slowing research to a crawl, and, by all means, every college campus and biotech and pharmaceutical company can commit to better training and better compliance reviews. The bigger challenge is not that scientists don't comply with standards but that our standards have not kept up with what we are learning about animals' welfare or the many ways we can replace animal experiments altogether and refine those we cannot yet replace.

An animal committee may self-inspect its animal rooms and find mice stereotypically flipping somersaults or dogs spinning in small cages and feel stymied. I once encouraged one of our nonscientists during a committee self-inspection to list her concern that many rats, a social species that will always pile into a group for sleep time, were singly caged with no clear explanation. In some of these cases, the head vet was also the vivarium director, tasked with keeping costs low for researchers. He explained to the committee that bigger cages might be nice, but they were not a legal requirement, and he would not be making changes. Raising our standards, not just because we thought we should but because our laws enforce it, can push us past such conflicts of interest in our self-regulating system.

The requirements in our two federal animal-welfare laws run a gamut from picky administrative matters to more expansive matters of real consequence for animals. Both are long overdue for updates in the face of massive amounts of new information in animal-welfare science. The last substantive update to the Animal Welfare Act was in 1991 (the USDA's final regulations for the 1985 congressional act). The NIH commissions the National Academies of Science, Engineering, and Medicine (NASEM) to manage and update the *Guide for the Care and Use of Laboratory Animals*. The *Guide* has broader impact, covering more species in more detail and serving as the main rule book, not just for the NIH's legal requirements but for the AAALAC accreditation program as well. An update to the Animal Welfare Act regulations is unlikely unless Congress were to amend the act again and call for substantive changes. NASEM updates the *Guide* only when and if the NIH and other organizations fund it to do so.

The *Guide* dates to the early 1960s, three years before the first Animal Welfare Act, when its veterinarian and scientist authors hoped to establish professional standards, bring second-rate research labs up to what the authors at top-tier labs were promoting, and in the process give scientists and vets, rather than government regulators, the chance to set standards. Though it acquired the force of law in the 1980s, insofar as all NIH grant recipients must follow it, NASEM has continued to rely on, or privilege, vets and scientists within the research community to write the rules they will obey. It pulls in leaders in their fields, including some animal-welfare scientists, who spend years poring through literature and debating standards, but all of them nonetheless are members of the animal-research pack.

In the 1980s *performance standards* became the buzzword, as the research lobbyists pushed for maximum flexibility while the animal activists wanted strict, clear rules. Performance standards offer flexibility in how to meet an outcome. For example, no one wants high-pressure water pipes to leak. Should the building code specify an engineering standard, a certain gauge of a certain material pipe, as the required way to prevent leaks? Or should it require a measurable standard, such as no leaks when tested at x pounds of pressure, that builders can meet with whatever state-of-the-art pipes they can demonstrate will hold the water? To work, performance standards need to specify the standard, such as no leaks at x pounds, not just state a platitude or aspiration, such as "leaky pipes should be avoided."

In the animal labs, our rules are a blend of flexible performance standards and detailed engineering standards, along with aspirations that lack specificity. Animal cage–size standards illustrate the difference between engineering and performance standards. The cost of animal caging makes any change in the regulations potentially quite costly. A standard-issue rhesus monkey like Eddie weighs somewhere around twenty pounds, a bit smaller than a beagle, though gangly. The law gives him four and a half square feet of floor space in a cage thirty inches high: a rigid engineering standard that an inspector with a tape measure can judge. The performance standard in the Animal Welfare Act requires that the cage allow him "normal postural adjustments with freedom of movement." The *Guide* adds a standard that the cage should be tall enough to allow "expression of species-typical behaviors."[10] Picture any monkey you've ever seen in a zoo or a video in a box that allows only enough freedom of movement to stand up, turn around, and sit down again without taking a full stride in any direction. The monkey cannot run or climb or even stand upright on his hind legs, as monkeys will do when they want to look around. They cannot climb, a "species-typical behavior" in virtually all monkeys, in such a box. Surely this fails our ethical standard that such confinement is a harm that calls for special justification. Surely too, a six-square-foot, low-ceiling cage, the engineering standard, fails to meet the USDA's performance standard of "freedom of movement." Yet a USDA vet inspector has no actual standard of "freedom of movement" by which to judge a monkey cage, only a tape measure and the monkey's weight records.

As for monkeys, so for mice: a cage should be large or complex enough, with barriers and the like, to allow the occupants "to escape aggression"; a conventional mouse cage does not approach that standard, and so I've seen many the mouse with a torn-up rump and tail when the aggressor's night-after-night attacks finally break skin and prompt us to separate them.[11] I've never seen an inspector or an ethics committee member call out a mouse or monkey cage for failing to meet these performance standards if the cage met the number of square inches in our rule book.

Many of the performance standards are way too squooshy to earn the word *standards*. Animals with restricted drinking water "should be closely monitored" to ensure their limited fluids are meeting their needs. Surgeries should entail "appropriate use of analgesics."[12] Scientists should "consider" alternatives to painful procedures. Staff and scientists conducting surgery or other procedures must be "qualified to perform their duties."[13] How thoroughly to pursue any of these calls for animal welfare the regulations do not say. Absent clear standards, research labs can vary widely on how well they minimize animal harm. Absent clear standards, ethics committees, vets, or scientists are on shaky ground pushing up on administrators to spend money for housing that exceeds the minimum engineering standards or training or other procedures the lead scientists find burdensome. A committee of vets, welfare scientists, and other experts spends years writing the *Guide*, then leaves it in the hands of hundreds of labs across the country. Few of them have the expertise of the *Guide* panel, and all of them face the potential conflict of putting restrictions on their colleague scientists, to judge what to score as "species-typical behavior" or what to monitor as evidence of excessive water restriction. Whoever writes the next edition of the *Guide* should test each "must" and "should" standard they prescribe against the question, "How could a lab or an auditor measure compliance with this standards?" Otherwise they leave too much up to local decision making in the hands of ethics committees, with a degree of flexibility that does not serve the animals well. They leave monkeys like

Eddie waiting a long time for someone to state in clear terms, "Do not take this monkey's food away until you have given him two months to show what he can learn through positive rewards."

Whereas some countries have rigid training and licensing requirements for scientists, staff, and students running animal experiments, the United States sticks with flexible performance standards. Here is the Animal Welfare Act: "Personnel conducting procedures on the species being maintained or studied will be appropriately qualified and trained on those procedures. . . . Training and instruction shall be made available, and the qualifications of personnel reviewed, with sufficient frequency to fulfill the research facility's responsibilities."[14] In many private labs, a staff of animal technicians and vet techs run all the animal experiments. They know what they're doing and what the rules are. In universities technicians are expensive, and, since students are on campus to learn, it's both cheaper and in their training interests to let them perform the animal procedures. Our legal requirements in the States for training animal workers are flexible enough that many students get permission to run animal experiments without the rigorous weeks-long mandatory training I see in foreign scientists who've come to our campus. Students need better training with a large enough cadre of vet techs, truly the unsung human heroes in animal labs, on hand to assist them until they're fully competent and can take pride in their animal skills. Absent such an explicit requirement in our US laws, I found myself on a succession of university task forces charged to find ways to reduce our researchers' "training burden," when every visit to a lab revealed yet another undertrained scientist.

From its birth in 1963, the *Guide* has doggedly hewed to the guiding principle that data and science can directly produce ethical animal-welfare standards untainted by politics or ideology. And sometimes they can. Data show that mice and guinea pigs, unlike most mammals, get sick without daily vitamin C supplements, so the unproblematic standard is that we must give them vitamin C every day. If every issue were that straightforward, it would not take a team of experts or fifteen-plus years to update the *Guide*. Even water requirements for monkeys like Eddie are complex. How much water does he need to stay alive? To grow to a normal adult weight? To be able to think about something other than where and when he can get another sip? The *Guide* may not be able to provide all those answers, but I promise you that individual campuses with but a handful of vets, no welfare experts on their ethics committee, and no water-balance physiologists are not equipped to make the determination. Where data are sparse, the *Guide* presumes that scientists and vets will supplement it with their own impartial expertise, but based on what?

Since the 1980s and 1990s, most industrialized countries with medical research programs have looked to ethics committees to fill in gaps in the rules and regulations, to apply updated knowledge to the animals who do not get a seat in the meeting room, and to ensure that no vet or behaviorist stands alone as the only voice for animal welfare. Rather than stating in law or the *Guide* how much water and fruit a monkey in Eddie's circumstances must receive, ethics committees around the country all individually make their own judgments, reflecting, supposedly, local conditions and regional standards. Because . . . water tastes different to lab monkeys in San Francisco and in New Jersey? Because they experience thirst differently? Because committee members in those states care more or less about thirsty monkeys in distress?

Almost every issue the *Guide* and the Animal Welfare Act cover are similarly complex. Housing standards are particularly fraught as, unlike telling individual scientists to give their monkey more water, improving high-tech animal-unfriendly housing would cost big dollars. Vets running vivaria and caging manufacturers focus on the "recommended minimum space" tables in the *Guide*, ignoring all the performance factors of sociality, warmth, lighting, and opportunities for exploration. Animals need some control over their own lives that a simplistic table with entries like sixteen square inches of floor space for a one-hundred-gram hamster do not capture. An animal-centric update to the *Guide* would shred those tables and clearly state what hamsters need in their enclosure for a decent quality of life, which may (or may not) require more than sixteen square inches. This does not fit so neatly into table. Regulators and ethics committees on their semiannual self-inspections would need to let go of the check-the-box approach to evaluating animal care and instead ask how much water or what kind of home an animal needs, in some biological sense, or deserves, in an ethical sense. And if that costs, in dollars or in slowed research, who decides what animals will get? Is that a decision best made at the national level or one by one in ethics committees?

In the previous chapter, "Flea," I've laid out my belief that science labs should expand their concerns to all sentient animals and expand their concept of harm to embrace lifelong animal flourishing. I've bemoaned that animal ethics reviews focus almost exclusively on harm reduction à la the Three Rs of alternatives, but far too little on the justification for harming animals. We have an ethical obligation to balance the costs in animal welfare against the odds that a proposed experiment will accomplish its aims and that those aims are important enough to justify harming animals. I want this to be a legal requirement, lest we continue to dodge this important task.

The obvious candidate to accept this duty on a project-by-project basis is the animal ethics committee. The animal ethicist Jerrold Tannenbaum has argued that such committees may actually violate the federal law if they take on that task. Tannenbaum, a lawyer and ethicist, cites congressional language and regulatory language that committees have no authority to interfere with research decisions, goals, or methods and must restrict themselves to reviewing the care and treatment of animals.[15] He may be correct. These committees do indeed "interfere" with a scientist's planned methods when they require analgesic treatments or set end points for removing animals from the experiment. The further step I advocate, of reviewing whether a particular project merits animal use and prohibiting studies that do not make the cut, may be pushing the legal authority. And absent a clear mandate in law that they do just that, any committee that goes rogue and starts applying a higher standard of justification will surely incite researcher rebellion, and administrators, at least the ones I've known, will quickly muzzle the committee.

I urge anyone, working for a research institution or not, trained as a scientist or not, to find an ethics committee that needs members. They have a legal requirement to include outsiders and nonscientists, and anyone who reads this book has shown their commitment. This is your job description. Ethics committees (Institutional Animal Care and Use Committee is the common name in the United States) spend little of their time in compliance work, whether disciplining errant researchers or inspecting vivaria run by vets and professionals who already know what they're doing to meet the animal-care standards. Overwhelmingly, the job is to review scientists' plans, their protocol, to run an animal experiment.

No scientist gets to touch an animal until the committee has approved. Small labs' committees meet monthly or quarterly to review a handful of protocols. On my huge campus, our committee met three times a month, reviewing a dozen or more animal protocols that ran up to a hundred pages, plus various modifications for which scientists needed approval. That's a big commitment, and we cannot really pay outsiders beyond a lunch and a parking pass, or then they wouldn't be "nonaffiliated." Scientists describe the experiments they want to run and why they want to run them and how many animals they plan to use and how they came up with that number. The level of detail may vary. Our committee had learned from experience to ask minute details, what size scalpel or sutures the animal surgeons would place, learning time and again that in the face of loose standards for training, even highly qualified MD surgeons can make a mess of a sheep or hamster surgery. The committee asks how the scientists know nonanimal models cannot completely replace animals; what databases did they search? Birth to death, the committee asks for every detail of the animals' service. What you likely won't do is vote on whether, after researchers have done all they can, working with their vets, to reduce animal harm, the projects' methods have scientific merit or, beyond that, whether the projects' goals are important enough to justify whatever harm the animals will endure.

The name "Institutional Animal Care and Use Committee" dodges the word *ethics* with good reason, as the expectation in our US system is that the committee performs a technical task rather than an ethical judgment. The committee's task boils down to oversight of the scientists' efforts to explore the Three Rs of animal alternatives. Have they read up on replacing the animals with nonsentient cells, computer modeling, and the like? Check. Will they use the right statistical tests to calculate the smallest necessary number of animals consistent with producing statistically robust data? Check. Have they explored refinements of the experiments' care and the animals? Check. In Eddie's case that would mean training without water restriction; testing without rigid confinement; pairing him up with compatible species-mates; avoiding surgery and, when that's not possible, using the best pain treatments; housing them in ways that exceed our Animal Welfare Act standards; and setting a cap on how many hours a day he spent doing the scientists' work. Often the checklist approach remains stuck and lets committees dodge deeper ethical questions.[16] My committee in San Francisco required scientists to provide keywords they used to investigate alternatives in their animal experiments, forcing them to name at least two different search engines and the date of their online searches. We did not ask them what their searches revealed, nor did we perform the occasional quality check using their search strategy. As long as the keywords looked good and included the word *alternative*, and they updated the date of the search for every triennial protocol review, they got a pass.

Our regulations call on committees to review scientists' justifications for how they want to run their experiments. Almost universally, *justification* means a technical justification, not an ethical one. Scientists explain why this experiment requires, or justifies, breeding or buying this many animals. They explain why a different project can work only if animals undergo two or three surgeries; the technical requirements of the experiment scientifically justify putting animals through the multiple surgeries. Social housing is the default expectation, but researchers and vets can give a technical justification for housing animals singly, if a data collection might require that, or if an animal is just too scrappy (or too easily bullied) for life with a roommate. Committees can review the scientific data that analgesics will affect research outcomes and invalidate or skew the project, and therefore untreated pain is justified. These various calls in our regulations for *justification* are really calls for technical explanations, more of a how-to rationale than an ethical justification, claims that a committee of technical experts should be able to evaluate. A technically oriented committee, in which scientists predominate as they presently do, can make these judgments just fine.

For many projects only the scientists involved simultaneously assess the amount of animal suffering, the quality of the experimental plans, and the importance of the questions the experiments are designed to answer. They alone put the three together to decide if projects are justified. In our current system, the NIH endorses a separation of the animal-welfare reviews and the merit reviews (which determine if experiments are well designed and can serve to advance an important goal). Expert panels that the NIH assembles to review proposals' merit and recommend funding have the deep knowledge to answer specialized questions. They can evaluate whether an experiment on how hormones affect monogamous behaviors in field voles is likely to yield results and whether there's evidence that those results would be relevant to human social behavior. The NIH does not expect such technical committees to evaluate the welfare aspects of each experiment in a large vole-research program. Nor does the NIH demand much cross talk between the scientific peer reviewers and the ethics committee. It does not expect ethics committee members to know details of what peer experts critiqued in the proposal or even that anyone performed a detailed verification that the scientific experts and the ethics committee reviewed the same project.[17] Remember too that the NIH jurisdiction is limited to NIH-funded projects, so its guidance does not extend to studies it declines to fund, to pilot studies, or to private, commercial labs.

I have no interest in slowing down important research or in adding to costs, but I want a more robust ethical review of animal-research projects, one that weeds out projects either too trivial or too farfetched to justify harming animals or, rather than outright rejection, a better system to modify projects to simultaneously improve welfare and science. Most other countries approach this with a requirement that ethics committees conduct an ethical harm-benefit review of animal projects. I've not yet seen that any country or individual campus or company meets my standards, but I have seen enough promise in the various approaches where I've visited to be able to envision my dream team.

My dream team includes experts in animal health and welfare, nonscientists who can bring fresh perspectives to issues vets and scientists take for granted, community representatives, animal protectionists, patient advocates, technicians on the front lines in the animal rooms and lab benches, and practicing research scientists in a range of disciplines. The team needs balance, with no constituency dominating the others. It needs safeguards against the various conflicts of interests to which an institution's own staff, whether scientists, vets, or nonscientists, are vulnerable.

Start with vets, my people. I'm proud of how vets were early leaders in improving laboratory animal care, well before we had laws compelling us to that role. And I'm saddened by how often vets have led the reaction against improving standards or tightening up regulations. The US system requires that a head vet (the term in the law is the "Attending Veterinarian") have a voice on the committee.[18] This is one of the most conflicted roles, especially when that vet is also the vivarium director struggling to keep costs down and income-generating animal rooms full. I know plenty of vets who feel they've been bullied by scientists, and scientists who feel the same about their vets. When I pushed for better pain medicines for one scientist's animals, I got little support from the committee chair or the university administration; they worried that I had a conflict of interest and suggested that I should recuse myself from meetings where that person's work would be on the agenda. It would have been better, perhaps, had that push for treating the monkeys' pain come from an outsider. And so I borrow the idea, from the Netherlands, that the committee should also have one or more vets and scientists who are not on the payroll of the institution and can speak independently on how projects may affect animals and whose paycheck does not depend on keeping the institution's animal labs buzzing, the cages full of rent-paying mice.

In the United States, a single community member (the jargon is "nonaffiliated member"), typically not a scientist by trade, may sit in a room with two dozen scientists and find themselves outnumbered, intimidated by all the PhD researchers at the table.[19] We had a community member, a retired insurance broker, on our UCSF committee. With no science training, it took him a year or more to find his voice, but, once he did, he had the courage to ask a roomful of researchers, "Can someone explain to me in lay terms why this project is worth doing?" Another, reviewing some project in its tenth year renewal, asked, "Do scientists here ever decide they're done with something and move on to other research?" After a short discussion, our committee chair would shut things down, saying, "Remember, our task is to focus on the animal welfare, not on the scientific merit." Different countries have different systems to balance committee membership so that animal researchers have representation but not overrepresentation. We should have more proportional representation. In Sweden half of the committee members are nonscientists. Other countries require fewer but set some cap on the monopoly of scientists so that nonaffiliated members and nonscientists do not feel intimidated. The Royal Society for the Prevention of Cruelty to Animals (RSPCA) in London has special outreach and support for community and lay members on committees.

I started my ethics committee work as a student representative on Cornell's nascent committee and believe that animal-care technicians, students, and veterinary technicians, the people who see animals all day every day, have valuable wisdom on how projects translate from paper to the animal cage. They're the ones who see which labs are skillful, which are attentive to their animals, and which should not be allowed to touch animals on their own without more training or assistance.

Ethics committees need senior experienced researchers, not only for the credibility they give the committee when their peers complain it's too slow or too picky but also for their expertise. In-house committees may lack the expertise that NIH's scientific review groups represent, but scientists who know that process can sniff out second-rate projects, even if not in their specialty, and work with their colleagues to increase the quality of the work. Our regulations started from the presumption that researchers revert to running cruel, repetitive, and wasteful experiments unless kept on a short leash. And some may. But all the technical innovations that allow for nonanimal alternative tests came from scientists themselves— and many of the critiques too. Henry Spira knew nothing about eye safety testing when he embarked on his quest to spare rabbits the suffering of the Draize test. Rather, he heard scientists themselves decrying the amount of animal suffering in a test they saw as unreliable. I've described scientists leading the way to better monkey diabetes experiments, to housing that does not stress animals, or to judging whether inflammation in mice sufficiently models human pathology to stay in the researcher's tool kit. Scientists really have the catbird seat for seeing what animal techniques are useless or distressing and to do the research to come up with better options. I want to acknowledge and capitalize on scientists' own contributions to more responsible animal use.

Committees need true animal-welfare scientists and specialists, a relatively young profession only now standardizing training and credentialing for such folks. These are the people who pushed vets to see that welfare is more than just the physical health of animals' bodies. Few but the largest research institutions have such a trained and qualified person on staff, and, of those who do, many are situated in the animal-care department, supervised (and, sadly, often squelched) by senior administrators and vets. Welfare experts need more independence. In Australia the state government reviews an institution's nominees for its animal-welfare specialist on the committee and decides whom to appoint.

For institutions focused on human medical care, clinicians and patient advocates bring wisdom. For many years I read animal protocols in which researchers declared that the human condition they were studying, multiple sclerosis, for example, was not painful in people and would not be in their lab animal models either. That would be news to the National Multiple Sclerosis Society, which claims that "pain syndromes are common," and to an MS patient as well.[20] Perhaps just the experience of being a patient with a chronic illness could bring perspectives about chronic diseases more generally, even if not the disease the particular ethics committee member is living with. I know of no country that requires a seat for representatives of patients, but what a valuable perspective for deciding what projects are important enough to warrant harming animals. Clinicians, even with no animal-research history in their background, likewise bring valuable knowledge. One of my favorite committee members at UCSF was a cardiac critical-care nurse who'd never worked in research. She brought a wealth of knowledge to her animal-protocol reviews, and, after decades of working with exalted surgeons and physicians, she had a fearlessness in calling them out on any posturing.

One huge challenge: What place should animal protectionists and advocates have on a research animal ethics committee? What place for sharp critics of the usefulness or the possibility of justification of animal testing? In Australia I've talked to lab vets about the expectation to appoint representatives (or nominees) of local humane societies, the sorts of people running animal shelters rather than protesting at the lab's doors. They are not often successful convincing such people to join. I tried getting the director of our local Society for the Prevention of Cruelty to Animals to consider a place with Cornell's ethics committee. She knew that the SPCA was the place people expected to bring their complaints when they drove past our horses and cows and, for one short-term project, sled dogs, penned outside through the Ithaca winter. "What would I tell them," she asked me, "that I was on the committee that approved this?" Yet what better person to have on the committee to represent a community's interest that scientists not run roughshod over animals' interests? And if the local SPCA itself will not join the committee, should we hold a space, and will they help with this, for someone the SPCA nominates and appoints? Our first community representative was a perfectly nice and diligent insurance salesperson, a friend of our vice-chancellor. He asked good questions, but how I'd have loved a real animal expert to sit alongside him.

Even more challenging, is there a role for representatives of the more strident animal liberation and animal rights groups? Trust is difficult, especially if labs invite animal activists in to sit on committees, review protocols, and come behind the science to join the committee's inspections. A local SPCA director will face allegations of colluding with enemies. Scientists will fear that activists are looking for that "gotcha" with which they can run to the press. More, they'll fear the threats of violence and of campaigns to persecute them into quitting science. Can activists commit to condemning violence against scientists? Can activists and the institution agree on how much of the ethics committee's business will be broadcast or kept in confidence? I recognize that including such activists will require careful scrutiny, but yes, some can. Behind the strong posturing, groups like PETA employ plenty of trained scientists and experienced staff who know how to work incrementally for progress toward their organization's proudly proclaimed vision that animal testing breathe its last breath. But isn't that the long-term goal all of us in animal labs should want?

While borrowing other countries' practices and aiming for some of my own, I'm also intrigued by how human subjects committees work to safeguard vulnerable study participants, be they pregnant women, prison inmates, or children. NIH rules require appointing a special advocate for research with children who are wards of the state and have no parent or guardian to protect their interests.[21] Animal rights lawyers have sued for guardianship of apes and elephants, and, round the world, countries are recognizing rights for natural entities such as rivers (teeming with sentient animals but not sentient in their own right) and appointing guardians to speak for what a river or mountain "wants."[22] Surely mice and monkeys are vulnerable subjects and sentient in a way a river is not. Alas, the most important stakeholders sit in cages while committees deliberate, as we have no shared language that could give them a voice, nor even a clear way for human guardians to advocate their interests. This may be the closest they get to representation, as we presently have no way of giving them their own voice on an ethics committee.

My dream team goes far beyond the NIH-required five-member committee (the USDA and the Animal Welfare Act require only three members!). On my campus we had a committee of some twenty-five members; my proposal here does not require more committee members so much as different members. I question how much work we've put on ethics committees by pushing so many decisions, like how thirsty is too thirsty, down to local reviews in the absence of clearer national-level standards. We should set our current *Guide* and regulations to the side and start fresh. It's time to get more specific about standards, to work with the complexity in animal-welfare science, and to understand that how we interpret that science and the rules we write following its evidence are ethical issues that demand a range of stakeholders.

In creating animal ethics committees in 1985, Congress demanded community representation at the level of institutional committees. Meanwhile, the USDA and NIH periodically update their rules and post updates on how they plan to interpret them, all without any significant citizen input. The *Guide* is the closest we get to a national ethical standard on lab animal care, but it comprises research insiders—most of them quite good people, some of them true animal-welfare experts—almost exclusively in its authoring committees. The National Academies of Science, Engineering, and Medicine's Institute for Laboratory Animal Research produces the *Guide* in its various editions, as well as numerous other reports on animal testing, but has never included broad representation on the *Guide* authoring committees. Since the 2011 *Guide*, the NASEM and NIH have called on nonscientists for its review of chimps in labs and on animal rights–organization representatives for reviews of dogs in Veterans Administration labs and in reviews for the National Toxicology Program. Lesson learned: animal advocates can work with welfare scientists, vets, and others in developing national standards for laboratory animal care and use. They can bring not only a broader range of values more representative of the country's populace but also insights on how what we know about animals should shape how we commit to treating them.

I want science and medical progress to continue, but in addition to a far higher standard of care for whatever animals may yet be necessary in labs, I want sufficient community support to let scientists do their work. That may require more (or better) external oversight rather than less. Dare I go further without alienating a research community that already feels over-regulated? Sure I will. If animal welfare is a community priority and responsibility, why not have city or county governing boards overseeing private and academic animal labs, as we have school boards, citizen police commissions, and zoning boards. Certainly, polarized school board meetings can heat up, and so will animal forums. Animal advocates, from the local SPCA to strong animal rights proponents, should have a place on these boards or in public forums.

Some years back Andrew Rowan, at Tufts University's Center for Animals and Public Policy and later at the Humane Society of the United States, joined with other science-trained animal-welfare proponents to spark dialogue between scientists and welfare advocates. Folks could mingle at conferences hosted by the Scientists Center for Animal Welfare, and animal-protection organizations could attend our lab animal science meetings. In the wake of the 1985 Animal Welfare Act and the creation of institutional ethics committees, scientists and welfare advocates seemed willing to find common ground. That's mostly broken down now, with little cross talk in the United States between reform-minded scientists and incrementalist animal-welfare advocates. People have retreated into their respective corners, fearing and vilifying one another as they dominate the discourse that people in the middle hear. Yet many people in science labs and in animal-advocacy organizations actually do want to set aside the rancor and find ways to fund research to replace more and more animals, trim experiments of low promise, and improve the welfare of those animals who remain in the labs.

Looking for inspiration and ideas to bring back to the United States, I traveled to a conference in London in 2022. I had a delightfully friendly dinner with representatives of the RSPCA and of the main UK research-advocacy group, Understanding Animal Research. Some fifteen years ago, people in the United Kingdom took stock of how animal rights vandalism and threats of violence had subsided but also, to the researchers' concern, how public support for animal research was trending down in polls. Understanding Animal Research spearheaded efforts to birth a Concordat on Openness in Animal Research, with input from the RSPCA and other stakeholders, to which numerous labs in the United Kingdom have signed on. Institutions that join the Concordat commit to increasing their transparency about their animal use.[23] The effort could lead to alliances with moderate animal-protection organizations, who, if they get a voice on animal use, might insulate lab scientists from the more strident badgering of animal rights activists. Moreover, when people feel researchers make the effort to show that their animal labs produce important data while making efforts to take good care of their animals, public support for animal experimentation increases. Member institutions produce annual reports for the Concordat, heralding whatever improvements they've made for animal care in the past year, as well as what research achievements they've accomplished. They commit to maintaining a public-facing website that documents how many animals they use and what problems they've had over the past year. Along with any self-congratulatory statements about how well they treat their animals, they commit to posting information about the limitations of animal labs. As my dinner mates caught up with one another, they explained to me that a progressive move like their Concordat really required buy-in from multiple actors, from protectionists to research advocates, so why not approach it collegially, listening rather than shouting?

Since well before the launch of the Concordat, the RSPCA has worked with scientists to develop best practices for various models in animal research, such as using animals in sepsis and septic shock research.[24] Complementing the Concordat's voluntary efforts at increased transparency, the UK Home Office, which licenses and inspects British research institutions, publishes nontechnical summaries of all the animal licenses it approves, though with limited transparency, revealing neither scientists' names nor even their institution.[25] We have nothing comparable in the United States. Imagine being able to look online to see what animal experiments a local college or company has in progress or to use these approval summaries for some homework on what publications are resulting from the experiments. Imagine seeing more than the blandest assurances that we follow all the rules or that we are accredited, if a college or company publicly admits to having animals in its labs.

I do not want to overromanticize the progress I see in the United Kingdom and the European Union, where plenty of enmity still lurks between science advocates and animal advocates. I do not want to disparage industry groups like the North American 3Rs Collaborative, working to advance lab animal welfare, though I would encourage them to welcome animal-protection groups as collaborators. And I likewise applaud the Animal Research Development Fund, an offshoot of the American Anti-vivisection Society, which enlists scientist and veterinary advisers to distribute research grants for developing nonanimal tests. Still, the British efforts should inspire us in the United States to do something different, to set aside enmity and celebrate diversity in the quest for something better for both animals and science.

Despite loud voices at the opposite poles of animal-research policy, many people believe, as I do, that animal research remains a necessary path to knowledge we want and that animals deserve better care. We know dialogue and cooperation could be powerful, but fear spurs us to label even peaceful protesters as "extremists," raise the barricades, and clamp down on the flow of information. As we draw the shades on what we are doing, we feed the temptation for outsiders to assume the worst and for animal-protection organizations to spread fear and anger among their membership, with questionable facts and inflammatory rhetoric, in which every scientist is cruel and every experiment pointless torture. In this environment reformist voices that accept scientists' claims that animal labs are still vital but who nonetheless want to push for higher standards of justification and animal welfare find themselves shut out. It need not be this way.

Throughout these chapters I've made my case that animal experiments are more useful than animal rights proponents would have you believe but less so than research advocates insist. Likewise, animals suffer more in labs than the research advocates claim, while the scenes of sadistic scientists torturing animals are similarly deceptive exaggerations. Scientists can bring us important knowledge for promoting human, animal, and environmental health, but not every experiment they envision is productive enough to warrant harming sentient animals.

Gorilla

Gorilla gorilla

BACK TO THE ZOO, SEARCHING FOR
A MORE HUMANE FUTURE

As I started work on this book, I returned to the Boston Zoo, where I launched my professional life with animals, for a reunion of zookeepers from my teen years. I was studying animal law at the time while living in Boston for a few months, and I reenacted my bike commute to work from fifty years earlier. Some old friends and I sat under an oak tree, debating which animals had lived beneath that tree so many years earlier. I said otters, but a friend remembered acorns hitting him on the head as he shoveled our dairy cow's manure. Who could dispute such a vivid memory? We watched a slideshow from the old days and shouted out the names of people and animals we had known. Only Aggie, our baby African elephant, could remotely have still been alive fifty years on from our time together, but one of the current zoo vets did some tracing for us and reported back the sad news: Aggie had died in her thirties in another zoo. I privately thought of what I'd heard of Winston, that as a young chimp getting too big for our small zoo, he'd found himself in a research lab—like me, but on the other side of the bars.

We toured the renovated zoo and enjoyed lunch with the current vet team and the gorilla caregivers. The kind of violent raid that had brought Winston and the other chimps of my zoo stint from the jungles of Africa had also brought the parents and grandparents of the Boston zoo's gorillas there. The current inhabitants, however, were all born in the United States and never knew the pleasures and challenges of life in the wild or suffered the violence that their forebears, and my Winston, had. The caregivers' gorilla love was strong, and they told us of the newest baby's antics and the time Little Joe climbed the enclosure walls, escaping down Blue Hill Avenue. Their baby gorilla did not wear diapers. The keepers who loved him so fiercely did not cuddle and kiss him. They did not pull him from his family for visitors to coo at the sight of zookeepers feeding him a bottle. Theirs was a hands-off love. They left his family group intact and let them give the baby his loving.

The Children's Zoo where I worked in 1971 was a short walk from the lion house that had been shuttered the year before. "We don't believe animals should be caged this way," a huge banner on the vacated building read. "We're working on a new zoo." People's views of animals and zoos were evolving, and lions behind bars were passé. Vets in zoos, labs, and just about every corner of veterinary practice have developed our craft through the ensuing decades. We pay much better attention to animal pain, for instance, and, happily, we have much better medicines available to help us help them. In zoos and in labs, we have come to see that animal health is more than just the physical health of the body, that animals have rich mental lives that deserve our respect.

The 1985 Animal Welfare Act called on both zoos and labs to mind the psychological well-being of their primates, a legal requirement that boosted an expanded ethical commitment to the other animals in our care. Many zoo goers and zookeepers felt a growing unease about the iron bars of old-school zoos. Lions pacing behind bars, shackled elephants rocking back and forth, and chimps and gorillas listlessly sitting in tiny cages provided evidence of zoo cruelty. The Animal Welfare Act's legal requirements for monkeys' and apes' mental health gave a term—*environmental enrichment*—to what zookeepers knew all along they should offer to all their animals. Score one for my friends and me biking around hunting tadpoles for our turtles' predatory inclinations.

Zoos have taken up the call to heed the mental well-being of their animals, far beyond the physical health of their bodies. They work to keep families together. They train their animals to cooperate with health exams, breaking up their day and making a visitor event out of the process, with posted times when a tiger will come to the grated fence and stand tall on hind legs while a vet technician with a stethoscope auscultates the heart. Whatever obstacles zoos and aquariums may face, I envy them. In the labs we get little chance to keep our animals in intact family groups, and we rarely take the time to train our animals for better lives in our hands. We cycle through animals so quickly. We use them in experiments and then dispatch them, maintaining large numbers that challenge any opportunity for human-animal bonding.

Conscientious vets and researchers in the 1960s managed vivaria in a country with no legal protections for the animals in labs. They launched an accreditation program, now called the Association for Assessment and Accreditation of Laboratory Animal Care International (AAALAC), to raise standards they and other labs would follow, with the side hope that their efforts at self-regulation would keep rigid laws at bay. A decade later the young Animal Welfare Act was regulating animal labs and zoos, but with the barest of requirements, such as the nebulous performance standard that enclosures be large enough for animals to "make normal postural adjustments."[1] The Association of Zoos and Aquariums decided that it was not sufficient to ensure a skeptical public that zoos were turning their backs on yesteryears' cruel confinement and launched its own accreditation program. Though I push for higher standards than what our current lab animal laws require, the AAALAC presently avoids going too far beyond the lab animal regulations—though it could. Legal requirements for zoos are far laxer than what labs face, yet the Association of Zoos and Aquariums shoots to go far beyond what the Animal Welfare Act mandates.

Zoo and aquarium professionals have set themselves a challenge, through their national and international accrediting standards: conduct annual welfare audits for all the animals in their possession, right down to the anemones and minnows in their largest tanks. As I started my work on this book in 2021, I made the rounds of some zoos and aquariums to ask how folks were meeting this new commitment. They didn't know all the details for sure but were eager to take up the challenge they had set for themselves with their accreditation processes. What a great model for us in the labs to follow. I'm adding that to my list of changes I want to see, starting now, in how we care for the animals in our labs. I am heartened by the number of scientists working to find nonanimal replacements in labs and finding ways to humanely refine animal experiments in their specialties. I applaud the various Three Rs organizations around the world for developing and promoting better animal care. I'm cautiously optimistic that technical innovations and artificial intelligence can improve animal monitoring to detect pain and suffering, while also leading to yet more ways to lower our reliance on animal studies for medical progress. Yet waiting for every lab and scientist to embrace the urgency to get animals out of labs and to treat the ones who remain better has frustrated me throughout my career. I have seen top-down laws make a difference, from instituting ethics committee oversight of animal experiments to mandating attention to animals' mental health and well-being. Our laws and their enforcement are due for an update, one I am skeptical that we will see. Scientists, vets, ethics committees, animal-welfare specialists, and animal activists, if allowed and willing to participate, can achieve much without waiting for our elected leaders to follow our lead.

Harming animals requires justification. In labs we currently exclude too many animals and ignore some clear harms in our harm-reduction efforts. Our regulatory requirements take the form of vague performance criteria, left to the jurisdiction of institutional self-regulation. As for justification, scientific experts score the value of proposed experiments and ethics committees focus on refining painful experiments, but with little cross-talk between the two. The tepid ethical harm-benefit analysis this produces does not give the animals what they deserve, given how much we take from them.

Throughout this book I've expressed my belief that, for all the progress scientists are making replacing animals in labs, the animals remain necessary if we want our current level of progress to continue. I've accepted the species bias most of us humans share, that potential benefits for present and future human patients can justify harming animals. I've argued that our current practices fail to give animals what many of us believe they deserve. I've asserted that we can go further in reducing the harm they experience and in restricting animal testing only to labs that can deliver high-quality data in pursuit of important research questions.

I believe that animals suffer less in labs than the continuous torture that animal activists allege and more in labs than the happy pronouncements made by research defenders. I find myself in the middle, too, about the usefulness of animal research: more than the naysayers claim but not as bountiful as scientists' lobbyists insist. From this middle ground, I have gathered my own and others' ideas for how to better the lives of the millions of animals still in labs while maintaining or even improving the science we derive from studying them so that we humans may live better lives.

And thus I offer my proposals to make the world a bit more humane for the animals in labs, my manifesto. Some of my proposals scientists and vets can adopt on their own. Some are for institutions, the universities and companies who run animal experiments, and the organization that accredits their animal programs. Knowing institutions will not change voluntarily if change could cost money or slow some research projects, we need changes in our governing laws, adopting some of the best from other countries' practices. Lab outsiders are crucial; they led to the first lab animal–welfare protections in England in the 1800s, and outside agitators and campaigns have been successful since then. Though acts of violence against people are exceedingly rare, all of us in the labs fear the more strident and destructive tactics, from shaming and intimidation campaigns to vandalism and threats of violence to scientists and their families. But not every bit of outsider inquiry and involvement is something we need fear.

I've discussed in prior chapters various barriers to reform. Scientists often fear that moves that might help their animals feel better could hurt the value of their experiments. Fair enough, but often those effects are too weak to worry about, or they actually lead to the contrary. In most situations happier, healthier animals make for better science. More than anything, money is the barrier to animal welfare, and some of my reforms, especially phasing out unfriendly animal housing, would cost money. I am not so naive as to believe federal funds for better animal welfare, possibly not even National Institutes of Health (NIH) funds to update the hopelessly outdated *Guide for the Care and Use of Laboratory Animals*, will come riding over the horizon any time soon. Far from a slush fund, universities' overhead and indirect costs that they receive with a scientist's grants are the funding for vet salaries and enrichment programs. When the federal government cuts research dollars, expect animal-welfare cuts as scientists cling to their ability to run experiments. That said, my former employer, the University of California–San Francisco (UCSF), has a

$10 billion annual operating budget. Harvard sits on a $50 billion endowment. Large drug companies take in upward of $50 billion annually. Even as the US disinvests in science, they can spend some tiny portion of these mounds of gold as some small recompense to the animals who make involuntary sacrifices so that we may live healthier lives.

1. Let All Sentient Animals Be Legal Animals

We have had animal ethics committees in most US labs for more than forty years, enough time to see the limits of the honor system that self-regulation entails. We need good laws, not just good intentions. Our laws and policies should protect all sentient animals, those creatures with sufficient mental capacity to be capable of suffering. I have explained that some animal labs receive no third-party accreditation or legal oversight and that, in labs that do, the Animal Welfare Act defines most animals away as "not animals." The overwhelming majority of lab animals are in the hands of scientists who know they will never, ever be subject to an unannounced inspection by the US Department of Agriculture (USDA). Informally, mice and rats' invisibility allows ethics committees to treat them as third-class citizens. Mice, rats, birds, and octopuses are all unequivocally sentient and deserve legal protections. Borderline animals like hermit crabs may not yet be due full legal protections, though we certainly should treat them with respect. Fleas, I'm afraid, will have to wait while we get caught up with moving highly sentient creatures into our zone of protection.

A caveat: while many or most labs include mice and rats in some sort of self-policing committee system, my belief that mice deserve the attention of Animal Welfare Act inspectors requires action from the same US Congress that voted in 2002 to permanently proclaim rats and mice not animals. I'm not sanguine about that, but the NIH could increase its own oversight and data collection on mice and rats for the subset of labs under its coverage. I don't know what resources the NIH Office of Laboratory Animal Welfare puts into responding to Freedom of Information requests, but, if it posted information from its annual reports online for all to see, it could spare itself those efforts. State and local governments could also partly address this lapse.

2. Enforce Animal-Welfare Standards Through One Law Only

Another task for Congress: move the best of the NIH lab animal–welfare oversight provisions—the broader range of animal species and the self-reporting mechanism for welfare failures—to the USDA. Then make the Animal Welfare Act, with its cadre of USDA veterinary inspectors, our only federal lab animal–welfare law. A separate law at the NIH that covers only animals that the government funds is superfluous, and the NIH office overseeing that separate law is redundant if the USDA and the Animal Welfare Act covered all animals. A law is necessary; red tape is optional and needs to go away. All sentient animals in all US labs deserve Animal Welfare Act protections, regardless of funding.

With a single, rigorously enforced animal-welfare law, independent accreditation through the AAALAC process would likely be superfluous. Alternatively, the AAALAC could remain relevant were it to develop standards that go beyond the Animal Welfare Act and verify that accredited labs truly do achieve animal care superior to the pitiful minimum standards of the law. The Association of Zoos and Aquariums could serve as a role model for setting and enforcing industry standards beyond what any law requires, if only for its new program of annual welfare audits in institutions that want to make the grade.

3. Enrich Animal Lives as the Standard of Care, Not an Optional Nicety

I've visited parks where zoo goers can contribute to an animal-enrichment fund. Is it so easily axed from the budget if zoo visitors don't ante up? I've surveyed labs' websites that promote basic nutrition supplements as "enrichments." Enrichment implies some extra little nicety we bestow on our animals when it's not too inconvenient or expensive. Enrichment implies something we add on top of lives that already give animals all they could need, that already meet the meager standards of the *Guide* and the Animal Welfare Act. Let's get rid of that word and get serious about animals' welfare.

We need to shift how we conceptualize animal welfare for confined animals. We need to recognize that virtually all confinement is a form of deprivation. Lab dogs may never have chased a ball on a beach. Lab monkeys may never have climbed a tree or joined the bustling social life of a troop. Woodchucks born in a cage have never sat in the sun outside the burrow they dug with their own four paws. These animals may not have a clear vision of what they are missing, but we, their keepers, do. When we weigh animal harm against the hoped-for benefits of research, these deprivations need to count.

Dogs racing on the beach show us unequivocally what they are capable of enjoying. Cats have different ideas of fun. For whatever species, animal-welfare science guides us toward meeting animals' needs and desires. Do captive woodchucks have a strong need to run down into a burrow? A desire to sit up and scan for predators? Do they want a burrow of their own, or would they rather pile all in together? We have ways of asking them what they prefer and how strong that preference is. Those strong preferences should inform what we consider the basic standard of care, not as an occasional enrichment opportunity but as daily care.

Animals deserve a rich environment, not a bare cell that we occasionally "enrich." Our current rule books, the Animal Welfare Act regulations and the *Guide for the Care and Use of Laboratory Animals*, allow for small, barren steel or plastic cages that are easily scrubbed and sanitized but that deprive animals the opportunity for digging, climbing, nesting, socializing, or getting some distance from other animals, all of which may be vital to their sense of well-being. It's time to send these cages to the trash heap and replace them with animal homes that allow us to fulfill more than the most basic of animal needs. And always the warning: this will cost tens of millions of dollars or more, so it's best to prioritize the worst of the worse, with sound data on what would serve the animals better.

4. Rethink Death as a Harm

Death deprives animals of the joys and sorrows that could await them, and, assuming they remain in human custody, their humans should strive to make the joys far outweigh any sorrows. Under our current practices and rules, killing animals is our obligation if they are suffering but otherwise it is considered value-neutral. If we need tissues or simply have no further use for the animal, we kill them as our default course of action. This makes sense in a country (like most) where a huge majority believe that killing animals for meat is ethical if the animals are fairly well cared for in their short lives. Monkey and ape retirement sanctuaries and dog and cat adoption laws reflect an alternate ethic, that at least some animals, maybe especially those from whom we've extracted whatever sacrifices our experiments entail, merit a different fate. As scientists weigh the harm and benefits of a proposed experiment, killing their lab animals, no matter how painlessly, should count in the harm column.

5. Unmask Invisible Pain

Too many animals suffer from too much pain in the laboratories. We know from living with dogs and cats that animal pain is often devilishly difficult to detect. In the labs we see most of our animals but a few minutes each day and almost never get to know them as individuals. Vets serve as the point people for pain management for lab animals and must not only promote the most effective pain treatments available for the species they care for but also adhere to the principle that we must presume that what we know to be painful to people is also painful to animals, unless we have strong evidence to the contrary. Administering analgesics for animal pain "as needed" rather than medicating preemptively before seeing (or often missing) signs of pain invariably undertreats that pain. Animals suffer, while their untreated pain undermines quality science and contributes to the crisis of reproducibility.

Scientists sometimes fear that pain medicines will ruin their animal experiments, disrupting immune function, cognitive function, cancer biology, and other aspects of the animal body. And, unfortunately, sometimes they do. Before any committee approves a scientist to run painful experiments without treating the pain, they and the scientist must also consider how untreated pain will similarly disrupt physical and mental function, to the detriment of both the animal and the data. Scientists and ethics committees must compare research data when pain medicines are deployed versus when the animal is left with untreated pain. If results differ, they must consider whether the animals with their medicines might in fact be the truer representative model of human biology. If so, or if the differences are trivial, animals deserve the benefit of the doubt and their pain meds.

Better yet: avoid performing painful procedures at all!

6. Be Transparent About the Value and the Limits of Animal Experiments

From childhood vaccines to later-life medical care, all of us have been touched by advances for which animals have been put in service. I run on my titanium knees. If my mother's macular degeneration that so blinded her late in life should strike me, I am grateful that I can tap into treatments that only came along later. Diabetes and HIV infections were once certain death sentences, but my friends and family now live with these conditions, while we hope for a permanent cure. I would call the lifesaving advances in cancer care modern miracles, but they are not miracles. They are the hard-earned result of many scientists' efforts, plus the involuntary service of countless animals. Look around in your own and your loved one's lives and see how medical breakthroughs have helped or might help you. Get your vaccines and thank the animals. I am sure they were in fact necessary for most of the medical advances we now enjoy. I don't expect my readers to know the details, but I do wish their doctors and nurses understood better how much we rely on animals. Around the globe some research centers sponsor a gratitude event for the lab animals they have relied on in the prior year. Alone that does nothing for the animals who have passed. But it does give clinical practitioners insight into where the medicines they prescribe have originated and may inspire lab workers to recommit to taking better care of their animal subjects.

The Concordat on Openness in Animal Research in the United Kingdom encourages research institutions to maintain public-facing websites about their animal programs. They should be able to post the good that has come from their animal labs, but they are also challenged to post the limitations and the downsides of animal testing as well. A student writing a term paper on animal testing should be able to find some honest information from labs in their area and not just the hyperbole from two poles of this polarized topic. They should be able to contact a university or company and get substance beyond the boilerplate assurances that "we follow all applicable regulations." I want a next generation of scientists who decide that good people can work with lab animals if and only if they treat the privilege as a sacred trust.

7. Teach Young Animal Researchers to See Their Animals Holistically

A monkey is more than a disembodied head full of neurons. A mouse is not just immune cells in the bloodstream. When scientists see their animals as whole creatures, thinking and feeling, with many moving parts that all affect one another, they can design better experiments. They can see how pain and stressors affect how animals' neurons and immune cells perform in an experiment. They can set aside any notion that ideal animals exist, which they could study if they could just eliminate all the environmental variability in the animals' lives. They need to examine the notion that if they could shrink their worlds down to a simple bare box with no social interactions and no chance to perform any behaviors whatsoever, that they are somehow a purer research subject. They could see that, instead of studying normal biology in such restricted environments, they are studying the pathology of sensory deprivation, frustration, and overall lack of stimulation. When scientists see that impoverished, developmentally stunted animals are poor research subjects, they will want better lives for their subjects for the scientist's own interest in better, more translatable data and, value added, for their animals' emotional well-being.

Young scientists in the United States receive on-the-job skills development, ranging from how to make babies from a female and male mouse to how to place electrodes in a monkey's motor cortex. Full-on coursework in laboratory animal biology and behavior, in how to design an animal experiment, in how the animal body responds to infection or social interaction or opportunities for exercise are rare in the United States. Before scientists tunnel down to the neuron or immune cell that will yield them their dissertation, they deserve a chance to learn about whole bodies, how animals sometimes do and sometimes do not yield data that translates to human health and illness. Scientists can run better experiments when they see their animals are not just stimulus-response machines or furry test tubes full of cells and chemicals. Seeing animals as more than the sum of their parts can improve experiments and should encourage young researchers to care how their experiments might feel to the animals.

8. Train Young Scientists to See Animals as Worthy of Ethical Treatment

The NIH requires that all trainee scientists receive training in "responsible conduct of research," with at least eight hours of instruction every four years.[2] Labs submit their syllabus for NIH approval and are free to exceed the eight-hour minimum, which is good, given they must cover plagiarism, conflicts of interest, lab safety, grant management, and, yes, use of human and animal research subjects. This allows time, an hour or two at the maximum, to give some basics of the laws for animal experiments, alternatives in animal testing, and the Three Rs of replacement, reduction, and refinement. Every medical scientist uses knowledge from animal labs in their research and so should get at least this much information, but, for those who will actually do animal experiments, they need more.

Students who will one day be running their own lab should engage all the issues I've covered in this book. They should learn how to evaluate the merit and necessity of an experiment and how animal pain and distress can disrupt a research project. They should understand that ethical justification means more than technical merit but includes values in the harm-benefit analysis. They should explore their own values in the class and engage with others who hold different values, who might be skeptical that their research holds as much worth as they believe it does. If such training for a lifetime's work takes more than fifteen hours, I would be surprised, but it can work only if they go back into a lab where the head scientist and the lab culture see animal use as a responsibility they must not take lightly.

9. Give Lab Vets Ethics and Welfare Training Too

Lab vets typically wield the scepter as the campus animal expert, ruling on matters of veterinary medicine, animal welfare, regulatory compliance, zoonotic diseases, and more. In all my years on ethics committees, never did we successfully recruit a campus ethics professor to join us, so we vets become the de facto animal ethicists, though so few have had the chances I had to study this in any formal way. Often vets are scientists' collaborators in planning an experiment and assuring the ethics committee that it meets standards and merits approval. Often, too, they manage the vivarium, with a responsibility to contain animal-care costs so research dollars go to staffing and equipment without diverting too much to animal care. Sometimes they are the voice against approving a scientist's plans, stepping up as the animals' best ally. The potential conflicts of interest abound.

We vets also face a professional conflict when our training in veterinary medicine and infection control makes us want to curb animal happiness in favor of sanitation and safety. We are the ones who developed the small sterilizable cages our animals live in, not trusting that a pig in mud or a dog running on a lawn could be as sanitary for an experiment as an animal on a steel grid floor, sanitized to the 180 degrees Fahrenheit our rules require. We distrust our behaviorists when they push us to risk two monkeys brawling as they work out who's boss and then settle into peaceful cohabitation. We vets will have to stitch up the loser in a monkey dominance fight, so why risk it? Why? Because a good monkey whisperer knows how to set the conditions to guide the animals through an introduction and let them share some companionship for months to come.

Vets need training and support for the ethically fraught leadership role we undertake in animal labs. We need to recognize our biases that place physical, bodily health over mental well-being and happiness. We need to understand that not every little medical fact a scientist might discover meets societal standards for ethically justifying harming innocent animals. Like trainee scientists, we will not develop as ethical practitioners in a bubble but instead should welcome chances to engage with people who see the world differently than we do.

10. Empower Animal Behaviorists and Welfare Specialists in the Lab

Since I first got into animal biz, animal welfare has professionalized. Visionary animal behaviorists plus the occasional vet created the science of measuring animal welfare and, by extension, the science of measuring, with all due caution and humility, how our efforts hurt animals from their own perspective. No research institution today runs an animal lab without a credentialed vet with authority over animal health. We now have a range of professionally trained and credentialed animal-welfare experts who should similarly be part of every animal lab and every animal ethics committee. Behaviorists often see animal behaviors, such as mice neurotically, stereotypically chewing up their chow without eating it, that we vets either do not see or do not take to be abnormal. They know how to go beyond vets' and scientists' quick look at the animals and glib assurances all is well, putting video recorders, for instance, on monkeys to see just how nervous they are when they hear scientists approaching from down the hall. And the good ones, the ones I've been honored to call colleagues, speak up for the animals and advocate for their welfare. Too bad such folks rarely have much authority in the animal lab. All forms of self-regulation are replete with potential conflicts of interest, but I must insist that welfare specialists should not be under the rule of vets or other administrators whose performance reviews score how well they trim animal-care costs. If colleges and companies have money enough to run animal labs, they have money to treat their animals better.

11. Send Our Rule Books to the Shredder and Start Fresh

The *Guide for the Care and Use of Laboratory Animals* serves in the United States as the standard of care for NIH oversight and AAALAC accreditation. What a heavy responsibility on the teams of vets, scientists, and others who periodically update it! Unfortunately, even in its eighth edition, published in 2011, it's a muddled mess. In part it's a user's manual for animal subjects just as lab equipment requires, with information on biology and physical needs of laboratory animals' bodies. In part it promotes a standard of humane animal care, which it defines as treating animals to "high ethical and scientific standards." To the extent it defines *ethics*, it seems mostly to limit itself, and therefore to limit our obligations, to the harm-reduction principles of the Three Rs. It claims that "members of the public should feel assured that adherence to the *Guide* will ensure humane care and use of laboratory animals."[3] Skeptic that I am, I am not so assured.

The *Guide*'s authors, reviewers, and sponsoring bodies skew very heavily toward animal-research insiders, though with some input from animal-welfare scientists. Progress comes very slowly to the *Guide*'s updates, including the present edition's addition that ethics committees are "obliged to weigh the objectives of the study against potential animal welfare concerns," a small step toward moving these committees from their technical focus to a genuine ethics-review group.[4] Nonetheless, it continues its previous editions' allowance of tiny lab cages and dodges giving guidance on how a committee should act when it matches up weak study objectives against real and present animal-welfare problems. Though claiming to be evidence based, it calls for an ethics committee with the same composition as what the NIH rules require, with at least one community representative and at least one nonscientist. It offers no evidence that one of each of those members is either necessary or sufficient for a quality review, especially given that it makes no call to limit the number of research insiders who could drown out other voices.

One clear fact about our *Guide for the Care and Use of Laboratory Animals* and its sister document, the Animal Welfare Act regulations, is that the NIH and USDA have excluded a real diversity of human voices in setting our national standard of care for animals in labs. The 2011 edition is way past due for an update, but I predict the same exclusion in the next edition as well. The *Guide* could not possibly contain all the necessary information to guide scientists to generating high-quality, reproducible, translatable animal data. If it is to have a function, it is to set ethical standards for when and how to justify allowing animal harm, but for that role it needs to include multiple stakeholders and to recognize that every *must* and *should* sentence is an ethical statement of what we owe animals.

12. Validate Animal Alternatives and Animal Models

Hooray for scientists' and regulators' accelerating efforts to replace animals in safety testing and other lab tests with nonanimal alternatives: they don't just develop new tests but must then validate how faithfully they model how a human patient will respond to a new drug or treatment. Scientists have mostly replaced rabbits in eye-safety tests and in screening medical products for endotoxins, but only by demonstrating that nonanimal replacements serve eye safety as well as rabbits and horseshoe crabs. Scientists' successes validating nonanimal methods in safety and toxicity labs is great news.

Meanwhile, we have no such validation for hundreds of animal models in current use in more basic research. How faithfully does a marmoset monkey or a mouse model human multiple sclerosis? Faithfully enough to keep inducing the animal version of that disease in animals? Are there mouse cancers we can cure but fail to replicate our success in their human counterparts? If so, does this mean some mouse models in cancer biology lack the validity they need to merit research funds or allotted animals? If so, let's get rid of those.

Ethics committees cannot make these technical assessments, nor need we have every committee around the country trying to gather the expertise to do so. Earlier I described one expert panel's work to see if genetic studies in inflammation in mice work at predicting human responses to similar injuries any better than a coin toss. If they don't, scientists must not receive grant funding using those animal models. Expert committees should come together and put these animal models through the rigorous validation process we are using to test potential animal replacements. When they do, they risk alienating peers in their own subspeciality, so let me request that they include community representation and specialists from other fields to hold them accountable.

13. Fund and Disseminate Alternatives

In the United States, the federal government and powerhouse pharmaceutical companies have funded most of the medical research in our animal labs. Science costs money, and money is finite. Funding in animal labs will go to testing a cancer drug, but scientists also need support for developing better ways to study cancer drugs, better imaging equipment or protein-purification machines or other infrastructure needed to run cutting-edge experiments. Some of the money goes to new tests that, almost by coincidence, will replace animals, but targeted funding specifically to replace animals and validate the testing to regulators' satisfaction has always lagged. We need better government support as well as industry support. A petition in London some years back called (unsuccessfully) for a per-use tax, in which labs pay into a fund supporting alternatives research based on the numbers of animals they use in a given year.

Certainly, I want to lower the financial barriers to finding nonanimal tests to end our reliance on live lab animals.

Scientists need to know what nonanimal tests, as well as what welfare refinements, could make their studies more humane. In updating the Animal Welfare Act, Congress funded a USDA Animal Welfare Information Center, hoping it would help scientists search for alternatives in and to animal experiments. Circa 1985 it was a good effort, though never great. Earnest staffers generated bibliographies on topics such as managing pain in lab animals, bibliographies that ran several pages but, even at that, could not capture all the then cutting-edge information on the topic. Scientists need to know (and have access to) nonanimal technologies and need to understand statistical study-design strategies to reduce the animals they enroll as subjects and the myriad refinements to animal housing, handling, running experiments, and pain and distress. Standard database searches are not up to this task; no wonder my ethics committee barely glanced at scientists' descriptions of their alternatives searches. Artificial intelligence should bring improvement on this. And, as I was reminded as I tried researching for this book once UCSF had closed the door on me and my library access, scientists may find dozens of articles to review in planning their experiments, but many of the best lie hidden behind journals' paywalls. What to do? Decide that weeks of delay or fifty dollars per article cannot pay for themselves and stick with reading the free-access literature? Better artificial-intelligence–powered tools for literature searches and financial support for scientists aiming for the best for their animals are essential.

14. Transform Care and Use Committees into Ethics Committees

What we in the United States call Institutional Animal Care and Use Committees (IACUCs) should commit to serving as ethics committees. The difference: they must move beyond the harm-reduction principles of the Three Rs and ask, given that almost all animal experiments confine and harm animals, does this project deserve the privilege of using animals? This requires scoring the scientific merit of the project, as only well-designed, competently performed experiments that test plausible hypotheses deserve the animals they require. Currently, not all IACUCs even do that much, trusting that an NIH scientific peer-review group has reviewed all the separate experiments contained in a large grant application and deemed them meritorious. That error itself needs correction.

Ethical review goes even further, asking not just whether a scientist can competently answer a scientific question with the least possible harm to the animals. It asks, "Is this question important enough to justify harm?" In Europe and Australia, ethics committees have borne the obligation for several years now to do an ethical (animal) harm to (human) benefit assessment. Neither in publications, nor in conversations with folks in those countries, have I seen evidence that anyone is satisfied they have yet figured out how to do this. I talk to scientists, vets, and committee members who report the conversation with the researcher is productive, but, as for weeding out or limiting projects whose benefits don't appear to justify harming animals, few if any are actually accomplishing that.

Granting agencies and pharma executives somehow decide that some projects warrant the finite pool of available funds and others do not. We do not look at the social license to harm animals as a similarly finite resource, but we should. What if an ethics committee decided that only the top two-thirds of its applications will receive approval in a given month? Or what if the granting agencies managed research dollars like a mutual-fund portfolio, capping animal experiments at x percent of the funds it would disburse, with the rest going to human or other nonanimal tests? Our research laboratories should also be laboratories of experimentation for how we balance necessary medical progress against the harm that progress does to animals.

15. Add More Guards for the Henhouse

Current committees in the United States tend to have far, far more scientists than other members, and these committees of course all skew heavily toward people with a higher acceptance of animal use. In a pluralistic, democratic society, we should have far broader representation reflecting the range of values within our communities—and all the more so as we do not hear the animals' voices. I don't believe we need bigger committees than many institutions now have, just more diversity of knowledge, values, and expertise.

I say, let's bring in more outsiders, both nonscientists but also vets and scientists not on the institution's payroll. Let's bring in more insiders, the students and techs with daily close contact with the animals and the humanities or engineering professors who may look with suspicion at biologists' assertions of the value of their work. In most current ethics committees, the scientists submit a project, then wait weeks for an impersonal letter with the results of a closed-door meeting. If local review and oversight is so integral to our system, why not invite the scientists in for a dialogue where they might teach and learn?

No animal ethics committee should meet without expert animal-welfare specialists, but we should also invite animal advocates, ranging from shelter workers from the Society for the Prevention of Cruelty to Animals (SPCA) to hardcore antivivisectionists. My warning to the strong animal rights folks: yours is not a majority opinion in our country. Our country is no more ready to liberate all the lab animals than it is ready to turn its back on meat eating. Nonetheless, we lab insiders should hear your voice, as should other nonscientists and outsiders on the committee. If you are willing to accept incremental progress, your participation may move the needle, resulting in fewer bad experiments and better animal welfare.

16. Bring Vets and Vet Professionals to Ethics Committees

Vet students I've counseled confess the feelings I had to stare down in starting in this career. How to take the joy and satisfaction of healing much-loved pets and bring that skill and commitment to the labs, where few animals' lives will end well? How can you look in the mirror when you've voted to approve experiments that will harm and kill innocent animals? How can you examine lab dogs, listen to their heart or look into their eyes, then put them back into a cramped box of a cage? Maybe you love animals too much to consider a job in research. If yes, the labs need you. If that's not for you, you might meet me halfway. We had a private-practice vet on our UCSF ethics committee for several years. I hated him second-guessing my case management, but he knew his stuff and brought a fresh set of veterinary eyes to our animal program. Contact colleges or companies in your area and see if they have a role for you, on the staff or on the ethics committee.

17. Take Action to Help Animals

Do you care about animals? Consider how your daily life impacts them. Eat less meat or none at all. Volunteer in an animal shelter. Adopt an elderly, abandoned dog.

For readers who believe me that animal labs still produce important discoveries, who want animal interests to weigh more heavily in science, and who feel my frustration that our government shows little interest in coming to their aid, can I convince you to try to get involved? The USDA's website lists registered research labs (if their animals are legally "animals"), including those in your area. Pubmed, Google Scholar, and other search engines will show you if your local college (add its name as a search term) is publishing animal experiments. Private research labs in your area may be harder to find, but, if they are accredited, they may be listed on AAALAC's website. Contact the institutions. Tell them you've learned of their legal requirement for community representatives on their animal committee. Ask if there's a role for you. You may not get a warm welcome, especially if they think you're an animal rights activist trying to infiltrate and blow the whistle. There's one way to find out. If they let you in, you may be pleasantly surprised at the diligence of the committee, the vets, and the researchers. If you've got animal expertise, as a dog trainer or even as a pet guardian, you'll bring valuable knowledge—so too if you are a patient with a condition the campus scientists are studying. If you have no science expertise to claim, that's fine. Don't be shy to ask. I've seen times when the nonscientist on the committee questions what the researcher is trying to accomplish, only to find that we vets and scientists likewise didn't understand but were embarrassed to ask. I have seen many occasions where fresh eyes spark conversations that inspire the researchers and vets to do better by their animals while making the experiment better too. The animals cannot thank you, not in words you understand, at least. But you will know what a contribution you are making, for the animals and for science.

IN CONCLUSION

I've spent more than forty years in animal labs. How can I ever honor all the animals? Most never even got a name. I remember best some of the ones who found a life after the laboratory. Dolly was an indefatigable beagle whom no fence could ever contain. How many times did the SPCA find her on her various excursions and deliver her home again to the woman who'd adopted her? Eddie, the rhesus monkey, went to a sanctuary in Oklahoma, and, though he and Wyatt had never looked like a compatible pair in the lab's monkey room, they bonded and lived close to twenty years together in retirement, having, I hope, more fun than a barrel of monkeys. Others we cared for while knowing their lives would not outlast their role in the lab. Harriet was a big, strong rat whose lab role required him (yes, him) to sport a large array of electrodes on his head as he demonstrated his sharp memory navigating a maze in search of chocolate. Stretch was a cat with a rare collagen disorder whose skin tore easily. He could never have survived in his human family's home; they donated him to the vet college, where he lived several years, never siring the offspring the scientists needed to try to model this still-uncurable condition, Ehlers-Danlos syndrome, in children. I never saw most of the unnamed animals my vivarium housed or my ethics committee approved.

As the vet and compliance officer, I mostly just saw the animals if they needed vet care, or the experiments were going awry. And rarely did I see the fruits of the animals' lab lives. Many experiments produced no data at all. Others added their small bit to the cumulative knowledge in a field. Some projects led to human clinical trials and some successes, though that too was difficult to see, as once a treatment is in the pipeline, it's in the hands of a drug company for final testing and marketing, not a college campus. Our marmoset monkeys were instrumental in developing ocrelizumab, the multiple sclerosis treatment. Larry (no relation!), Goliath, and Octopus received stem-cell treatments in their brains for their parkinsonism, a monkey treatment that then went on for human studies. The lab they inhabited also used rats and monkeys to develop techniques to target cancer chemotherapies directly into a patient's brain tumor, a technique much more effective and less toxic than intravenous injections. Paraplegic mice contributed to a potential therapy (MMP-9 inhibitors) for dogs and humans with disc disease and spinal injuries. Always, on my job, there were just enough animals I could help and just enough exciting science coming from the labs that I stayed on the inside, walking past the occasional protesters locked out of our labs and naturally imagining the worst for our hidden animals.

I am out of the animal labs now. I'm grateful for what the scientists have accomplished and grateful for my colleagues' commitment to the animals. I've been vexed by indifferent scientists and vets. I've been frustrated with myself that I do not know any discipline in enough depth to assure a reader that the animal experiments really and truly are necessary for the scientific progress we are looking for. I've worked hard to improve animal care on the campuses where I worked and to push my colleagues around the country to push for higher standards of animal welfare in the labs. This book is my parting contribution to this field, both my effort to let lab outsiders get a rare look inside the animal labs and to push the needle of progress as I hand over this work to another generation.

Acknowledgments

In 2019, never dreaming a pandemic was bearing down on the world, my husband, David Takacs, urged me to come to Australia with him. We went not just to watch the wallabies; he taught and studied biodiversity law as a visiting professor, and I launched this, my second book. I reluctantly left my work and my colleagues in my university's lab animal–welfare program and headed off to Tasmania. I said goodbye to some great scientists, many dedicated animal caregivers, and a few hundred thousand animals. Being a lab vet is emotionally and ethically exhausting, and I felt such relief to step away from the day-to-day animal care—but not entirely, as I still had much to say about the sad reality that we cannot have the medical progress we want without some continued use of lab animals and the equally sad reality that our treatment of them falls so far short of what it should be.

From emu watching in the Outback to our wide-ranging conversations on ethics and animals and the natural world, David has been my companion, my muse, my editor, my travel buddy, and my unflagging support. Every word in this book, every punctuation mark, has benefited from his wisdom and patience.

As COVID lockdowns passed, I grabbed the opportunity for a visiting fellowship at the Brooks McCormick Jr. Animal Law and Policy Program at Harvard Law School. My thanks go to the staff and fellows of the program for encouraging me to weave my own stories into what I want to show of the hidden lab animals on whom we rely and from whom we take so much.

In these pages I cover the issues in laboratory animal care and research that I know intimately from my decades of practice. On other fronts, safety and toxicity testing in particular, I am more of an outsider. That broad coverage has required tapping ever so many scientists, animal-welfare activists, animal behaviorists, and vets for fact checks and insights. This broad range of experts could not possibly all agree with what I present here, as they do not all agree with one another, not on the facts of animal research and not on what we should do with those facts. I hope they can all see my effort to honor their perspective and present it fairly. Pulling up fifty years of memories of animals I've known has been a fun journey. My thanks to folks I've known through those years who've gently corrected my recollection of our shared stories and forgiveness for any words I've put in your mouths in retelling them.

For their time in writing or meeting with me to answer my questions, challenge my assumptions, or otherwise help with this effort, I thank (with an ever-looming fear of missing people in my public gratitude) Jamie Austin, Melissa Bain, Eric Baitchman, Jaco Bakker, Allen Basbaum, Marc Bekoff, Chris Bonar, Thom Boyce, Linda Brovarney, Gordon Burghardt, Kristian Cantens, Don Casebolt, Alka Chandna, Leigh Clayton, Amy Clippinger, Douglas Cohn, Craig Cooper, Robyn Crook, Gail Davies, Yvette DeBoer, David DeGrazia, Louis DiVicenti, Jane Dunnett, Mandy Errington, Gerard van Essen, David Favre, Andrew Fisher, Dan Flanagan, Andy Flies, Malcolm French, Joe Garner, Brianna Gaskill, John Gluck, Huw Golledge, Jenny Gray, Chris Green, David Grimm, Luce Guanzini, Thomas Hartung, Stephen Hauser, Penny Hawkins, Crystal Heath, Patricia Hedenqvist, Karen Hirsch, Pru Hobson-West, Jim Hunter, John Inns, Wendy Jarrett, Paulin Jirkof, Lisa Jones-Engel, Soraya Juarbe-Diaz, Melanie Kaplan, Jean Kazez, Geoff Kerr, Matt Kessler, Nicole Kleinstreuer, Wendy Koch, Megan LaFollette, Garet Lahvis, Rafael Landea, Dale Langford, Matt Leach, Sue Leary, Patrick Lester, Jennie Lofgren, John Ludders, Dev Manoli, Joe Martinez, James O. Marx, Georgia Mason, David Mellor, Joy Mench, Mike Mendl, Stephen Menne, Jeff Mogil, Lis Moses, Steve Niemi, Terri O'Lonergan, Steve Osofsky, Daniel Pang, Hester Parker, Monique Perron, Pandora Pound, Ruth Pye, Barney Reed, Dario Ringach, Cliff Roberts, Amy Robinson-Junker, Margaret Rose, Valerie Ross, Johnny Roughan, Andrew Rowan, Bill Satterfield, Ray Schreiber, Cathy Schuppli, Carine Serageldine, Dan Shapiro, Sally Sherwen, Adam Shriver, Peter Singer, Robert Sorge, Autumn Sorrells, Joe Spinelli, Bill Stokes, Bernie Unti, Nelleke Verhave, Greg Vicino, Jason Watters, Rebbecca Wilcox, Kate Willett, Axel Wolff, Rochelle Woods, and Joanne Zurlo.

I also thank my agent, Rachel Sussman, and my tireless University of California editor, Chloe Layman, for their help over the years of writing, as well as many good folk who took it on themselves to read whole chapters or more and share their feedback. And so I send my gratitude to Toni Andres, Chad Attenborough, Deborah Blum, Richard Entlich, Melanie Graham, Brian Hare, Hal Herzog, Amy Johnson, Alicia Karas, Jose Peralta, Anita Piccoli, Eileen Pollack, Kate Pritchett-Corning, Anne Quain, David Robertson, Susan Silver, and Dan Weary.

From the start I've faced the challenge of bringing to life animals I've known, knowing how few of those lives ended well. Thanking them for a sacrifice they had forced on them is too little, too late. I've seen and benefited from incredible medical progress in my lifetime, and I know how animals contributed—or were forced to contribute. We need to move past animal experiments with nonanimal replacements and the sooner, the better. Until that day, what can I say in appreciation other than this call for more humane care and concern for all the animals we rely on? I can only express my appreciation by demanding more humane care and concern for the animals our society depends on—until we finally embrace alternatives that will set them free.

Notes

PYTHON

1. Some grainy vestiges of *Wild Cargo* still survive on sites such as YouTube; see "*Wild Cargo #1* with Arthur Jones," posted March 26, 2016, by Cyberpump!, YouTube, www.youtube.com/watch?v=dNW0Zp4uOeE.

2. For a look at the complicated ethics of modern zoos, I recommend Jenny Gray's *Zoo Ethics* (2017).

3. The article on worst jobs in science humorously plumbed the ickiness of being an anal-wart researcher and the sad realities of what nurses must endure. Charles Leroux, "Think You've Gota Lousy Job? Try Tick Dragging," *Chicago Tribune*, November 9, 2004, www.chicagotribune.com/2004/11/09 /think-youve-gota-lousy-job-try-tick-dragging/.

4. Bernie Rollin became a friend and mentor through the years. His memoir, *Putting the Horse Before Descartes* (2011), gives an overview of the issues he cared about and a feel for his irreverent norm shattering. *The Unheeded Cry* (1989) is the Rollin book that most influenced my own work, exploring how, in their pursuit of purer data, behavioral scientists such as B. F. Skinner banned talk of animals' inner lives, shaping the science they produced while also giving themselves permission to perform painful and distressful animal experiments.

WOODCHUCK

1. In the 1970s the newly identified human hepatitis B virus was the only known member of its virus family, the *Hepadnaviridae*. Next came the woodchuck hepatitis virus, ground squirrel hepatitis virus, woolly monkey hepatitis B virus, and duck hepatitis virus. There are now close to twenty that have been identified in various animal species. The other human hepatitis viruses, labeled A, C, D, and E, are in other virus families.

2. The Italian team published their findings in *Lancet*; see Villa et al. (1982).

3. Scientists are quick to publish their positive findings, such as successfully infecting chimpanzees with hepatitis viruses. Negative findings rarely see the light of day in journals. Thus, when I find old articles describing the successful infection of chimps, I can only deduce that mice, rabbits, and other animals had failed as hepatitis models. No scientist would have jumped to chimps—large, scarce, expensive—if initial (but unpublished) attempts to infect smaller animals had succeeded.

4. Roderick Murray and other National Institutes of Health scientists conducted the experiments on decontaminating blood supplies for transfusions, writing, "The service rendered by these volunteers is gratefully acknowledged" (1954; 1955, 8).

5. As Saul Krugman launched his hepatitis experiments, no one knew for sure that a virus lurked as the cause, though that was the strong suspicion. Evidence also hinted that two versions of contagious hepatitis might in fact be symptomatically similar but distinct viral diseases. Krugman's work contributed to identifying that, indeed, the hepatitis A virus caused "infectious hepatitis." The common version spread via infected food or spiked chocolate milk was unrelated to the hepatitis B virus, or "serum hepatitis," which woodchucks were able to mimic. See Krugman (1976) for a version of some of this history.

6. Medical, military, and psychiatric researchers through the twentieth century ran human experiments constrained mostly only through limitations in available funding and scientists' informal ethical standards. Several books cover the Tuskegee abuses and other harmful experiments. Activism for protecting human subjects sometimes went hand in hand with protecting animals, all under the rubric of antivisectionism, though ultimately, greater protections for human subjects often shifted the onus onto animals, continuing the Nuremberg precedent that animal studies should precede human research.

7. The International Union for the Conservation of Nature lists animals it judges to be endangered, but the US Fish and Wildlife Service determines a species's legal status in the United States. The International Union listed long-tailed macaques (*Macaca fascicularis*), both for habitat loss and for overcollection for medical research. Importers buy captive-born monkeys from breeders in Asia, but we lack good systems for knowing how much breeders capture monkeys to

supplement their captive breeding core, putting pressure on the local, wild populations. Were the United States to adopt the International Union's determination of their status, imports of both wild-caught and captive-reared animals could become illegal (Gamalo et al. 2024).

8. Animal experiments in the University of California–San Francisco that have led to Nobel Prizes include the discovery in chickens that viruses can cause cancer; studies in rodents elucidating the role of prion proteins in neurological diseases such as mad cow disease; tests performed on mice and worms on chromosomes and aging; studies of heat- and pain-sensing nerves in mice, rattlesnakes, and other animals; and the reactivation of a mouse's stem cells so they can develop into a full range of cells and tissues.

9. The European Union, along with Norway, updates their annual statistics and posts them for public access. They report data by species, country, degree of severity, and use. The UK data were included in the past, but since Brexit they no longer are. "Alures—Animal Use Reporting—EU System: EU Statistics Database on the Use of Animals for Scientific Purposes Under Directive 2010/63/Eu," European Commission, accessed November 11, 2024, https://webgate.ec.europa.eu/envdataportal/content/alures/section1_number-of-animals.html.

10. Three vole biologists have summarized the work in *Scientific American*: Steven Phelps, Zoe Donaldson, and Dev Manoli; see "Monogamous Prairie Voles Reveal the Neurobiology of Love," *Scientific American*, February 1, 2023, 40.

MARMOSET

1. For European data of how many lives COVID vaccines have saved, see Mesle et al. (2024); for global estimates, see John P. A. Ioannidis, Angelo Maria Pezzullo, Antonio Cristiano, and Stefania Boccia, "Global Estimates of Lives and Life-Years Saved by COVID-19 Vaccination During 2020–2024," MedRxiv, December 17, 2024, https://doi.org/10.1101/2024.11.03.24316673.

2. The UCSF neurologist and marmoset scientist, Dr. Stephen Hauser, led Genentech's animal and human tests that won the drug its Food and Drug Administration approval. He has written about his quest for a useful treatment, including his conviction that, while scientists can induce experimental autoimmune encephalomyelitis, the most common lab model of MS in both mice and monkeys, and produce similar MS-like symptoms, the immune cells involved are different enough (T lymphocytes in mice, B lymphocytes in monkeys and people) that only the marmosets could have served to develop this drug that targets B cells. Greenfield and Hauser (2018); Hauser (2015); Nina Bai, "New Multiple Sclerosis Drug, Backed by 40 Years of Research, Could Halt Disease," March 28, 2017, www.ucsf.edu/news/2017/03/406296/new-multiple-sclerosis-drug-backed-40-years-research-could-halt-disease.

3. The issue of twins is interesting biology but only a tangent, and very little experimental autoimmune encephalomyelitis research includes the transmission via transfusion these scientists had been thinking would be useful. Marmoset twins' placentas fuse with each other during gestation, and so siblings end up immunologically tolerant and will readily accept a transfusion from each other. UCSF and Harvard scientists found they could establish experimental autoimmune encephalomyelitis in one sib and transfuse immune cells to the other, with resulting symptoms of an immune attack on the nerves. Once established as the primate model for MS, marmosets have continued that fate, even though the twin issue is irrelevant to most studies (Massacesi et al. 1995).

4. Chemie Grünenthal and its associated Grünenthal Foundation are one source of information on the thalidomide tragedy. They describe the criminal trial against company leaders, which ended without a verdict, and the settlement they reached with patients, with ongoing compensation payments sixty years after they removed the drug from the market. "The History of the Thalidomide Tragedy," Grünenthal Foundation, accessed November 28, 2023, www.thalidomide-tragedy.com/en/the-history-of-the-thalidomide-tragedy#Development.

5. To read more about why Darwinian evolution results in species differences that can make animal-lab data misleading for human health applications, start with Pandora Pound's *Rat Trap: The Capture of Medicine by Animal Research—and How to Break Free* (2023) or Hugh LaFollette and Niall Shanks's *Brute Science* (1996).

6. The NASEM report on primates in NIH-funded research is available on its website; see National Academies of Sciences (2023).

7. The House of Representatives Subcommittee on Cybersecurity, Information Technology, and Government Innovation heard from three speakers, from the White Coat Waste Project, from the Physicians Committee for Responsible Medicine, and from the Johns Hopkins School of Public Health. "Transgender Rats and Poisoned Puppies: Oversight of Taxpayer-Funded Animal Cruelty," Committee on Oversight and Government Reform, February 26, 2025, www.congress.gov/119/chrg/CHRG-119hhrg58804/CHRG-119hhrg58804.pdf.

8. See Seok et al. (2013).

9. See Ineichen and peers' (2024) primer on systematic reviews to improve the use of lab animals. Hooijmans and colleagues (2012) published the work on probiotics for man and mouse pancreatitis, while Perel and colleagues (2007) compared mouse laboratory findings with human clinical trials.

10. Roche and Genentech scientists, as well as unaffiliated researchers, have published a number of assessments of ocrelizumab, both for MS and for another autoimmune disease, rheumatoid arthritis. For the published phase 2 and phase 3 clinical trials and the prescribing information that mentions safety studies in pregnant monkeys, see Kappos et al. (2011); "Highlights of Prescribing Informa-

tion," Genentech, last revised June 2024, www.gene.com/download/pdf /ocrevus_prescribing.pdf; and Hauser et al. (2017).

11. After a drug has cleared preclinical studies, not just in animals but in whatever in vitro or other tests, it enters phase 1 trials in a few dozen volunteers. Healthy volunteers or patients with the disease the drug is targeting serve to catch safety concerns and to establish dosing regimens. The FDA says that about 70 percent of such drugs move on to phase 2, where larger cohorts of patients yield fuller information on side effects and efficacy, with about a third of those moving on to phase 3. The FDA, the agency that reviews and approves drugs and so should know, reports that 25 to 30 percent survive through phase 3, where a few thousand patients produce more efficacy data and details of adverse reactions. Phase 4 trials come after the drug is already approved and in use, helping the company and the FDA fine-tune the safety data and, on rare occasions, calling for removal of an approved drug from the market, either voluntarily or by FDA decree, the fate thalidomide would have met were it in widespread use in the United States. FDA guidance may shift with time, especially now that Congress has given it greater freedom to waive animal data if drug companies present solid in vitro safety data. "Step 3: Clinical Research," US Food and Drug Administration, accessed December 6, 2024, www .fda.gov/patients/drug-development-process/step-3-clinical-research#Clinical _Research_Phase_Studies.

12. Several scientists and data analysts are working with whatever clinical-trial data they can access to determine rates of success and failures in the quest for assuring that only the most promising preclinical data lead to expansive three-phase human testing. See Hay et al. (2014) and Wong et al. (2019).

13. For arguments that basic science is less likely to bear fruit than experiments further along in the drug pipeline and the suggestion that this fact should weigh against some animal experiments, see Ioannidis (2007); Greek and Greek (2010); and Munafo et al. (2017).

14. "The Experimental Design Assistant: Helping Researchers Worldwide Design Robust and Reliable Experiments," National Centre for the Replacement, Refinement and Reduction of Animals in Research, accessed December 15, 2024, https://eda.nc3rs.org.uk/.

15. Bailoo et al. (2014); Garner et al. (2017).

16. Garner and his colleagues (2017) show how to incorporate animal-subject diversity into animal experiments to better model the diversity of a human-patient population; see also Richter et al. (2010) and Garner (2014).

17. Katie Thomas, "F.D.A. Approves First Drug to Treat Severe Multiple Sclerosis," *New York Times*, March 28, 2017, sec. B, p. 8, www.nytimes.com/2017/03 /28/health/fda-drug-approved-multiple-sclerosis-ocrevus.html.

DOG

1. For this chapter I rely heavily on my research on the Animal Welfare Act for my first book, *What Animals Want: Expertise and Advocacy in Laboratory Animal Welfare Policy* (2004), as well as the National Academies of Sciences, Engineering, and Medicine report *Necessity, Use, and Care of Laboratory Dogs at the U.S. Department of Veterans Affairs* (2020).

2. For a description of the melee at Magnan's demonstration of alcohol and absinthe, see Eric Michael Johnson, "Charles Darwin and the Vivisection Outrage," *Scientific American*, October 11, 2011. For more on Darwin's role following the Magnan episode, see Feller (2009) and Susan Hamilton, "On the Cruelty to Animals Act, 15 August 1876," February 2013, https://branchcollective .org/?ps_articles=susan-hamilton-on-the-cruelty-to-animals-act-15-august -1876.

3. Darwin wrote about dog vivisection in both editions of *The Descent of Man and Selection in Relation to Sex* (1871, 1874); see also Herzog (2002).

4. For history of the Brown Dog Affair and the dog's statue in Battersea, see Mason (1997) and Lansbury (1985).

5. Since 1971 the USDA's enforcement of the Animal Welfare Act has included collating facilities' annual reports of the numbers of dogs and other act-covered animals they have used. Despite some shifts in USDA's reporting expectations, they reflect rough trends in animal use, including the decline in lab dogs and cats since the start of data collection. While the National Institutes of Health makes the statistics it collects available only through Freedom of Information requests, the USDA proactively posts its data and is working to make its site more user-friendly. Animal and Plant Health Inspection Service, "Animal Care: Animal Welfare Act and Animal Welfare Regulations," July 2023, www.aphis .usda.gov/sites/default/files/ac_bluebook_awa_508_comp_version.pdf.

6. Congress directed the NIH to commission the National Academies of Science to "critically examine the general desirability and necessity of using random source dogs and cats in NIH-funded research, and the specific necessity of using dogs and cats from Class B dealers for such research," and the National Academies of Science published the results; see Committee on Scientific and Humane Issues (2009).

7. A few times lab cats have indeed been the focus of animal-activist campaigns, most notably a campaign in the 1970s to stop the American Museum of Natural History's behind-the-scenes cat experiments. Researchers at the museum were running neurology experiments with cats, studying sexual behavior after spinal cord injury. Before he turned to animals in cosmetics testing, the activist Henry Spira launched a successful campaign to stop the museum's experiments on how spinal cord injuries affect cats' sexual performance, a scenario ready-made for an activist's exposé; see Spira (1985).

8. Both the guidelines and the accreditation program have had a few name changes since the early 1960s. The guidelines are now called the *Guide for the Care and Use of Laboratory Animals*; see Animal Care Panel (1963); National Research Council (2011); and "About: What Is AAALAC?," AAALAC International, accessed February 20, 2021, www.aaalac.org/about/what-is-aaalac/.

9. For this important article in the history of laboratory animal–welfare laws, see Coles Phinizy, "The Lost Pets That Stray to the Labs," *Sports Illustrated*, November 27, 1965, 36–49.

10. Though unnamed, the dog featured in the *LIFE* magazine article gave a face to the early efforts to pass the 1966 Animal Welfare Act. Michel Silva, "Concentration Camps for Dogs," *LIFE*, February 4, 1966, 22–29.

11. Carbone et al. (2003).

12. For two versions of laboratory animal adoptions with opposite stances on the justification for using dogs in laboratories, see "Our Mission: Educate. Legislate. Liberate," Beagle Freedom Project, accessed March 21, 2025, https://bfp.org/mission/; and "Homes for Animal Heroes," National Animal Interest Alliance, accessed March 21, 2025, https://homesforanimalheroes.com/about-us/mission/.

13. The USDA published its update for vets in the *Journal of the American Veterinary Medical Association* (Jones 1968).

14. "State Pound Seizure Laws," Humane Society of the United States, accessed March 21, 2025, www.humanesociety.org/sites/default/files/docs/state-pound-seizure-laws-HSUS.pdf.

15. I am glossing over the scientific rationales for using random-source or purpose-bred dogs in labs. Random-source dogs bring a degree of genetic variability that is hard to match in a lab colony and that may be useful for some experiments. For most uses, such as in safety-testing compounds and medicines, a uniform cohort of beagles in the experimental and control groups can yield clearer data with fewer animals. The National Academies of Science examined the use of random-source dogs in NIH-funded research, which does cover dogs in private, for-profit safety-testing and toxicology labs; see Committee on Scientific and Humane Issues (2009).

16. For some coverage of the closure of the Envigo kennel, its adoption of close to four thousand beagles to willing homes, and the fine it paid, see Martin Weil, "Dog Breeder to Pay Record $35M Fine After Surrendering Thousands of Beagles," *Washington Post*, June 4, 2024, www.washingtonpost.com/dc-md-va/2024/06/04/envigo-beagles-rescue-breeder-fine/. For a smart and sensitive account of one adopted lab beagle's life, read Kaplan (2025).

17. The museum shut the lab within a year of Spira launching his protests. This campaign had no effect that I know of on cat research more generally but did lay the strategic groundwork for the campaign to end eye-safety tests in rabbits (see Spira 1985).

18. To read more about the legal case to let California students opt out of dissecting frogs in their biology classes, see Howard Rosenberg, "Apple Computer's 'Frog' Ad Is Taken Off the Air." *Los Angeles Times*, November 10, 1987, www.latimes.com/archives/la-xpm-1987-11-10-ca-20057-story.html; and Orlans et al. (1998).

RABBIT

1. "Santaland Diaries" is in Sedaris's story collection *Barrel Fever* (1994).

2. WellBeing International maintains a copy of Spira's ad targeting Revlon on its WBI Studies Repository. "Revlon Advertisement," Millennium Guild, 1979, www.wellbeingintlstudiesrepository.org/cgi/viewcontent.cgi?article=1001&context=dratcam. WellBeing International also posts Spira's quite extraordinary earlier notes on the campaign and his goal to "pressure the cosmetic trade association to tax their members one hundredths of 1% of the industry's $11 billion gross" to fund research into finding an alternative to the Draize test that regulators would accept. Henry Spira, "Draize: A Blueprint for Change," WellBeing International, August 23, 1979, www.wellbeingintlstudiesrepository.org/cgi/viewcontent.cgi?article=1000&context=dratcam.

3. I owe much of what I know about Henry Spira and his campaign to Singer (1998) and Rowan (1984), along with Parascandola's 1991 history of the Draize test.

4. The Thackeray Museum of Medicine includes the *Vanity Fair* lithograph of Pasteur with two rabbits. Rabbits easily injure their spines if held incorrectly, but Pasteur shows good technique in his portrait. "Chromolithograph Cartoon from 'Vanity Fair' of Louis Pasteur Holding Two White Rabbits, Captioned 'Hydrophobia,' 1887," Thackeray Museum of Medicine, accessed March 21, 2025, https://collections.thackraymuseum.co.uk/object-959-025.

5. All the Tropes is a community-edited site on the model of Wikipedia, with an entry on cultural references to rabbits in pregnancy diagnosis. "The Rabbit Died," All the Tropes, last modified September 25, 2024, https://allthetropes.org/wiki/The_Rabbit_Died.

6. For a review of the Draize test's scientific challenges and the evaluation of slaughterhouse chicken eyes as a replacement, see Prinsen et al. (2017). The authors include a sample illustration of the disturbing eye inflammation toxicologists score in the Draize test, as once standardized (and now largely replaced) by the FDA.

7. Two issues in animal rights campaigns, cosmetics testing and fur, resonate for their juxtaposition of cruelty with the pursuit of vanity. Both are primarily associated with female vanity, however, and I always wonder about this. Is it because something primarily associated with women can be coded as too trivial

to justify? Leather wearing and meat eating are less gendered but have some claim to acceptability to the extent that leather is mostly seen as a by-product of meat production and meat eating is often seen as important or even necessary for health. Or is it that a campaign that targets women's products intentionally targets the people most likely to care about animals and act on the issue? Men have received prominent coverage as the leaders of the modern animal-protection movement, but, as Emily Gaarder writes in her book *Women and the Animal Rights Movement* (2011), from Victorian times to the present, the majority of animal rights advocates have been women.

8. *Lab Animal* started as a trade magazine, a place to advertise a company's rabbits and other animals, along with all the newest equipment, but Lynne Harriton as editor elevated its quality of journalism (see Harriton 1981).

9. See Singer's (1998) biography of Spira.

10. Andrew Rowan describes the eclipse of guinea pigs for tuberculosis diagnostics, citing Marks (1972) from the Tuberculosis Reference Laboratory in Cardiff, Wales.

11. As the Leaping Bunny Program notes, there is no standard for labeling products "cruelty free" or even "not tested in animals." See "About Leaping Bunny," accessed January 8, 2024, www.leapingbunny.org/about/about-leaping-bunny.

12. Several US government agencies contribute staff and expertise to the Interagency Coordinating Committee on the Validation of Alternative Methods, with the National Toxicology Program in the lead role (Interagency Coordinating Committee 2010).

13. Rebecca Skloot has written a celebration of Henrietta Lacks, with documentation of the ethical and policy controversies surrounding her HeLa cells (see Skloot 2011).

14. For a very technical overview of the various assays that combine to replace the Draize test, I recommend PETA Science Consortium International's 2021 review. This paper is also my source for evidence that, as of 2018 at least, some labs were still submitting Draize data to the EPA (Clippinger et al. 2021).

15. For press coverage of the 2024 Nobel Prize for AI-designed proteins, see "David Baker," Department of Biochemistry, University of Washington, accessed February 6, 2025, https://sites.uw.edu/biochemistry/faculty/david-baker/; and Sneha Khedkar, "Nobel Prize in Chemistry for Work on Proteins," *Scientist*, October 9, 2024, www.the-scientist.com/nobel-prize-in-chemistry-for-work -on-proteins-72232.

16. Bayer maintains a website on glyphosate (Roundup) litigation, castigating the "litigation industry"; see "The Challenge with Glyphosate Litigation," Bayer Global, last updated October 17, 2024, www.bayer.com/en/roundup -litigation?gad_source=1&gbraid=0AAAAAo3bpgODE-yv4DsrZZm-3akIh2kNi &gclid=CjwKCAjwnPS-BhBxEiwAZjMF0i4H9nq6lXX2Mb82opTT9AowGhgrk

ORCoyMYldTeV-E3R2PUfNSB2BoCv7UQAvD_BwE. Miller and Lois are litigators in several Roundup cases, with updates on Roundup lawsuits; see Ronald V. Miller Jr., "Monsanto Roundup Lawsuit Update," Lawsuit Information Center, April 29, 2025, www.lawsuit-information-center.com/roundup-lawsuit.html.

17. For more on industry funding as a source of bias, see Munafo et al. (2017); Lundh et al. (2018); and Cochrane, "Cochrane: 30 Years of Evidence," December 15, 2023, www.cochrane.org. See also the Animal Research: Reporting of *In Vivo* Experiments guidelines for reporting animal experiments, which call for authors to disclose sponsorship and financial interests in their work. "New ARRIVE Guidelines 2.0 Released," National Centre for the Replacement, Refinement and Reduction of Animals in Research, July 14, 2020, https://arriveguidelines.org/news/new-arrive-guidelines.

18. Rovida and Hartung (2009), Knight et al. (2023), and Rovida et al. (2023) have analyzed the animal-use implications of the European Union's Registration, Evaluation, Authorization, and Restriction of Chemicals program to increase toxicology data on chemicals in use in Europe.

19. Luckily for companies, the FDA doesn't wait until the end of a years-long study to disapprove it but instead reviews plans up front to tell a company on a one-to-one private basis what data it will accept before allowing a human clinical trial to commence (National Academies of Sciences 2023).

20. For more on the Medical Research Modernization Act 2.0, see Adashi et al. (2023). See also U.S. Food and Drug Administration, "FDA Announces Plan to Phase Out Animal Testing Requirement for Monoclonal Antibodies and Other Drugs," April 10, 2025, www.fda.gov/news-events/press-announcements/fda-announces-plan-phase-out-animal-testing-requirement-monoclonal-antibodies-and-other-drugs; and Asher Mullard, "Is the FDA's Plan to Phase Out Animal Toxicity Testing Realistic?," *Nature Reviews Drug Discovery*, May 15, 2025, www.nature.com/articles/d41573-025-00087-x.

21. Andrew Rowan tracks the EU and UK statistics for insights into trends in those places and, by extrapolation, in the United States. "Replacing Laboratory Animals in Safety Testing," WellBeing International, December 31, 2021, https://wellbeingintl.org/replacing-laboratory-animals-in-safety-testing/.

CHICKEN

1. The Brambell Committee morphed into the Farm Animal Welfare Committee, whose restatement of the Five Freedoms is the most common version in use:

Freedom from hunger and thirst, by ready access to water and a diet to maintain health and vigour.

Freedom from discomfort, by providing an appropriate environment.

Freedom from pain, injury and disease, by prevention or rapid diagnosis and treatment.

Freedom to express normal behaviour, by providing sufficient space, proper facilities and appropriate company of the animal's own kind.

Freedom from fear and distress, by ensuring conditions and treatment, which avoid mental suffering.

I do not find the Five Freedoms rubric all that helpful, but I appreciate its historical significance. Obviously, there are more than five—freedom from pain *and* injury *and* disease need not be lumped as one freedom, while freedom from discomfort should include a cozy bed, shelter from the weather, a floor that's comfortable for walking, and more. And as Bekoff and Pierce point out, actual freedom itself, not "freedom to" or "freedom from," is nowhere in the list. "FAWC Report on Farm Animal Welfare in Great Britain: Past, Present and Future," Animal Welfare Committee, October 12, 2009, www.gov.uk/government /publications/fawc-report-on-farm-animal-welfare-in-great-britain-past-present -and-future; *Report of the Technical Committee to Enquire into the Welfare of Animals Kept Under Intensive Livestock Husbandry Conditions*, Wageningen University and Research eDepot, December 1965, https://edepot.wur.nl/134379; Bekoff and Pierce (2017).

2. Hughes and Black (1973) published their experiment about six years after the Brambell Committee report. Oversimplifying history a wee bit, modern-day ethologists nonetheless credit them with establishing preference studies as the foundation of animal-welfare science.

3. Professor Mench later became my postdoctoral mentor at the University of California Center for Animal Welfare in Davis. In addition to her expertise in poultry welfare, she has written extensively on laboratory animal welfare as well (see Mench et al. 1992; Mench 1998; and Brown et al. 2006).

4. Morton and colleagues described critical anthropomorphism at a workshop at the Hasting Center, a bioethics research institute (see Morton et al. 1990).

5. In addition to numerous technical publications of her animal-welfare studies, Grandin has described how her insights as an autistic person, such as her aversion to sensory overload or to the comfort of being closely held, lead to testable hypotheses about animals' welfare (see Grandin 1995).

6. Lahvis has written of how his frustrations in trying to study normal mouse behavior (as controls in his autism studies) ultimately led him to step away from that work. Garet Lahvis, "Freefall into Darkness," Aeon, June 2, 2022, https:// aeon.co/essays/what-do-caged-animals-really-tell-us-about-our-mental-lives.

7. See Manser et al. (1996).

8. Sherwin has written extensively on how to ask animals what they most want; see Sherwin et al. (2004); and Chris M. Sherwin, "Validating Refinements to Laboratory Housing: Asking the Animals," National Centre for the Replacement, Refinement and Reduction of Animals in Research, September 2007, https:// nc3rs.org.uk/sites/default/files/2022-01/CSherwinValidatingRefinementsto LaboratoryHousing.pdf.

9. AVMA Panel on Euthanasia, *AVMA Guidelines for the Euthanasia of Animals*, 2020, www.avma.org/sites/default/files/2020-02/Guidelines-on-Euthanasia-2020.pdf. I worked on the Laboratory Animals Working Group for the 2013 and 2020 editions.

10. Carbon dioxide for killing animals has received plenty of coverage and experimental investigation, with researchers divided on whether to condemn its use or find a way to fine-tune it; see Raj and Gregory (1995); Kirkden et al. (2008); Hawkins et al. (2016); and Turner et al. (2020).

11. See Peckmezian and Taylor (2017).

12. See Mellor (2019).

13. See the Rygula et al. (2012) study of cognitive bias in rats and the LaFollette et al. (2017) study and promotion of rat tickling in labs; see also "Rat Tickling Background and FAQ," 3Rs Collaborative, accessed April 1, 2024, https://3rc.org/rat-tickling-faq/.

14. The New York group focuses on the type of consciousness that we also call *sentience*, or the ability to have what they call "subjective experiences." Choose either word, consciousness or sentience, but remember that we are talking about which animals are capable of having pleasant or unpleasant feelings. "The New York Declaration on Animal Consciousness," New York University, April 19, 2024, https://sites.google.com/nyu.edu/nydeclaration/declaration?authuser=0. See also Phillip Low's earlier "Cambridge Declaration on Consciousness," paper presented at the Francis Crick Memorial Conference on Consciousness in Human and Non-human Animals, University of Cambridge, July 7, 2012.

15. On various kinds of crabs, including shell-swapping hermit crabs, see Appel and Elwood (2009) and Elwood (2022).

16. Elwood has summarized his findings in scientific manuscripts and in works for more general audiences. He has also teamed up with mammal and fish experts to explore which animals have the capacity to experience pain as an emotionally unpleasant sensation beyond the physical phenomenon of simple nociception (see Elwood 2022; Sneddon et al. 2014; and Elwood 1991).

17. See Giancola-Detmering and Crook (2024).

18. See Singer's *Animal Liberation Now* (2023) for his current assessment. Scientists have published behavioral data on oysters' shell-shutting responses to loud noises or evidence of oyster-eating crabs. In giving readers guidance on what to think of oysters on a menu, Singer reminds us that oyster farming may be much better environmentally than most vegetable farming, including all the collateral damage to sentient animals who are harmed by pest-control measures. But do read carefully, as he makes a compelling case for why scallop harvesting should keep those mollusks off your plate even if you're comfortable shucking some dubiously sentient oysters (see also Carroll and Clements 2019; and Ledoux et al. 2023).

19. I review the dog exercise and guinea pig issues in more depth in my book *What Animals Want*. For the actual studies that the USDA relied on, see Carbone (2004); White et al. (1989); and Hughes et al. (1989).

20. The later study of guinea pigs' space utilization started with better background knowledge of wild and domestic guinea pigs' natural behaviors, noting that such animals rarely dart about very far from sheltered spaces (Byrd et al. 2016).

21. In an article claiming that "most published research findings are false," the epidemiologist John Ioannidis warned of the risks I see in the studies of guinea pig cages and dog exercise. In fields where study designs are not standardized or where financial stakes are high, expect a high risk of false findings. If only a single group is studying an issue, such as cage sizes for guinea pig harems, expect problems (see Ioannidis 2005).

22. For Dr. Melcher's assessments of the USDA's handling of the Animal Welfare Act, see Stephen Labaton, "Animal Advocates Win Court Ruling," *New York Times*, February 26, 1993, A12; and John Melcher, "The Mental Health of Primates: We're Still Needlessly Cruel to Research Animals in Our Labs," *Washington Post*, September 7, 1991, www.washingtonpost.com/archive/opinions /1991/09/08/the-mental-health-of-primates/08aebb35-f1c6-4a2a-b00c-37e0a 406a812/.

23. For critiques of how ventilated cages damage animals' welfare and psychological health, see Lahvis (2022); Baumans et al. (2002); Ahlgren and Voikar (2019); David et al. (2013); Mineur and Crusio (2009); Garner et al. (2017); and Beura et al. (2016).

24. Most primate researchers would report simply that "blood was collected," with the dreaded passive voice conferring no mention of restraint, sedation, or cooperation. Graham, on the other hand, developed and has very actively campaigned among scientists and vets to train animals to cooperate voluntarily with research procedures (see Graham et al. 2012).

25. Most of Hare's published studies allow dogs or apes to participate in various mental and behavioral challenges. See, for example, Hare et al. (2007) and Salomons et al. (2024).

26. For the campaigns to end tail restraint of lab mice, see "Refined Mouse Handling Overview," 3Rs Collaborative, accessed April 1, 2024, https://3rc.org /refined-mouse-handling-overview/; and "Mouse Handling," National Centre for the Replacement, Refinement and Reduction of Animals in Research, accessed April 1, 2024, www.nc3rs.org.uk/3rs-resources/mouse-handling.

CHIMPANZEE

1. Enos, the second chimp in space, suffered painfully when his testing equipment malfunctioned. He continued receiving electric shocks whether he

responded correctly, incorrectly, or not at all. Among the numerous articles that cover Ham's story, many are space enthusiasts' hagiographies. For a more nuanced treatment, I recommend Henry Nichols, "Ham the Astrochimp: Hero or Victim?," *The Guardian*, December 16, 2013, www.theguardian.com/science /animal-magic/2013/dec/16/ham-chimpanzee-hero-or-victim.

2. "Taxidermy Is the Wrong Stuff," *Washington Post*, January 26, 1983, www .washingtonpost.com/archive/politics/1983/01/27/taxidermy-is-the-wrong-stuff /db9b8616-54a9-4649-80d8-9d701d64ba62/.

3. One of Harlow's students, John Gluck, continued his own version of this work as a professor in New Mexico, until the ethical ramifications of causing so much suffering with too little benefit, in his eyes, led him to shut down his lab and raise his voice for more constraints on how we experiment on animals (see Gluck 2016).

4. I have covered the two exposés in more detail in my book *What Animals Want* (2004), as has Deborah Blum in her book *The Monkey Wars* (1994). PETA maintains the video, *Unnecessary Fuss*, on its website, with a narrative that makes a disturbing video look even worse. When a scientist spills some sort of liquid (I'd have guessed water or saline solution, as I viewed it) on an animal, the narrator muses without evidence, "maybe it's acid." *Unnecessary Fuss*, People for the Ethical Treatment of Animals, accessed March 8, 2025, www.peta.org /videos/unnecessary-fuss/. To read more about the Silver Spring monkeys, go to the horse's mouth, Alex Pacheco, for his tale of gaining access to the monkey lab and later bringing in the police (Pacheco and Francione 1985), as well as Caroline Fraser, "The Raid at Silver Spring," *New Yorker*, April 19, 1993, 66–84.

5. Warren G. Magnuson, "Statement Before Committee on Commerce, United States Senate," *Animal Dealer Regulation*, March 25, 1966, https://babel .hathitrust.org/cgi/pt?id=uc1.$b642732&seq=7.

6. See John Melcher, "The Mental Health of Primates: We're Still Needlessly Cruel to Research Animals in our Labs," *Washington Post*, September 7, 1991, www.washingtonpost.com/archive/opinions/1991/09/08/the-mental-health-of -primates/08aebb35-f1c6-4a2a-b00c-37e0a406a812/; and *Boston Globe*, "Animal Houses," *Chicago Tribune*, updated August 8, 2021, www.chicagotribune .com/1988/09/28/animal-houses/.

7. The amended Animal Welfare Act was passed in 1985, embedded in the Farm Bill, which comes for renewal every ten years. Food Security Act of 1985, Pub. L. No. 99–198, Title XVII, Subtitle F: Animal Welfare.

8. In an article for *Science*, "Experts Ponder Simian Well-Being," Constance Holden quotes De Waal and other leading primatologists of the day (see Holden 1988).

9. Melcher went to work for an animal-protection organization, the Animal Welfare Institute, that had been influential when the Animal Welfare Act was having its update. In that role he wrote an op-ed piece criticizing how the USDA

had thwarted his and Congress's intentions with their weak rule for primate welfare; see John Melcher, "The Mental Health of Primates: We're Still Needlessly Cruel to Research Animals in Our Labs," *Washington Post*, September 7, 1991, www.washingtonpost.com/archive/opinions/1991/09/08/the-mental-health-of-primates/08aebb35-f1c6-4a2a-b00c-37e0a406a812/.

10. Here is the text of the animal-welfare regulations, finalized in 1991 after five years of back-and-forth lobbying by the researchers' lobbies and the animal rights groups: *"Environmental enrichment.* The physical environment in the primary enclosures must be enriched by providing means of expressing noninjurious species-typical activities. Species differences should be considered when determining the type or methods of enrichment. Examples of environmental enrichments include providing perches, swings, mirrors, and other increased cage complexities; providing objects to manipulate; varied food items; using foraging or task-oriented feeding methods; and providing interaction with the care giver or other familiar and knowledgeable person consistent with personnel safety precautions." See Animal and Plant Health (1991); and Animal and Plant Health Inspection Service, "Animal Care: Animal Welfare Act and Animal Welfare Regulations," July 2023, www.aphis.usda.gov/sites/default/files/ac_bluebook_awa_508_comp_version.pdf.

11. For pharmacologists' review of the value of mouse burying as an anxiety test, see De Boer and Koolhaas (2003), and for the dismissal of marbles as a fun enrichment for mice, see Würbel and Garner (2007).

12. I make no product endorsements, but Bio-Serv's catalog is full of examples of enrichment equipment it sells for animals, and the cage and aquarium manufacturer sports an enrichment device for zebra fish, a picture of riverbed rocks to put under the clear fish tank, which apparently they prefer as an alternative to looking down through clear plastic; see "Enrichment Solutions," Bio-Serv, accessed July 15, 2024, www.bio-serv.com/enrichment.html; and Techniplast, "Zebrafish Housing," accessed June 8, 2025, https://aquaticsolutions.it/products/zebrafish-housing.html#zpark. For some of the animal-welfare science supporting this tank modification, see Schroeder et al. (2014).

13. "Laboratory Animal Refinement and Enrichment Forum (LAREF)," Animal Welfare Institute, accessed March 15, 2025, https://awionline.org/content/refinement-forum-laref.

14. Markowitz and Spinelli (1986); Markowitz (2011).

15. Male bias is a scientific quality issue in research and much less an issue of murine quality of life, though male mice can certainly fight quite aggressively. Scientists have traditionally avoided female study subjects because of the hormonal fluctuations of the female reproductive cycle. Scientists, feminists, and the NIH have found that limiting a study population to males only can lead to serious oversights in using the animal data to apply to human health. NIH now expects labs to have a balance of male and female (human and nonhuman) test

subjects in their experiments or provide a scientific explanation for why they do not. The justification would mostly be that in the human, one sex or the other is disproportionately affected. X-linked Duchenne muscular dystrophy is a disease in which female mouse or human patients are quite rare. For Mason's critique of CRAMPED mice in unenriched lab housing, see Cait et al. (2022).

16. I read Hediger's (1969) work voraciously in my zookeeper days, which may have inspired me to go tadpole hunting for the turtles. Burghardt (1999) added the comparison between additive and controlled deprivation perspectives to captive animal welfare.

17. Watters and Krebs detail their Three Needs approach in their how-to book for zoo workers, *Managing Zoo Animal Welfare* (2025). Another good overview of zoo animal–welfare assessment comes from the Animal Welfare Education Center at the veterinary college in Barcelona (Tallo-Parra et al. 2023).

18. For a write-up of Lahvis's ideas, see Lahvis (2017) and Garet Lahvis, "Freefall into Darkness," Aeon, June 2, 2022, https://aeon.co/essays/what-do -caged-animals-really-tell-us-about-our-mental-lives.

19. Committee on the Use (2011).

20. Not all NIH chimps immediately went to sanctuaries, especially older infirm animals that vets had deemed unfit for travel. Animal-welfare organizations have brought legal action to finish this process; see Francis S. Collins, "NIH Will No Longer Support Biomedical Research on Chimpanzees," National Institutes of Health, November 17, 2015, www.reginfo.gov/public/do/DownloadDoc ument?objectID=132266801; US Fish and Wildlife Service, "U.S. Fish and Wildlife Service Finalizes Rule Listing All Chimpanzees as Endangered Under the Endangered Species Act," June 12, 2015, www.fws.gov/press-release /2015-06/us-fish-and-wildlife-service-finalizes-rule-listing-all-chimpanzees; Committee on the Use (2011); and Rachel Dobkin, "Retired Research Chimps Moving to Louisiana Sanctuary," *Newsweek*, November 8, 2024, www.newsweek .com/retired-research-chimps-moving-new-mexico-louisiana-sanctuary-1983112.

RAT

1. See Wilks (1881) and Chapple and Hadwen (1914).

2. Octopus anesthesiology is where mammal anesthesiology was decades ago. Scientists are identifying various chemicals and medicines that appear to anesthetize the animals and trying to determine if they indeed induce the octopus version of unconsciousness or just some sort of immobilization (Crook 2021; Roumbedakis et al. 2020; Deutsch et al. 2023).

3. The American Veterinary Medical Association has updated its Veterinarian's Oath since I first swore it, emphasizing vets' responsibility for animal welfare: "Being admitted to the profession of veterinary medicine, I solemnly swear

to use my scientific knowledge and skills for the benefit of society through the protection of animal health and welfare, the prevention and relief of animal suffering, the conservation of animal resources, the promotion of public health, and the advancement of medical knowledge. I will practice my profession conscientiously, with dignity, and in keeping with the principles of veterinary medical ethics. I accept as a lifelong obligation the continual improvement of my professional knowledge and competence." See "Veterinarian's Oath," accessed July 19, 2021, www.avma.org/resources-tools/avma-policies/veterinarians-oath.

4. For the current International Association for the Study of Pain, see Raja et al. (2020).

5. In apparently anesthetized human and nonhuman animals, consciousness and, as a consequence, pain are possible. Anesthetists use paralyzing drugs to prevent reflexes during surgery, but those drugs can also mask evidence that the immobilized patient may feel what the surgeon is doing. The concern is real, but vets and human doctors do have methods to minimize this occurrence, and my concern in this chapter remains the pain of fully conscious animals.

6. A bit of vet trivia: "hardware disease" from swallowed nails perforating the stomach is common enough that farmers have their cows swallow a magnet. Swallowed nails and bits of wire stick to the magnet inside the reticulum, the first of their four stomachs, rather than straying to the stomach wall and puncturing it. James R. Mosely, "Animal Welfare: Politics or Facts?," paper presented at the Food Animal Well-Being conference, Purdue University, Lafayette, IN, April 13–15, 1993.

7. "1970 Amendments to AWA, House Report No. 91-1651," Animal Legal and Historical Center, December 7, 1970, www.animallaw.info/statute/us-awa-1970-amendments-awa-house-report-no-91-1651.

8. See Roughan and Flecknell (2001).

9. Paul Flecknell, presentation for the Fifty-Fourth National Meeting of the American Association for Laboratory Animal Science, Seattle, October 12–16, 2003.

10. Alicia Karas, the animal pain specialist at the vet college at Tufts University, and her colleagues developed their TINT (Time to Incorporate to Nest Test) assay to efficiently survey postsurgical pain in large groups of mice. They distributed a measured ten grams of cotton nesting material to each cage and scored whether mice had started using it for nest construction within ten minutes (Rock et al. 2014; Gallo et al. 2020).

11. Langford worked with neonatologist Kenneth Craig to translate human studies of "pain faces" to the mouse lab (see Langford et al. 2010; and Craig et al. 1994).

12. See, as an example, Roughan et al. (2014), a study of place preference in mice in bladder-cancer experiments. Cognitive-bias tests (turning optimistic animals into pessimists) have worked on farms and should be useful in labs too (Neave et al. 2013).

13. See Danbury et al. (2000).

14. Column A is where the lab lists the species of animals, and column B lists animals—newly purchased, breeders, or young being reared—that it has on-site but has not yet used in an experiment.

15. I traced the *Guide*'s treatment of the issue of withholding pain treatments shortly after the eighth edition came out in 2011. In some of the middle editions, the *Guide* was clear about the occasional researcher request to withhold pain medicines, but, in the seventh and eighth editions, the issue was submerged a bit (see Carbone 2012).

16. For our full manuscript on scientists' (non)publication of their animals' pain management, see Carbone and Austin (2016).

17. For scientific publications on how different opioid analgesics affect experimental cancer metastasis in rats, see Franchi et al. (2007) and Page et al. (2001).

18. Paulin Jirkof is a veterinarian and pain expert working in Switzerland. She has published a concise review, aimed at vets and ethics committee members, of some of the ways that pain medicines and untreated pain can affect the outcomes of experiments (see Jirkof 2017).

MOUSE

1. I based my estimate of animal numbers on information I gathered from one US lab that was willing to share data and from fifteen whose numbers I obtained through open records requests. Others have tried estimating US numbers through extrapolation from European statistics (where lab mice are legally animals) or through numbers of animals in published scientific manuscripts. See Carbone (2021); Hannah Sparks, "More Than 110 Million Rats, Mice Used in US Labs: Report," *New York Post*, January 18, 2021, https://nypost.com/2021/01/18/more-than-100-million-rats-mice-used-in-us-labs-report/amp/; and Taylor and Alvarez (2019).

2. The Animal Welfare Act defined these as its initial legal "animals": dogs, cats, rabbits, hamsters, guinea pigs, and all species of nonhuman primates, from lemurs to gorillas.

3. The 1970 congressional Animal Welfare Act explicitly excluded agricultural animals in food and fiber research and implicitly included pigs and sheep in medical research labs. The act is silent about mice and rats, but the USDA vet overseeing enforcement took that silence as flexibility to enforce the law for mice or not. I've covered much of this history in detail in my book *What Animals Want* (2004). See also Animal Welfare Act of 1970, Pub. L. No. 91-579; and Schwindaman et al. (1973).

4. Schwindaman (1973, 1274).

5. Animal and Plant Health Inspection Service, "Animal Care: Animal Welfare Act and Animal Welfare Regulations," July 2023, www.aphis.usda.gov /sites/default/files/ac_bluebook_awa_508_comp_version.pdf, 6.

6. The National Association for Biomedical Research is a source for some of this information, though it gives no evidence for its claim that 90 percent of mice and rats were already being covered by other regulations or by the AAALAC. "2001 Year-End Animal Research Policy Summary," January 16, 2002, https:// apps.fass.org/fass/fass-science/policysummaryYearEnd20011.pdf. That would require knowing how many mice are covered in AAALAC's confidential accreditation processes, how many report to the NIH (which does not publish a tally of animal numbers reported to it), and how many uncounted mice are in private labs with no accreditation and no federal funding.

7. There are dozens of rodents whose English names include rat or mouse, such as the jumping mouse, wood rat, and grasshopper mouse. The genera *Rattus* and *Mus* include only the common house mouse, the Norway rat, and the black rat (aka the ship rat), their lab-bred descendants, and a few closely related species.

The definition of *animal* in the Animal Welfare Act (as passed by Congress) and the "Animal Welfare Regulations" (as written and enforced by USDA) is on the USDA's website and is anything but reader-friendly:

> The term "animal" means any live or dead dog, cat, monkey (nonhuman primate mammal), guinea pig, hamster, rabbit, or such other warm-blooded animal, as the Secretary may determine is being used, or is intended for use, for research, testing, experimentation, or exhibition purposes, or as a pet; but *such term excludes (1) birds, rats of the genus Rattus, and mice of the genus Mus, bred for use in research* [emphasis added], (2) horses not used for research purposes, and (3) other farm animals, such as, but not limited to livestock or poultry, used or intended for use as food or fiber, or livestock or poultry used or intended for use for improving animal nutrition, breeding, management, or production efficiency, or for improving the quality of food or fiber. With respect to a dog, the term means all dogs including those used for hunting, security, or breeding purposes. (Animal and Plant Health, "Animal Care")

8. Various people and groups, such as the National Association for Biomedical Research, put the number of mice and rats in US labs at around 95 percent of all warm-blooded (i.e., mammal and bird) animals in research. Animal-protection groups such as the Animal Legal Defense Fund use this figure as well. In my research I estimated 97 percent. Factoring in fish, similarly excluded from Animal Welfare Act protections and from requirements for counting them, the mice, rats, and fish presently excluded from the Animal Welfare Act have got to be well in excess of 99 percent of all lab animals. Animal Legal Defense Fund, "Federal Laws and Agencies Involved with Animal Testing," accessed June 12, 2025, https://aldf .org/article/federal-laws-and-agencies-involved-with-animal-testing/; "The Importance of Animal Research," National Association for Biomedical Research,

accessed May 8, 2020, www.nabr.org/biomedical-research/importance-biomedical
-research; Carbone (2021).

9. While I've had dozens of off-the-record conversations about problem or
difficult PIs (i.e., principal investigators who head their own labs and grants),
I've found only a single in-print mention of this phenomenon; see Schuppli
(2004).

10. For the National Academies' examinations of dogs, chimps, and monkeys
in NIH and Veterans Administration research, see National Academies of Sciences (2023, 2020) and Committee on the Use of Chimpanzees in Biomedical
and Behavioral Research (2011).

11. The American Fancy Rat and Mouse Association is my source for this history. Nicole Royer, "The History of Fancy Mice," last updated February 23, 2015,
www.afrma.org/historymse.htm.

12. See these histories of the Jackson Labs and of Abbie Lathrop: "Our History," Jackson Laboratory, accessed March 22, 2025, www.jax.org/about-us
/history; and Steensma et al. (2010).

13. All animal species, humans included, can get cancer. A minuscule fraction
of these cancers is contagious, where animals catch cancer from others of their
kind. Dogs have a transmissible venereal tumor, unique to their species and
passed dog to dog. Tasmanian devils are an Australian marsupial with not just
one but two contagious cancers, Devil facial tumor disease types 1 and 2, both
spread by fighting and biting. Tassie devils are a threatened species in low
number, so scientists have relied on mice as lab models (Kreiss et al. 2011; Murchison 2008). Much of animal-based cancer research relies on cancer cells' ability to grow in a body when deliberately injected. Human cancers and Tasmanian
devil cancers can grow as transplants in immune-suppressed mice, though they
are not then contagious among the mice sharing a cage.

14. Tristan Free and Jenny Straiton write that inbreeding did indeed end
some European royal lineages and note that the highly inbred families of the
Game of Thrones television series somehow escaped such consequences suspiciously easily. "Family, Duty, Honor: Looking at Historical Parallels to the
Inbreeding in Europe and *Game of Thrones*," *BioTechniques*, April 12, 2019,
www.biotechniques.com/general-interest/family-duty-honor-looking-at-historical
-parallels-of-the-royal-incest-in-game-of-thrones/.

15. To read some history of the OncoMouse, see Hanahan et al. (2007).

16. CRISPR is an acronym for the *clustered regularly interspaced short palin-
dromic repeats* in bacterial cells. It's part of the bacteria's defense against bacteria-eating viruses (Mojica and Montoliu 2016).

17. The Food and Drug Administration's material on drugs it approves is
online. For Casgevy they review the mouse data for pharmacology and safety
testing, noting there's no good animal model of sickle cell disease in which the
company could test the drug's efficacy. US Food and Drug Administration,

"Approval History, Letters, Reviews, and Related Documents: CASGEVY," January 5, 2024, www.fda.gov/media/175235.

18. CRISPR gene editing is crucial to efforts to genetically engineer pigs to serve as organ donors for humans, retooling the genes that would stimulate organ rejection and modifying just how large a pig organ can grow in a human recipient (Ryczek et al. 2021). For two ethicists' take on the implications of how CRISPR makes large-animal research more feasible, I recommend Walker and Eggel (2020).

19. Yet darker are the possibilities to genetically engineer humans, not for programmed depression and anxiety but for what, better docility? To "correct" any flaws and deviations from ideal height, eye color, or gender identity? In 2019, following a Chinese scientist's news that he had used CRISPR to edit some human babies' genomes, two of the CRISPR inventors joined with other scientists to press for a moratorium on modifying sperm, ova, or embryos to make genetically modified children (Lander et al. 2019). This situation is in flux legally, jurisdiction by jurisdiction, with no emergent international consensus at this time. He Juankui received a three-year jail sentence for forging his ethics approval and exceeding Chinese ethics and practice rules. Dennis Normile, "Chinese Scientist Who Produced Genetically Altered Babies Sentenced to 3 Years in Jail," *Science*, December 30, 2019, https://doi.org/10.1126/science.aba7347.

20. Harlow and Suomi (1971). For a deeper discussion of Harlow's work, see Blum (1994).

FLEA

1. I recommend reading Bernard Rollin's general books on animal ethics, as well as his most research-focused effort, *The Unheeded Cry* (1989), along with his memoir of his improbable and inspirational life, *Putting the Horse Before Descartes* (2011).

2. Anyone wishing to read more about animal ethics in general should start with Peter Singer's *Animal Liberation Now* (2023), now in its third edition; Tom Regan's seminal *The Case for Animal Rights* (1983); and David DeGrazia's *Taking Animals Seriously* (1996). In addition to a wealth of authors with book-length treatments of animal ethics, several anthologies and encyclopedias include a range of shorter contributions from diverse thinkers that can serve as an introduction for deeper exploration (Fischer 2019; Botzler and Armstrong 2016; Beauchamp and Frey 2011). For work more focused on animals in laboratories, start with Beauchamp and DeGrazia (2020); Rollin (1989, 1995); and me (Carbone 2004).

3. These percentages are very, very rough, as different geneticists score genetic similarity differently. Some compare only the genes that encode proteins. Others

cast a wider net, with DNA genes that just make RNA or that may not encode for anything at all. No matter the precise number, chimps are far and away closer relatives than mice, dogs, or chickens.

4. The Great Ape Project began with two philosophers, Paola Cavalieri and Peter Singer, and has inspired chimpanzee sanctuaries and political movements in several countries. Lawyers have brought lawsuits on behalf of individual chimps, elephants, and orcas that would grant those individuals legal rights hitherto reserved exclusively for us *Homo sapiens*, with the express expectation that success would set a legal precedent that others of their species have the same legal rights (Cavalieri and Singer 1993).

5. Cher's rescue of Kaavan the elephant is the subject of a documentary, *Cher and the Loneliest Elephant* (see Finnigan 2021).

6. Bentham presaged Ryder's and Singer's urge to extrapolate from the evils of racism and slave-holding to think about the evils of speciesism. The fuller Bentham passage lays out his case:

> The day has been, I grieve to say in many places it is not yet past, in which the greater part of the species, under the denomination of slaves, have been treated by the law exactly upon the same footing as, in England for example, the inferior races of animals are still. The day *may come*, when the rest of animal creation may acquire those rights which never could have been withholden from them but by the hand of tyranny. The French have already discovered that the blackness of the skin is no reason why a human being should be abandoned without redress to the caprice of a tormentor. It may come one day to be recognized that the number of the legs, or the villosity of the skin, or the termination of the *os sacrum*, are reasons equally insufficient for abandoning a sensitive being to the same fate. What else is it that should trace the insuperable line? Is it the faculty of reason, or, perhaps, the faculty of discourse? But a full-grown horse or dog is beyond comparison a more rational, as well as a more conversable animal, than an infant of a day, or a week, or even a month, old. But suppose they were otherwise, what would it avail? the question is not, Can they *reason*? nor, Can they *talk*? but, Can they *suffer*? (Bentham 1789, 310–11)

7. Richard Ryder's essays appear in several anthologies of animal ethics (see Ryder 1985).

8. A panel reviewing the use of dogs in Veterans Administration research agrees with me on the unpopular belief in animal-welfare arguments in favor of keeping dogs in labs rather than replacing them with pigs. See Forster et al. (2010) and National Academies of Sciences (2020).

9. Hal Herzog's book *Some We Love, Some We Hate, Some We Eat* (2021) catalogs some of the many conflicting feelings people have about the various animals in our world.

10. The World Animal Protection group updated its list of countries that explicitly reference sentience in their laws in 2023. "Encouraging Animal Sentience Laws Around the World," April 27, 2023, www.worldanimalprotection .org/latest/blogs/encouraging-animal-sentience-laws-around-world/. The list of

countries protecting live conscious lobsters from the pot is growing. Joseph Prezioso, "Cooking a Lobster Alive? Whether This Is Legal Depends on Where in Europe You Are Cooking," Belga News Agency, December 24, 2022, www .belganewsagency.eu/cooking-a-lobster-alive-whether-this-is-legal-depends-on -where-in-europe-you-are-cooking.

11. Arluke (1994); Sharp (2018).

12. Carbone (2004, 213).

13. Reflecting the central place the Three Rs have achieved among laboratory animal people, numerous writers have hitched their agendas to them and have proliferated a nearly endless list of "fourth Rs," including respect, rehoming, rehabilitation, responsibility, refusal (of fruitless protocols), relaxation, reproducibility, reuse, relationship, repayment, relevance, recourse, and remembering, with surely more Rs to come.

14. The Animal Welfare Information Center could not work miracles with its allotted funding and staffing but still holds much of value. "Animal Welfare Information Center (AWIC)," National Agricultural Library, accessed March 15, 2025, www.nal.usda.gov/programs/awic.

15. Russell and Burch (1959, 14).

16. A caveat to the premise that animal subjects do not gain from their service is research, especially clinical trials, of therapies for individual animals. Studies of pets' or zoo animals' cancers or surgeries may help the afflicted dog or rhinoceros, along with producing knowledge relevant for their own or other species, including humans. Research, such as studies that zoo vets often conduct, that may help the species but not the research subjects themselves calls for a different ethical review. Absent data collection or reliable statistics, as far as anyone knows, such animals are a tiny fraction of the large numbers of animals living as models in labs devoted to human and veterinary health and safety.

17. Beauchamp and DeGrazia (2020) organize six principles in two categories, the "Principles of Social Benefit" and the "Principles of Animal Welfare." The three "Principles of Social Benefit" are "The Principle of No Alternative Method," "The Principle of Expected Net Benefit," and "The Principle of Sufficient Value to Justify Harm." Animal-welfare principles comprise the "Principle of No Unnecessary Harm," the "Principle of Basic Needs," and the "Principle of Upper Limits to Harm."

18. Bateson published his cube in a brief article in *New Scientist*, and lab animal ethicists have modified and relied on it since. Even his three-dimensional structure is an oversimplification, as it condenses all possible kinds of harm to all kinds of animals into a single axis. See his "When to Experiment on Animals," February 20, 1986, 30–32.

19. Seok et al. (2013, 3507).

20. Aronson and Green's (2020) history and analysis of "me-too" pharmaceuticals does not delve into the role of animals, but it is a helpful summary of why

and when companies pursue duplicative drugs, including within the same company.

21. Tannenbaum (2017). The Golden Goose Awards are not just for animal studies, so look to Speaking of Research to annually highlight the animal experiments that look pointless but come with big payoffs. For a fuller read on Bang and Levin's horseshoe crab award, start with the Golden Goose press release: "2019: The Blood of the Horseshoe Crab," Golden Goose Award, accessed June 14, 2019, www.goldengooseaward.org/01awardees/horseshoe-crab-blood; see also "The Golden Goose Awards," Speaking of Research, April 2012, https://speakingofresearch.com/2012/04/30/the-golden-goose-awards/; and "The Golden Goose Award: History," Golden Goose Award, accessed June 14, 2024, www.goldengooseaward.org/history.

22. John Ioannidis explains this more fully in his essay "Why Most Published Research Findings Are False" (2005).

RHESUS MONKEY

1. Singer writes of using hundreds of monkeys to develop deep-brain stimulation for tens of thousands of human patients, "If suffering was minimized, then this research does appear to bring greater benefits than it causes harm, even when we give this harm no less weight than we would give to similar harm inflicted on a human being with cognitive capacities no greater than those of the monkeys used." See Singer (2023).

2. My university, the University of California–San Francisco, was in the crosshairs of Stop Animal Exploitation NOW! for many years. Their posting on UCSF monkey use is disturbing reading and exactly what we in the labs tried to minimize. I can state with confidence that the pictures they used were not our animals. "Primate Experimentation in the U.S.—UCSF," Stop Animal Exploitation NOW!, accessed June 20, 2025, https://saenonline.org/fact-primate-ucsf.html.

3. As I describe in the chapter "Python," the US Animal Welfare Act covers all labs but excludes most animals—that is, the mice, rats, fish, birds, and frogs. NIH rules (based on the 1985 Health Research Extension Act) cover all vertebrates and octopuses, but only in federally funded labs, not in private companies or small colleges. AAALAC is the body that accredits lab animal programs if they seek and pay for its imprimatur. A handful of states and cities have lab animal-welfare laws as well. Curiously, after thirty years of dual federal laws, Congress passed the Twenty-First Century Cures Act in 2016. In terms of animals, the goal was to reduce regulatory burden and bureaucracy but, rather than combine the NIH and Animal Welfare Act laws into one comprehensive law overseen by a single agency, it simply called on the NIH, the USDA, and the Food and Drug

Administration to work together to harmonize their requirements to the extent possible.

4. For coverage of the USDA closure and fine of the Envigo dog-breeding kennel in Virginia, see Martin Weil, "Dog Breeder to Pay Record $35M Fine After Surrendering Thousands of Beagles," *Washington Post*, June 4, 2024, www.washingtonpost.com/dc-md-va/2024/06/04/envigo-beagles-rescue-breeder-fine/. Readers can freely find Envigo's inspection reports on the USDA's website, though navigating that site has always been cumbersome and confusing. The People for the Ethical Treatment of Animals coverage includes links to the USDA's inspection reports, so readers can more easily see the specific infractions, plus see what these sorts of inspection reports look like. "Deprivation, Despair, and Death at Envigo: A PETA Undercover Investigation; Federal Officials' Findings," People for the Ethical Treatment of Animals, accessed March 15, 2025, https://investigations.peta.org/dog-beagle-breeding-mill-envigo/usda-findings/.

5. Various organizations make use of the Foundation for Biomedical Research's materials; see, for example, Indiana University, "Fact vs. Myth: The Essential Need for Animals in Biomedical Research," accessed June 9, 2025, https://research.iu.edu/compliance/animal-care/lab-animal-resources/about/fact-vs-myth.html; and Walter Jessen, "Animal Research: Animal Welfare vs. Animal Rights," Highlight Health, September 22, 2010, www.highlighthealth.com/biomedical-research/animal-research-animal-welfare-vs-animal-rights/. For a collection of similar statements in the United States and Canada and a legal refutation, see Black (2011).

6. For opposing half truths about animal research and its regulations, see the research lobbyists at the National Association for Biomedical Research ("Crisis Management Guide," Fall 2013, www.nabr.org/view_file/6116/1893/8941/CMG-NABR.pdf) and the activists at People for the Ethical Treatment of Animals ("Don't Existing Laws Ensure That Animals in Labs Are Treated Humanely?," accessed April 29, 2025, www.peta.org/about-peta/faq/dont-existing-laws-ensure-that-animals-in-labs-are-treated-humanely/).

7. Reflecting this ambivalent view of the USDA's handling of its Animal Welfare Act responsibilities, the Animal Law Clinic at Harvard University filed two lawsuits, one to get the USDA to rewrite its overly broad (in the Harvard clinic's view) standards for primate well-being and the other to stop USDA plans to institute a merit-based system in which labs with good track records get reduced coverage in inspections to devote resources for more focus on labs with bad performance. Jo B. Lemann and Neil H. Shah, "District Court Judge Rules in Favor of HLS Animal Law Clinic in Suit Against USDA," *Harvard Crimson*, April 7, 2023, www.thecrimson.com/article/2023/4/7/animal-law-clinic-wins/.

8. The UCSF student newspaper wrote up the news of the 1998 San Francisco City Hall hearing, then reprised the story twenty years later, without updating

any information. "This Date in UCSF History: SF Supervisors Debate UCSF Animal Experiments," Synapse, November 16, 2018, https://synapse.ucsf .edu/articles/2018/11/16/date-ucsf-history-sf-supervisors-debate-ucsf-animal -experiments.

9. Scoring how well any inspection or accreditation system works is challenging, though PETA tried that by comparing USDA inspection reports of accredited versus nonaccredited facilities and declaring accredited facilities worse. Though critical of the AAALAC's lack of transparency, PETA never published the dataset on which it based its analysis or shared it with AAALAC, who responded critically to PETA's publication. Readers can see the article and rebuttal and its coverage in *Science* magazine (Goodman et al. 2015; Grimm 2014; Newcomer 2015).

10. Animal and Plant Health Inspection Service, "Animal Care: Animal Welfare Act and Animal Welfare Regulations," July 2023, www.aphis.usda.gov /sites/default/files/ac_bluebook_awa_508_comp_version.pdf, 209.

11. National Research Council (2011, 55).

12. National Research Council (2011, 31, 114).

13. Animal and Plant Health, "Animal Care," 77, 71.

14. Animal and Plant Health, "Animal Care," 68, 71.

15. See Tannenbaum (2017) for the argument on why the individual scientist, not an animal-ethics committee, should conduct the ethical harm-benefit analysis of a project's justification.

16. Landi (a lab vet and ethicist) teamed up with bioethicists Shriver and Mueller to argue for moving committee reviews from a check-the-box mentality to a more engaged ethical deliberation (see Landi et al. 2015).

17. The NIH posted its guidance against ethics committees (it uses the name Institutional Animal Care and Use Committee, or IACUC) reviewing the scientific merit of a project, as well as the minimal comparison it requires to ensure that the proposal the experts review and the more detailed one the ethics committee reviews are in fact identical and congruent; see "Frequently Asked Questions: Is the IACUC Responsible for Judging the Scientific Merit of Proposals?," NIH Office of Laboratory Animal Welfare, last updated February 15, 2023, https://olaw.nih.gov/faqs#/guidance/faqs?anchor=questionhttps://olaw.nih.gov /guidance/faqs. I've critiqued the NIH approach more thoroughly in Beauchamp and DeGrazia's book *Principles of Animal Research Ethics* (Carbone 2020).

18. National Research Council (2011); Animal and Plant Health, "Animal Care."

19. US Department of Agriculture, "Animal Welfare Inspection Guide," January 2021, https://nationalawa.org/files/Animal_Care_Inspection_Guide _Jan_2021.pdf.

20. "Pain and Itching: Pain in Multiple Sclerosis," National Multiple Sclerosis Society, accessed March 11, 2025, www.nationalmssociety.org/understanding -ms/what-is-ms/ms-symptoms/pain-itching.

21. See the Office for Human Research Protections Part D rules for human subjects protections: "Subpart D: Additional Protections for Children Involved as Subjects in Research," US Department of Health and Human Services, last reviewed November 13, 2024, www.hhs.gov/ohrp/regulations-and-policy /regulations/45-cfr-46/common-rule-subpart-d/index.html.

22. For more on habeas corpus rights and guardianship for great apes and elephants, see "Our Clients: Their Stories and Cases," Nonhuman Rights Project, accessed March 11, 2025, www.nonhumanrights.org/our-clients/. For a review on legal rights for nonsentient natural entities, see Takacs (2021).

23. Animal rights groups in the United Kingdom are not universally supportive of the Concordat transparency effort, calling it propaganda, not transparency (Jarrett 2016). It far eclipses anything in the United States.

24. For a range of examples of UK animal-protection groups collaborating with scientists to develop best practices for animal models of septic shock, rodent euthanasia, and octopus pain and distress, see Lilley et al. (2015); Hawkins et al. (2016); and Andrews et al. (2013).

25. For the semitransparent, nontechnical summaries the UK Home Office publishes, see "Non-technical Summaries of Projects Granted Under ASPA," Home Office, January 23, 2025, www.gov.uk/government/collections/non -technical-summaries-of-projects-granted-under-aspa.

GORILLA

1. Animal and Plant Health Inspection Service, "Animal Care: Animal Welfare Act and Animal Welfare Regulations," July 2023, www.aphis.usda.gov /sites/default/files/ac_bluebook_awa_508_comp_version.pdf, 137.

2. National Institutes of Health, "Update on the Requirement for Instruction in the Responsible Conduct of Research," accessed June 6, 2011, https://grants .nih.gov/grants/guide/notice-files/not-od-10-019.html.

3. National Research Council (2011, 4).

4. National Research Council (2011, 27).

Works Cited

Adashi, Eli Y., Daniel P. O'Mahony, and I. Glenn Cohen. 2023. "The FDA Modernization Act 2.0: Drug Testing in Animals Is Rendered Optional." *American Journal of Medicine* 136 (9): 853–54. www.ncbi.nlm.nih.gov /pubmed/37080328.

Ahlgren, Johanna, and Vootele Voikar. 2019. "Housing Mice in the Individually Ventilated or Open Cages: Does It Matter for Behavioral Phenotype?" *Genes, Brain and Behavior* 18 (7): e12564. www.ncbi.nlm.nih.gov/pubmed /30848040.

Andrews, Paul L. R., Anne-Sophie Darmaillacq, Ngaire Dennison, et al. 2013. "The Identification and Management of Pain, Suffering and Distress in Cephalopods, Including Anaesthesia, Analgesia and Humane Killing." Special issue, *Journal of Experimental Marine Biology and Ecology* 447:46–64. www.sciencedirect.com/science/article/pii /S0022098113000622?via%3Dihub.

Animal and Plant Health Inspection Service. 1991. "Environment Enhancement to Promote Psychological Well-Being." *Federal Register* 56 (32): 6426–505. www.govinfo.gov/content/pkg/CFR-2022-title9-vol1/pdf /CFR-2022-title9-vol1-sec3-81.pdf.

Animal Care Panel. 1963. *Guide for Laboratory Animal Facilities and Care.* Washington, DC: Public Health Service.

Appel, Mirjam, and Robert W. Elwood. 2009. "Motivational Trade-Offs and Potential Pain Experience in Hermit Crabs." *Applied Animal Behaviour*

Science 119 (1–2): 120–24. www.sciencedirect.com/science/article/pii /S0168159109001038.

Arluke, Arnold. 1994. "The Ethical Socialization of Animal Researchers." *Lab Animal* 23 (6): 30–35. www.researchgate.net/publication/269987131_The _Ethical_Socialization_of_Animal_Researchers.

Aronson, Jeffrey K., and A. Richard Green. 2020. "Me-Too Pharmaceutical Products: History, Definitions, Examples, and Relevance to Drug Shortages and Essential Medicines Lists." *British Journal of Clinical Pharmacology* 86 (11): 2114–22. www.ncbi.nlm.nih.gov/pubmed/32358800.

Bailoo, Jeremy D., Thomas S. Reichlin, and Hanno Würbel. 2014. "Refinement of Experimental Design and Conduct in Laboratory Animal Research." *ILAR Journal* 55 (3): 383–91. https://pubmed.ncbi.nlm.nih.gov/25541540/.

Baumans, Vera, Freek Schlingmann, Marlice Vonck, and Hein A. van Lith. 2002. "Individually Ventilated Cages: Beneficial for Mice and Men?" *Contemporary Topics in Laboratory Animal Science* 49 (1): 13–19. https:// pubmed.ncbi.nlm.nih.gov/11860253/.

Beauchamp, Tom, and David DeGrazia. 2020. *Principles of Animal Research Ethics.* New York: Oxford University Press.

Beauchamp, Tom L., and Raymond G. Frey, eds. 2011. *The Oxford Handbook of Animal Ethics.* New York: Oxford University Press.

Bekoff, Marc, and Jessica Pierce. 2017. *The Animals' Agenda: Freedom, Compassion, and Coexistence in the Human Age.* Boston: Beacon.

Bentham, Jeremy. 1948. *An Introduction to the Principles of Morals and Legislation.* New York: Hafner.

Beura, Lalit K., Sara E. Hamilton, Kevin Bi, et al. 2016. "Normalizing the Environment Recapitulates Adult Human Immune Traits in Laboratory Mice." *Nature* 532 (7600): 512–16. www.ncbi.nlm.nih.gov/pubmed /27096360.

Black, Vaughan. 2011. "A Regulated Regard: Comparing the Governance of Animal and Human Experimentation." *Revue Québécoise de Droit International* 24 (1): 237–48. www.persee.fr/doc/rqdi_0828-9999_2011_num _24_1_1225.

Blum, Deborah. 1994. *The Monkey Wars.* New York: Oxford University Press.

Botzler, Richard G., and Susan J. Armstrong, eds. 2016. *The Animal Ethics Reader.* London: Routledge.

Brown, Marilyn, Larry Carbone, Kathleen M. Conlee, et al. 2006. "Report of the Working Group on Animal Distress in the Laboratory." *Lab Animal* 35 (8): 26–30. www.ncbi.nlm.nih.gov/pubmed/16943790.

Buhot-Averseng, Marie-Christine. 1981. "Nest-Box Choice in the Laboratory Mouse: Preferences for Nest-Boxes Differing in Design (Size and/or Shape) and Composition." *Behavioural Processes* 6 (4): 337–84. www.ncbi.nlm.nih .gov/pubmed/24925866.

Burghardt, Gordon M. 1999. "Deprivation and Enrichment in Laboratory Animal Environments." *Journal of Applied Animal Welfare Science* 2 (4): 263–66. www.ncbi.nlm.nih.gov/pubmed/16363931.

Byrd, Charles P., Christina Winnicker, and Brianna N. Gaskill. 2016. "Instituting Dark-Colored Cover to Improve Central Space Use Within Guinea Pig Enclosure." *Journal of Applied Animal Welfare Science* 19 (4): 408–13. www.ncbi.nlm.nih.gov/pubmed/27223319.

Cait, Jessica, Alissa Cait, R. Wilder Scott, Charlotte B. Winder, and Georgia J. Mason. 2022. "Conventional Laboratory Housing Increases Morbidity and Mortality in Research Rodents: Results of a Meta-analysis." *BMC Biology* 20 (1): 15. www.ncbi.nlm.nih.gov/pubmed/35022024.

Carbone, Larry. 2004. *What Animals Want: Expertise and Advocacy in Laboratory Animal Welfare Policy.* New York: Oxford.

———. 2012. "Pain Management Standards in the Eighth Edition of the *Guide for the Care and Use of Laboratory Animals.*" *Journal of the American Association for Laboratory Animal Science* 51 (3): 1–5.

———. 2020. "The Potential and Impacts of Practical Application of Beauchamp and DeGrazia's Six Principles." In Beauchamp and DeGrazia 2020, 45–60.

———. 2021. "Estimating Mouse and Rat Use in American Laboratories by Extrapolation from Animal Welfare Act–Regulated Species." *Scientific Reports* 11:493. www.nature.com/articles/s41598-020-79961-0#article-comments.

Carbone, Larry, and Jamie Austin. 2016. "Pain and Laboratory Animals: Publication Practices for Better Data Reproducibility and Better Animal Welfare." *PLoS One* 11 (5): e0155001. www.ncbi.nlm.nih.gov/pubmed/27171143.

Carbone, Larry, Luce Guanzini, and Cary McDonald. 2003. "Adoption Options for Laboratory Animals." *Lab Animal* 32 (9): 37–41. www.nature.com/articles/laban1003-37.

Carroll, John M., and Jeff C. Clements. 2019. "Scaredy-Oysters: In Situ Documentation of an Oyster Behavioral Response to Predators." *Southeastern Naturalist* 18 (3): N21–22. https://bioone.org/journals/southeastern-naturalist/volume-18/issue-3/058.018.0303/Scaredy-Oysters--In-Situ-Documentation-of-an-Oyster-Behavioral/10.1656/058.018.0303.short.

Cavalieri, Paola, and Peter Singer, eds. 1993. *The Great Ape Project: Equality Beyond Humanity.* New York: St. Martin's Press.

Chapple, Walter A., and Walter R. Hadwen. 1914. *The Dogs Bill: A Verbatim Report of Debate Between Dr. W.A. Chapple, M.P., M.D., and Dr. Walter R. Hadwen, J.P., M.D.* London: British Union for the Abolition of Vivisection.

Clippinger, Amy J., Hans A. Raabe, David G. Allen, et al. 2021. "Human-Relevant Approaches to Assess Eye Corrosion/Irritation Potential of

Agrochemical Formulations." *Cutaneous and Ocular Toxicology* 40 (2): 145–67. www.ncbi.nlm.nih.gov/pubmed/33830843.

Committee on Scientific and Humane Issues in the Use of Random Source Dogs and Cats in Research. 2009. *Scientific and Humane Issues in the Use of Random Source Dogs and Cats in Research*. Washington, DC: National Academies Press.

Committee on the Use of Chimpanzees in Biomedical and Behavioral Research. 2011. *Chimpanzees in Biomedical and Behavioral Research: Assessing the Necessity*. Edited by Bruce M. Altevogt, Diana E. Pankevich, Marilee K. Shelton-Davenport, and Jeffrey P. Kahn. Washington, DC: National Academies Press. https://pubmed.ncbi.nlm.nih.gov/22514816/.

Craig, Kenneth D., Heather D. Hadjistavropoulos, Ruth V. E. Grunau, and Michael F. Whitfield. 1994. "A Comparison of Two Measures of Facial Activity During Pain in the Newborn Child." *Journal of Pediatric Psychology* 19 (3): 305–18. www.ncbi.nlm.nih.gov/pubmed/8071797.

Crook, Robyn J. 2021. "Behavioral and Neurophysiological Evidence Suggests Affective Pain Experience in Octopus." *iScience* 24 (3): 102229. https://doi.org/10.1016/j.isci.2021.102229.

Danbury, T. C., C. A. Weeks, J. P. Chambers, A. E. Waterman-Pearson, and S. C. Kestin. 2000. "Self-Selection of the Analgesic Drug Carprofen by Lame Broiler Chickens." *Veterinary Record* 146 (11): 307–11. www.ncbi.nlm.nih.gov/entrez/query.fcgi?cmd=Retrieve&db=PubMed&dopt=Citation&list_uids=10766114.

Darwin, Charles. 1871. *The Descent of Man and Selection in Relation to Sex*. London: Murray.

———. 1874. *The Descent of Man and Selection in Relation to Sex*. 2nd ed. London: Murray.

David, John M., Scott Knowles, Donald M. Lamkin, and David B. Stout. 2013. "Individually Ventilated Cages Impose Cold Stress on Laboratory Mice: A Source of Systemic Experimental Variability." *Journal of the American Association for Laboratory Animal Science* 52 (6): 738–44. www.ncbi.nlm.nih.gov/pubmed/24351762.

De Boer, Sietse F., and Jaap M. Koolhaas. 2003. "Defensive Burying in Rodents: Ethology, Neurobiology and Psychopharmacology." *European Journal of Pharmacology* 463 (1–3): 145–61. www.ncbi.nlm.nih.gov/pubmed/12600707.

DeGrazia, David. 1996. *Taking Animals Seriously*. Cambridge: Cambridge University Press.

Deutsch, Skyler, Rachel Parsons, Jonathan Shia, et al. 2023. "Evaluation of Candidates for Systemic Analgesia and General Anesthesia in the Emerging Model Cephalopod, *Euprymna berryi*." *Biology* 12 (2): 201. www.ncbi.nlm.nih.gov/pubmed/36829480.

Elwood, Robert W. 1991. "Ethical Implications of Studies on Infanticide and Maternal Aggression in Rodents." *Animal Behaviour* 42:841–49.

———. 2022. "Hermit Crabs, Shells, and Sentience." *Animal Cognition* 25:1241–57. https://link.springer.com/article/10.1007/s10071-022-01607-7.

Feller, David A. 2009. "Dog Fight: Darwin as Animal Advocate in the Antivivisection Controversy of 1875." *Studies in History and Philosophy of Science* 40 (4): 265–71. www.ncbi.nlm.nih.gov/pubmed/19917485.

Finnigan, Jonathan, dir. 2021. *Cher and the Loneliest Elephant*. Los Angeles: Paramount.

Fischer, Bob, ed. 2019. *The Routledge Handbook of Animal Ethics*. London: Routledge.

Forster, Roy, Gerd Bode, Lars Ellegaard, and Jan-Willem van der Laan. 2010. "The RETHINK Project on Minipigs in the Toxicity Testing of New Medicines and Chemicals: Conclusions and Recommendations." *Journal of Pharmacological and Toxicological Methods* 62 (3): 236–42. https://doi.org/10.1016/j.vascn.2010.05.008.

Franchi, Silvia, Alberto E. Panerai, and Paola Sacerdote. 2007. "Buprenorphine Ameliorates the Effect of Surgery on Hypothalamus-Pituitary-Adrenal Axis, Natural Killer Cell Activity and Metastatic Colonization in Rats in Comparison with Morphine or Fentanyl Treatment." *Brain, Behavior, and Immunity* 21 (6): 767–74. www.ncbi.nlm.nih.gov/pubmed/17291715.

Gaarder, Emily. 2011. *Women and the Animal Rights Movement*. New Brunswick, NJ: Rutgers University Press.

Gallo, Miranda S., Alicia Z. Karas, Kathleen Pritchett-Corning, Joseph P. Garner Guy Mulder, and Brianna N. Gaskill. 2020. "Tell-Tale TINT: Does the Time to Incorporate into Nest Test Evaluate Postsurgical Pain or Welfare in Mice?" *Journal of the American Association for Laboratory Animal Science* 59 (1): 37–45. www.ncbi.nlm.nih.gov/pubmed/31862018.

Gamalo, Lief Erikson, Kurnia Ilham, Lisa Jones-Engel, et al. 2024. "Removal from the Wild Endangers the Once Widespread Long-Tailed Macaque." *American Journal of Primatology* 86 (3): e23547.

Garner, Joseph P. 2014. "The Significance of Meaning: Why Do over 90% of Behavioral Neuroscience Results Fail to Translate to Humans, and What Can We Do to Fix It?" *ILAR Journal* 55 (3): 438–56. www.ncbi.nlm.nih.gov/pubmed/25541546.

Garner, Joseph P., Brianna N. Gaskill, Elin M. Weber, Jamie Ahloy-Dallaire, and Kathleen R. Pritchett-Corning. 2017. "Introducing Therioepistemology: The Study of How Knowledge Is Gained from Animal Research." *Lab Animal* 46 (4): 103–13. www.ncbi.nlm.nih.gov/pubmed/28328885.

Giancola-Detmering, Sarah E., and Robyn J. Crook. 2024. "Stress Produces Negative Judgement Bias in Cuttlefish." *Biology Letters* 20 (10): 20240228. https://royalsocietypublishing.org/doi/abs/10.1098/rsbl.2024.0228.

Gluck, John P. 2016. *Voracious Science and Vulnerable Animals: A Primate Scientist's Ethical Journey*. Chicago: University of Chicago Press.

Goodman, Justin R., Alka Chandna, and Casey Borch. 2015. "Does Accreditation by the Association for Assessment and Accreditation of Laboratory Animal Care International (AAALAC) Ensure Greater Compliance with Animal Welfare Laws?" *Journal of Applied Animal Welfare Science* 18 (1): 82–91. www.ncbi.nlm.nih.gov/pubmed/25174609.

Graham, Melanie L., Eric F. Rieke, Lucas A. Mutch, et al. 2012. "Successful Implementation of Cooperative Handling Eliminates the Need for Restraint in a Complex Non-human Primate Disease Model." *Journal of Medical Primatology* 41 (2): 89–106. www.ncbi.nlm.nih.gov/pubmed/22150842.

Grandin, Temple. 1995. *Thinking in Pictures*. New York: Doubleday.

Gray, Jenny. 2017. *Zoo Ethics: The Challenges of Compassionate Conservation*. Ithaca, NY: Comstock/Cornell University Press.

Greek, Ray, and Jean Greek. 2010. "Is the Use of Sentient Animals in Basic Research Justifiable?" *Philosophy, Ethics, and Humanities in Medicine* 5:14. www.ncbi.nlm.nih.gov/pubmed/20825676.

Greenfield, Ariele L., and Stephen L. Hauser. 2018. "B-Cell Therapy for Multiple Sclerosis: Entering an Era." *Annals of Neurology* 83:13–26. https://doi.org/10.1002/ana.25119.

Grimm, David. 2014. "Scientific Community: Animal Welfare Accreditation Called into Question." *Science* 345 (6200): 988. www.ncbi.nlm.nih.gov/pubmed/25170127.

Hanahan, Douglas, Erwin F. Wagner, and Richard D. Palmiter. 2007. "The Origins of Oncomice: A History of the First Transgenic Mice Genetically Engineered to Develop Cancer." *Genes and Development* 21 (18): 2258–70. www.ncbi.nlm.nih.gov/pubmed/17875663.

Hare, Brian, Alicia P. Melis, Vanessa Woods, Sara Hastings, and Richard Wrangham. 2007. "Tolerance Allows Bonobos to Outperform Chimpanzees on a Cooperative Task." *Current Biology* 17 (7): 619–23. https://doi.org/10.1016/j.cub.2007.02.040.

Harlow, Harry F., and Stephen J. Suomi. 1971. "Production of Depressive Behaviors in Young Monkeys." *Journal of Autism and Childhood Schizophrenia* 1 (3): 246–55.

Harrison, Ruth. 1964. *Animal Machines*. London: Stuart.

Harriton, Lynne. 1981. "Conversation with Henry Spira: Draize Test Activist." *Lab Animal* 10:16–22. www.wellbeingintlstudiesrepository.org/hensint/7/.

Hauser, Stephen L. 2015. "Beating MS: A Story of B Cells, with Twists and Turns." *Multiple Sclerosis* 21 (1): 8–21. https://pubmed.ncbi.nlm.nih.gov/25480864/.

Hauser, Stephen L., Amit Bar-Or, Giancarlo Comi, et al. 2017. "Ocrelizumab Versus Interferon Beta-1a in Relapsing Multiple Sclerosis." *New England*

Journal of Medicine 376 (3): 221–34. www.ncbi.nlm.nih.gov/pubmed /28002679.

Hawkins, Penny, Mark J. Prescott, Larry Carbone, et al. 2016. "A Good Death? Report of the Second Newcastle Meeting on Laboratory Animal Euthanasia." *Animals* (Basel) 6 (9): 50. https://pmc.ncbi.nlm.nih.gov/articles/PMC5035945/.

Hay, Michael, David W. Thomas, John L. Craighead, Celia Economides, and Jesse Rosenthal. 2014. "Clinical Development Success Rates for Investigational Drugs." *Nature Biotechnology* 32 (1): 40–51. www.ncbi.nlm.nih.gov /pubmed/24406927.

Hediger, Heini. 1969. *Psychology and Behaviour of Animals in Zoos and Circuses*. New York: Dover.

Herzog, Hal. 2002. "Ethical Aspects of Relationships Between Humans and Research Animals." *ILAR Journal* 43 (1): 27–32. www.ncbi.nlm.nih.gov /pubmed/11752728.

———. 2021. *Some We Love, Some We Hate, Some We Eat: Why It's So Hard to Think Straight About Animals*. 2nd ed. New York: Harper Perennial.

Holden, Constance. 1988. "Experts Ponder Simian Well-Being." *Science* 241 (4874): 1753–55. www.ncbi.nlm.nih.gov/pubmed/3175616.

Hooijmans, Carlijn R., Rob B. M. de Vries, Maroeska M. Rovers, Hein G. Gooszen, and Merel Ritskes-Hoitinga. 2012. "The Effects of Probiotic Supplementation on Experimental Acute Pancreatitis: A Systematic Review and Meta-analysis." *PLoS One* 7 (11): e48811. www.ncbi.nlm.nih.gov/pubmed /23152810.

Hughes, Barry, and A. J. Black. 1973. "The Preference of Domestic Hens for Different Types of Battery Cage Floor." *British Poultry Science* 14:615–19.

Hughes, Howard C., Sarah Campbell, and Cheryl Kenney. 1989. "The Effects of Cage Size and Pair Housing on Exercise of Beagle Dogs." *Laboratory Animal Science* 39 (4): 302–5.

Ineichen, Benjamin Victor, Ulrike Held, Georgia Salanti, Malcolm Robert Macleod, and Kimberley Elaine Wever. 2024. "Primer: Systematic Review and Meta-analysis of Preclinical Studies." *Nature Reviews Methods Primers* 4:72. https://doi.org/10.1038/s43586-024-00347-x.

Interagency Coordinating Committee on the Validation of Alternative Methods. 2010. *ICCVAM Test Method Evaluation Report: Recommendations for Routine Use of Topical Anesthetics, Systemic Analgesics, and Humane Endpoints to Avoid or Minimize Pain and Distress in Ocular Safety Testing*. Triangle Park, NC: National Toxicology Program.

Ioannidis, John P. A. 2005. "Why Most Published Research Findings Are False." *PLoS Medicine* 2 (8): e124. www.ncbi.nlm.nih.gov/pubmed/16060722.

———. 2007. "Why Most Published Research Findings Are False: Author's Reply to Goodman and Greenland." *PLoS Medicine* 4 (6): e215. www.ncbi .nlm.nih.gov/pubmed/17593900.

Jarrett, Wendy. 2016. "The Concordat on Openness and Its Benefits to Animal Research." *Lab Animal* 45 (6): 201–2. www.nature.com/articles/laban.102.

Jirkof, Paulin. 2017. "Side Effects of Pain and Analgesia in Animal Experimentation." *Lab Animal* 46 (4): 123–28. www.ncbi.nlm.nih.gov/pubmed /28328895.

Jones, Earl M. 1968. "The Laboratory Animal Welfare Act: Regulations, Standards, Enforcement, and Progress." *Journal of the American Veterinary Medical Association* 153 (12): 1874–77.

Kaplan, Melanie D. G. 2025. *Lab Dog: A Beagle and His Human Investigate the Surprising World of Animal Research*. New York: Basic Books.

Kappos, Ludwig, David Li, Peter A. Calabresi, et al. 2011. "Ocrelizumab in Relapsing-Remitting Multiple Sclerosis: A Phase 2, Randomised, Placebo-Controlled, Multicentre Trial." *Lancet* 378 (9805): 1779–87. www.ncbi.nlm .nih.gov/pubmed/22047971.

Kirkden, Richard D., Lee Niel, Gary Lee, I. Joanna Makowska, Marianne J. Pfaffinger, and Daniel M. Weary. 2008. "The Validity of Using an Approach-Avoidance Test to Measure the Strength of Aversion to Carbon Dioxide in Rats." *Applied Animal Behaviour Science* 114 (1–2): 216–34. www.sciencedirect.com /science/article/pii/S0168159108000695.

Knight, Jean, Thomas Hartung, and Costanza Rovida. 2023. "4.2 Million and Counting . . . : The Animal Toll for REACH Systemic Toxicity Studies." *ALTEX* 40 (3): 389–407. www.ncbi.nlm.nih.gov/pubmed/37470350.

Kreiss, A., C. Tovar, D. L. Obendorf, K. Dun, and G. M. Woods. 2011. "A Murine Xenograft Model for a Transmissible Cancer in Tasmanian Devils." *Veterinary Pathology* 48 (2): 475–81. www.ncbi.nlm.nih.gov/pubmed/20861503.

Krugman, Saul. 1976. "Viral Hepatitis: Overview and Historical Perspectives." *Yale Journal of Biology and Medicine* 49 (3): 199–203. www.ncbi.nlm.nih .gov/pubmed/785825.

LaFollette, Hugh, and Niall Shanks. 1996. *Brute Science: Dilemmas of Animal Experimentation*. London: Routledge.

LaFollette, Megan R., Marguerite E. O'Haire, Sylvie Cloutier, Whitney B. Blankenberger, and Brianna N. Gaskill. 2017. "Rat Tickling: A Systematic Review of Applications, Outcomes, and Moderators." *PLoS One* 12 (4): e0175320. www.ncbi.nlm.nih.gov/pubmed/28384364.

LaFollette, Megan R., Megan C. Riley, Sylvie Cloutier, Colleen M. Brady, Marguerite E. O'Haire, and Brianna N. Gaskill. 2020. "Laboratory Animal Welfare Meets Human Welfare: A Cross-Sectional Study of Professional Quality of Life, Including Compassion Fatigue in Laboratory Animal Personnel." *Frontiers in Veterinary Science* 7:114. www.ncbi.nlm.nih.gov /pubmed/32195275.

Lahvis, Garet P. 2017. "Unbridle Biomedical Research from the Laboratory Cage." *Elife* 29 (6): e27438. www.ncbi.nlm.nih.gov/pubmed/28661398.

Lander, Eric S., Françoise Baylis, Feng Zhang, et al. 2019. "Adopt a Moratorium on Heritable Genome Editing." *Nature* 567 (7747): 165–68. www.ncbi.nlm .nih.gov/pubmed/30867611.

Landi, Margaret S., Adam J. Shriver, and Anne Mueller. 2015. "Consideration and Checkboxes: Incorporating Ethics and Science into the 3Rs." *Journal of the American Association for Laboratory Animal Science* 54 (2): 224–30. www.ncbi.nlm.nih.gov/pubmed/25836970.

Langford, Dale J., Andrea L. Bailey, Mona Lisa Chanda, et al. 2010. "Coding of Facial Expressions of Pain in the Laboratory Mouse." *Nature Methods* 7 (6): 447–49. www.ncbi.nlm.nih.gov/pubmed/20453868.

Lansbury, Coral. 1985. *The Old Brown Dog: Women, Workers and Vivisection in Edwardian England.* Madison: University of Wisconsin Press.

Ledoux, Tamara, Jeff C. Clements, Luc A. Comeau, et al. 2023. "Effects of Anthropogenic Sounds on the Behavior and Physiology of the Eastern Oyster (*Crassostrea virginica*)." *Frontiers in Marine Science* 10:1104526. www .frontiersin.org/journals/marine-science/articles/10.3389/fmars .2023.1104526/full.

Lilley, Elliot, Rachel Armstrong, Nicole Clark, et al. 2015. "Refinement of Animal Models of Sepsis and Septic Shock." *Shock* 43 (4): 304–16. https:// journals.lww.com/shockjournal/Fulltext/2015/04000/Refinement_of _Animal_Models_of_Sepsis_and_Septic.2.aspx.

Lundh, Andreas, Joel Lexchin, Barbara Mintzes, Jeppe B. Schroll, and Lisa Bero. 2018. "Industry Sponsorship and Research Outcome: Systematic Review with Meta-analysis." *Intensive Care Medicine* 44 (10): 1603–12. www.ncbi.nlm.nih.gov/pubmed/30132025.

Manser, Caroline E., H. Elliot, T. H. Morris, and Donald M. Broom. 1996. "The Use of a Novel Operant Test to Determine the Strength of Preference for Flooring in Laboratory Rats." *Laboratory Animals* 30 (1): 1–6. https:// pubmed.ncbi.nlm.nih.gov/8709567/.

Markowitz, Hal. 2011. *Enriching Animals' Lives.* Pacifica, CA: Mauka.

Markowitz, Hal, and Joseph S. Spinelli. 1986. "Environmental Engineering for Primates." In *Primates: The Road to Self-Sustaining Populations*, edited by Kurt Benirschke, 489–98. New York: Springer.

Marks, J. 1972. "Ending the Routine Guinea-Pig Test." *Tubercle* 53 (1): 31–34. www.ncbi.nlm.nih.gov/pubmed/4556758.

Mason, Peter. 1997. *The Brown Dog Affair: The Story of a Monument That Divided the Nation.* London: Two Sevens.

Massacesi, Luca, Claude P. Genain, David Lee-Parritz, Norman L. Letvin, Donald Canfield, and Stephen Hauser. 1995. "Active and Passively Induced Experimental Autoimmune Encephalomyelitis in Common Marmosets: A New Model for Multiple Sclerosis." *Annals of Neurology* 37:519–30. https:// onlinelibrary.wiley.com/doi/epdf/10.1002/ana.410370415?saml_referrer.

Mellor, David J. 2019. "Welfare-Aligned Sentience: Enhanced Capacities to Experience, Interact, Anticipate, Choose and Survive." *Animals* 9 (7): 440. www.ncbi.nlm.nih.gov/pubmed/31337042.

Mench, Joy A. 1998. "Thirty Years After Brambell: Whither Animal Welfare Science?" *Journal of Applied Animal Welfare Science* 1 (2): 91–102. www.ncbi.nlm.nih.gov/pubmed/16363974.

Mench, Joy A., Stephen J. Mayer, Lee Krulisch, and Scientists Center for Animal Welfare. 1992. *The Well-Being of Agricultural Animals in Biomedical and Agricultural Research: Proceedings from a SCAW-Sponsored Conference, Agricultural Animals in Research, Held September 6–7, 1990 in Washington, D.C., with Additional Material Provided by the Authors.* Bethesda, MD: Scientists Center for Animal Welfare.

Mesle, Marguax M. I., Jeremy Brown, Piers Mook, et al. 2024. "Estimated Number of Lives Directly Saved by COVID-19 Vaccination Programmes in the WHO European Region from December, 2020, to March, 2023: A Retrospective Surveillance Study." *Lancet Respiratory Medicine* 12 (9): 714–27. www.ncbi.nlm.nih.gov/pubmed/39127051.

Mineur, Yann S., and Wim E. Crusio. 2009. "Behavioral Effects of Ventilated Micro-Environment Housing in Three Inbred Mouse Strains." *Physiology and Behavior* 97 (3–4): 334–40. www.ncbi.nlm.nih.gov/pubmed/19281831.

Mojica, Francisco J. M., and Lluis Montoliu. 2016. "On the Origin of CRISPR-Cas Technology: From Prokaryotes to Mammals." *Trends in Microbiology* 24 (10): 811–20. www.ncbi.nlm.nih.gov/pubmed/27401123.

Morton, David B., Gordon M. Burghardt, and Jane A. Smith. 1990. "Critical Anthropomorphism, Animal Suffering, and the Ecological Context." Special issue, *Hastings Center Report* 20 (3): 13–19.

Munafo, Marcus R., Brian A. Nosek, Dorothy V. M. Bishop, et al. 2017. "A Manifesto for Reproducible Science." *Nature Human Behaviour* 1 (1): 0021. www.ncbi.nlm.nih.gov/pubmed/33954258.

Murchison, Elizabeth P. 2008. "Clonally Transmissible Cancers in Dogs and Tasmanian Devils." Supplement, *Oncogene* 27 (S2): S19–30. www.ncbi.nlm.nih.gov/pubmed/19956175.

Murray, Roderick, John W. Oliphant, John T. Tripp, et al. 1955. "Effect of Ultraviolet Radiation on the Infectivity of Icterogenic Plasma." *Journal of the American Medical Association* 157 (1): 8–14. www.ncbi.nlm.nih.gov/pubmed/13211300.

Murray, Roderick, Frank Ratner, William C. L. Diefenbach, and Herman Geller. 1954. "Effect of Storage at Room Temperature on Infectivity of Icterogenic Plasma." *Journal of the American Medical Association* 155 (1): 13–15. www.ncbi.nlm.nih.gov/pubmed/13151878.

National Academies of Sciences, Engineering, and Medicine. 2020. *Necessity, Use, and Care of Laboratory Dogs at the U.S. Department of Veterans Affairs.* Washington, DC: National Academies Press.

———. 2023. *Nonhuman Primate Models in Biomedical Research: State of the Science and Future Needs*. Washington, DC: National Academies Press. https://nap.nationalacademies.org/catalog/26857/nonhuman-primate-models-in-biomedical-research-state-of-the-science.

National Research Council. 2011. *Guide for the Care and Use of Laboratory Animals*. 8th ed. Washington, DC: National Academies Press.

Neave, Heather W., Rolnei R. Daros, João H. Costa, Marina A. G. von Keyserlingk, and Daniel M. Weary. 2013. "Pain and Pessimism: Dairy Calves Exhibit Negative Judgement Bias Following Hot-Iron Disbudding." *PLoS One* 8 (12): e80556. www.ncbi.nlm.nih.gov/pubmed/24324609.

Newcomer, Christian. 2015. "A Defense of Animal Welfare Accreditation." *Science* 347 (6219): 243. www.ncbi.nlm.nih.gov/pubmed/25593178.

Orlans, F. Barbara, Tom L. Beauchamp, Rebecca Dresser, David B. Morton, and John P. Gluck. 1998. *The Human Use of Animals: Case Studies in Ethical Choice*. New York: Oxford University Press.

Pacheco, Alex, and Anna Francione. 1985. "The Silver Spring Monkeys." In Singer 1985, 135–47.

Page, Gayle Giboney, Wendy P. Blakely, and Shamgar Ben-Eliyahu. 2001. "Evidence That Postoperative Pain Is a Mediator of the Tumor-Promoting Effects of Surgery in Rats." *Pain* 90 (1–2): 191–99. www.ncbi.nlm.nih.gov/entrez/query.fcgi?cmd=Retrieve&db=PubMed&dopt=Citation&list_uids=11166986.

Parascandola, John. 1991. "The Development of the Draize Test for Eye Toxicity." *Pharmacy in History* 33 (3): 111–17.

Peckmezian, Tina, and Phillip W. Taylor. 2017. "Place Avoidance Learning and Memory in a Jumping Spider." *Animal Cognition* 20 (2): 275–84. www.ncbi.nlm.nih.gov/pubmed/27796659.

Perel, Pablo, Ian Roberts, Emily Sena, et al. 2007. "Comparison of Treatment Effects Between Animal Experiments and Clinical Trials: Systematic Review." *BMJ* 334 (7586): 197. www.ncbi.nlm.nih.gov/pubmed/17175568.

Pound, Pandora. 2023. *Rat Trap: The Capture of Medicine by Animal Research—and How to Break Free*. Leicestershire, UK: Troubador.

Prinsen, Menk K., Coenraad F. M. Hendriksen, Cyrille A. M. Krul, and Ruud A. Woutersen. 2017. "The Isolated Chicken Eye Test to Replace the Draize Test in Rabbits." *Regulatory Toxicology and Pharmacology* 85:132–49. www.ncbi.nlm.nih.gov/pubmed/28192172.

Raj, A. B. Mohan, and Neville G. Gregory. 1995. "Welfare Implications of the Gas Stunning of Pigs 1: Determination of Aversion to the Initial Inhalation of Carbon Dioxide or Argon." *Animal Welfare* 4:273–80.

Raja, Srinivasa N., Daniel B. Carr, Milton Cohen, et al. 2020. "The Revised International Association for the Study of Pain Definition of Pain: Concepts,

Challenges, and Compromises." *Pain* 161 (9): 1976–82. www.ncbi.nlm.nih
.gov/pubmed/32694387.

Regan, Tom. 1983. *The Case for Animal Rights*. Berkeley: University of
California Press.

Richter, S. Helene, Joseph P. Garner, Corinna Auer, Joachim Kunert, and
Hanno Würbel. 2010. "Systematic Variation Improves Reproducibility of
Animal Experiments." *Nature Methods* 7 (3): 167–68. www.ncbi.nlm.nih
.gov/pubmed/20195246.

Rock, Meagan L., Alicia Z. Karas, Katherine B. Gartrell Rodriguez, et al. 2014.
"The Time-to-Integrate-to-Nest Test as an Indicator of Wellbeing in
Laboratory Mice." *Journal of the American Association for Laboratory
Animal Science* 53 (1): 24–28. www.ncbi.nlm.nih.gov/pubmed/24411776.

Rollin, Bernard. 1989. *The Unheeded Cry: Animal Consciousness, Animal Pain
and Science*. Oxford: Oxford University Press.

———. 1995. *The Frankenstein Syndrome: Ethical and Social Issues in the
Genetic Engineering of Animals*. Cambridge: Cambridge University Press.

———. 2011. *Putting the Horse Before Descartes: My Life's Work on Behalf of
Animals*. Philadelphia: Temple University Press.

Roughan, John V., Claire A. Coulter, Paul A. Flecknell, Huw D. Thomas, and
Kenneth J. Sufka. 2014. "The Conditioned Place Preference Test for Assessing Welfare Consequences and Potential Refinements in a Mouse Bladder
Cancer Model." *PLoS One* 9 (8): e103362. www.ncbi.nlm.nih.gov/pubmed
/25100208.

Roughan, John V., and Paul A. Flecknell. 2001. "Behavioural Effects of Laparot-
omy and Analgesic Effects of Ketoprofen and Carprofen in Rats." *Pain* 90
(1–2): 65–74. www.ncbi.nlm.nih.gov/pubmed/11166971.

Roumbedakis, Katina, Marina N. Alexandre, José A. Puch, Maurício L. Martins,
Cristina Pascual, and Carlos Rosas. 2020. "Short and Long-Term Effects of
Anesthesia in *Octopus maya* (Cephalopoda, Octopodidae) Juveniles."
Frontiers in Physiology 11:697. www.ncbi.nlm.nih.gov/pubmed/32695019.

Rovida, Costanza, Francois Busquet, Marcel Leist, and Thomas Hartung. 2023.
"REACH Out-Numbered! The Future of REACH and Animal Numbers."
ALTEX 40 (3): 367–88. www.ncbi.nlm.nih.gov/pubmed/37470349.

Rovida, Costanza, and Thomas Hartung. 2009. "Re-evaluation of Animal
Numbers and Costs for In Vivo Tests to Accomplish REACH Legislation
Requirements for Chemicals: A Report by the Transatlantic Think Tank for
Toxicology (t(4))." *ALTEX* 26 (3): 187–208. www.ncbi.nlm.nih.gov/pubmed
/19907906.

Rowan, Andrew N. 1984. *Of Mice, Models, and Men: A Critical Evaluation of
Animal Research*. Albany: State University of New York Press.

Russell, William M. S., and Rex L. Burch. 1959. *The Principles of Humane
Experimental Technique*. London: Methuen.

Ryczek, Natalia, Magdalena Hryhorowicz, Joanna Zeyland, Daniel Lipinski, and Ryszard Slomski. 2021. "CRISPR/Cas Technology in Pig-to-Human Xenotransplantation Research." *International Journal of Molecular Sciences* 22 (6): 3196. www.ncbi.nlm.nih.gov/pubmed/33801123.

Ryder, Richard D. 1985. "Speciesism in the Laboratory." In Singer 1985, 77–88.

Rygula, Rafal, Helena Pluta, and Piotr Popik. 2012. "Laughing Rats Are Optimistic." *PLoS One* 7 (12): e51959. www.ncbi.nlm.nih.gov/pubmed /23300582.

Salomons, Hannah, Jordan Sokoloff, and Brian Hare. 2024. "Companion Dogs Flexibly and Spontaneously Comprehend Human Gestures in Multiple Contexts." *Animal Cognition* 27:78. https://link.springer.com/article/10.1007 /s10071-024-01901-6.

Schroeder, Paul, Soffia Jones, Iain S. Young, and Lynne U. Sneddon. 2014. "What Do Zebrafish Want? Impact of Social Grouping, Dominance and Gender on Preference for Enrichment." *Laboratory Animals* 48 (4): 328–37. https://pubmed.ncbi.nlm.nih.gov/24939904/.

Schuppli, Catherine Anne. 2004. "The Role of the Animal Ethics Committee in Achieving Humane Animal Research." PhD diss., University of British Columbia. https://open.library.ubc.ca/soa/cIRcle/collections/ubctheses/831 /items/1.0099790.

Schwindaman, Dale, Mark Conner, Charles McPherson, et al. 1973. "The Use of Animals in Medical Research and Experimentation." In *Research Animals in Medicine*, edited by Lowell T. Harmison, 1271–93. Washington, DC: US Department of Health, Education and Welfare.

Sedaris, David. 1994. *Barrel Fever: Stories and Essays.* Boston: Back Bay Books.

Seok, Junhee, H. Shaw Warren, Alex G. Cuenca, et al. 2013. "Genomic Responses in Mouse Models Poorly Mimic Human Inflammatory Diseases." *Proceedings of the National Academy of Sciences of the United States of America* 110 (9): 3507–12.

Sharp, Lesley A. 2018. *Animal Ethos: The Morality of Human-Animal Encounters in Experimental Lab Science.* Oakland: University of California Press.

Sherwin, Chris M., E. Haug, N. Terkelsen, and M. Vadgama. 2004. "Studies on the Motivation for Burrowing by Laboratory Mice." *Applied Animal Behaviour Science* 88 (3–4): 343–58. www.sciencedirect.com/science/article/pii /S0168159104000735?via%3Dihub.

Singer, Peter, ed. 1985. *In Defense of Animals.* Oxford: Blackwell.

———. 1998. *Ethics into Action: Henry Spira and the Animal Rights Movement.* Lanham, MD: Rowman and Littlefield.

———. 2023. *Animal Liberation Now: The Definitive Classic Renewed.* New York: Harper Perennial.

Skloot, Rebecca. 2011. *The Immortal Life of Henrietta Lacks.* New York: Crown Books.

Sneddon, Lynne U., Robert W. Elwood, Shelley A. Adamo, and Matthew C. Leach. 2014. "Defining and Assessing Animal Pain." *Animal Behaviour* 97:201–12. www.sciencedirect.com/science/article/pii/S0003347214003431 ?via%3Dihub.

Spira, Henry. 1985. "Fighting to Win." In Singer 1985, 194–208.

Steensma, David P., Robert A. Kyle, and Marc A. Shampo. 2010. "Abbie Lathrop, the 'Mouse Woman of Granby': Rodent Fancier and Accidental Genetics Pioneer." *Mayo Clinic Proceedings* 85 (11): e83. www.ncbi.nlm.nih .gov/pubmed/21061734.

Takacs, David. 2021. "We Are the River." *University of Illinois Law Review* 2021 (2): 545–606. https://heinonline.org/HOL/LandingPage?handle=hein .journals/unilllr2021&div=19&id=&page=.

Tallo-Parra, Oriol, Marina Salas, and Xavier Manteca. 2023. "Zoo Animal Welfare Assessment: Where Do We Stand?" *Animals* 13 (12): 1966. https:// doi.org/https://doi.org/10.3390/ani13121966.

Tannenbaum, Jerrold. 2017. "Ethics in Biomedical Animal Research: The Key Role of the Investigator." In *Animal Models for the Study of Human Disease*, edited by P. Michael Conn, 1–44. Cambridge, MA: Academic Press.

Taylor, Katy, and Laura Rego Alvarez. 2019. "An Estimate of the Number of Animals Used for Scientific Purposes Worldwide in 2015." *Alternatives to Laboratory Animals (ATLA)* 47 (5–6): 196–213. https://pubmed.ncbi.nlm .nih.gov/32090616/.

Turner, Patricia V., Debra L. Hickman, Judith van Luijk, et al. 2020. "Welfare Impact of Carbon Dioxide Euthanasia on Laboratory Mice and Rats: A Systematic Review." *Frontiers in Veterinary Science* 7:411. www.ncbi.nlm .nih.gov/pubmed/32793645.

Villa, Erica, Tiziana Barchi, A. Grisendi, et al. 1982. "Susceptibility of Chronic Symptomless HBsAg Carriers to Ethanol-Induced Hepatic Damage." *Lancet* 2 (8310): 1243–44. www.ncbi.nlm.nih.gov/pubmed/6128548.

Walker, Rebecca L., and Matthias Eggel. 2020. "From Mice to Monkeys? Beyond Orthodox Approaches to the Ethics of Animal Model Choice." *Animals* 10 (1): 77. www.ncbi.nlm.nih.gov/pubmed/31906319.

Watters, Jason V., and Bethany L. Krebs. 2025. *Managing Zoo Animal Welfare: A Behavior-Based Approach*. Hoboken, NJ: Wiley-Blackwell.

White, William J., Melvin W. Balk, and C. Max Lang. 1989. "Use of Cage Space by Guineapigs." *Laboratory Animals* 23:208–14.

Wilks, Samuel. 1881. "Vivisection: Its Pains and Its Uses." *Nineteenth Century* 10:936–48.

Wong, Chi Heem, Kien Wei Siah, and Andrew W. Lo. 2019. "Corrigendum: Estimation of Clinical Trial Success Rates and Related Parameters." *Biostatistics* 20 (2): 366. www.ncbi.nlm.nih.gov/pubmed/30445524.

Würbel, Hanno, and Joseph P. Garner. 2007. "Refinement of Rodent Research Through Environmental Enrichment and Systematic Randomization." *NC3Rs* 9:1–9. www.nc3rs.org.uk/sites/default/files/documents /Refinementenvironmentalenrichmentandsystematicrandomization.pdf.

Yong, Ed. 2022. *An Immense World: How Animal Senses Reveal the Hidden Realms Around Us*. New York: Random House.

Index

Abee, Christian, 106

adoption of laboratory animals, 13, 73, 82–83, 87–88, 218–19, 224, 277, 301n12, 301n16

agriculture: industrialized, 4, 110–11; pesticides for use in, 35. *See also* Harrison, Ruth

alcohol consumption, 12, 20, 22–25, 28, 32–33, 75; and exacerbation of liver infections, 115

alternative nonanimal experimental methods, 14, 40, 45, 54–58, 68, 88, 90, 95, 199, 209–12, 228–31, 259, 262; as complement to animal experiments, 101; funding and dissemination of, 284–85; and regulatory reform, 97; in safety and toxicology tests, 66–67, 95, 100; umbilical blood of aborted human fetuses in, 99; validation of, 283–84. *See also* experiments

alternatives in animal research. *See* animal ethics; reduction of animal numbers as an animal-research alternative; refinement as an animal-research alternative; replacement as an animal-research alternative

Alternatives Research and Development Fund, 196

American Anti-vivisection Society, 268

American Humane Association, 80

American Museum of Natural History, 300n7

American Veterinary Medical Association: Guidelines for the euthanasia of animals of the, 121; Veterinarian's Oath of the, 310n3

amyotrophic lateral sclerosis (ALS), 242

analgesics, 48, 165, 170–75, 178–89, 214, 254, 257, 277. *See also* aspirin; buprenorphine; carprofen; ibuprofen; morphine; opioids

anesthetics, 48, 168, 179–80, 182, 214, 311n5; for animal experiments, 75–77, 165, 175, 310n2; for animal surgeries, 169, 178, 184–86; chloroform, 76, 165, 184. *See also* pain

animal-care technicians. *See* technicians

animal ethics, 2, 10–11, 84, 217–20, 280, 315n2; broadening the range of voices in decisions on, 19, 235, 239; justification for harming animals in, 257, 272, 317n16; legal requirements and, 214–15, 240; as minimizing harm to animals, 44, 272; principles of, 33, 213–14; replacement, reduction, and refinement (Three Rs) in, 40, 61, 226, 228–30, 233–34, 240, 252, 257–58, 280, 282, 285, 317n13; sentience and, 217, 220–21; suffering and, 217, 219, 225. *See also* animal welfare; ethics committees

animal ethics committees. *See* ethics
  committees
Animal Legal Defense Fund, 136, 313n8
Animal Liberation Front, 148
Animal Machines. See Harrison, Ruth
animal-protection organizations, 56, 96,
  165, 247, 266–67, 321n24
Animal Research Development Fund, 268
animal-research facilities (vivaria), 4–6, 21,
  67, 78–82, 117, 139, 141; veterinarians
  as directors of, 79, 252, 256, 261, 281;
  for mice, 194, 220
animal rights activists, 11, 13, 48–51, 58,
  61, 74–84, 130–31, 173, 196, 212–16,
  237, 253, 264–67; campaigns of, 13, 16,
  72–73, 75–78, 89, 242, 300n7, 302n7;
  laboratory exposés of, 148, 245; as members of animal ethics committees, 287;
  messaging and images of, 20, 89–92,
  194; threats from, 31, 49, 212, 264, 266;
  watchdog groups of, 11, 16–18, 207,
  219, 250–51. *See also* animal welfare
animal testing: as distinguished from basic
  research, 13–14, 35–36, 103–4, 108,
  182–83, 209, 237–38, 283–84
animal welfare: advances in monkey, 244;
  advances in mouse, 207; environmental
  enrichment for, 152–61, 193, 225, 270,
  275–76, 309n10; "Five Freedoms" of the
  Brambell Report on, 111–12, 304n1;
  laws and regulations in different countries
  regarding, 16, 65, 129, 170, 212, 224,
  226, 231, 255, 260; policing of, 16, 49,
  65, 246; science of, 18, 35, 62, 64, 90,
  109–42, 152–55, 160–61, 166–67, 173,
  177, 188, 208, 215, 232, 252, 262, 265.
  See also animal ethics; animal rights
  activists; Animal Welfare Act (1966);
  Animal Welfare Act (1985); mental
  health
Animal Welfare Act (1966), 4–5, 13–18,
  72–73, 79–84, 96, 130, 148–49, 162,
  172–73, 195–99, 208–21, 245, 255–56,
  271, 275, 301n10, 312n2, 318n3;
  amendment (1970) of, 81, 173, 178–79,
  195, 312n3; amendment (2002) of, 274;
  definition of animal in, 274, 313n7;
  exclusion of mice and rats in, 16, 32, 162,
  182, 195–98, 212, 218, 274, 313n7,
  318n3. *See also* animal welfare; United
  States Department of Agriculture (USDA)
Animal Welfare Act (1970), 173, 178–79,
  311n7, 312n3

Animal Welfare Act (1985), 2, 9, 14–18, 32,
  84–88, 108, 113, 144, 149–55, 162,
  182, 191–99, 208, 212–21, 228, 245,
  266, 275, 308n7; final regulations for,
  252; standards for dog cages as defined
  by, 130–31, 133, 180; standards for dog
  exercise as defined by, 136, 195, 218–19,
  222, 225; standards for guinea pig cages
  as defined by, 130–31; standards for psychological well-being for primates as
  defined by, 195, 207, 213–15, 222, 270.
  See also animal welfare; ethics committees; Health Research Extension Act
  (1985); United States Department of
  Agriculture (USDA)
Animal Welfare Act (2002), 191, 274
Animal Welfare Institute, 308n9
animal-welfare science, 14, 111–13, 119,
  129, 305n2; role of in writing regulations, 131–37, 208, 252, 265, 276. *See
  also* measurements of animal welfare
anthropomorphism, 90, 114–17, 130, 141,
  146, 189; creative power of empathy and,
  116; critical, 115–17, 170, 305n4
antibiotics, 9, 10, 48, 71, 85, 104, 189, 202,
  204
antivivisectionists, 81, 164–65, 168, 287,
  296n6
aquariums, 161, 271
Arluke, Arnold, 224
arthritis, 68–69, 169, 183–84, 298n10
artificial-intelligence modeling, 55–56, 69,
  285; and cell-culture studies, 101. *See
  also* computer modeling
aspirin, 53, 172, 184, 223
Association for Assessment and Accreditation
  of Laboratory Animal Care International
  (AAALAC), 16, 79, 112, 191–93, 196–
  98, 208, 244–45, 248–52, 271, 275,
  282, 288, 313n6, 318n3, 320n9
Association of Zoos and Aquariums, 271, 275
Austin, Jamie, 185
Australia, 36, 202, 262–63, 314n13; ethics
  committees in, 286
autism, 115, 138, 305n6
autoimmune diseases, 45–46. *See also* multiple sclerosis (MS)
Avon, 95–96

bacteria, 7, 67, 238; immune-defense mechanisms of, 205; MRSA, 25
baldness: female, 236; male, 13, 231, 236
Bang, Fred, 238, 318n21

Founded in 1893,
UNIVERSITY OF CALIFORNIA PRESS
publishes bold, progressive books and journals
on topics in the arts, humanities, social sciences,
and natural sciences—with a focus on social
justice issues—that inspire thought and action
among readers worldwide.

The UC PRESS FOUNDATION
raises funds to uphold the press's vital role
as an independent, nonprofit publisher, and
receives philanthropic support from a wide
range of individuals and institutions—and from
committed readers like you. To learn more, visit
ucpress.edu/supportus.